Software Designers in Action

A Human-Centric Look at Design Work

Chapman & Hall/CRC Innovations in Software Engineering and Software Development

Series Editor

Richard LeBlanc

Chair, Department of Computer Science and Software Engineering, Seattle University

AIMS AND SCOPE

This series covers all aspects of software engineering and software development. Books in the series will be innovative reference books, research monographs, and textbooks at the undergraduate and graduate level. Coverage will include traditional subject matter, cutting-edge research, and current industry practice, such as agile software development methods and service-oriented architectures. We also welcome proposals for books that capture the latest results on the domains and conditions in which practices are most effective.

PUBLISHED TITLES

Software Development: An Open Source Approach
Allen Tucker, Ralph Morelli, and Chamindra de Silva

Building Enterprise Systems with ODP: An Introduction to Open Distributed Processing
Peter F. Linington, Zoran Milosevic, Akira Tanaka, and Antonio Vallecillo

Software Engineering: The Current Practice
Václav Rajlich

Fundamentals of Dependable Computing for Software Engineers
John Knight

Introduction to Combinatorial Testing
D. Richard Kuhn, Raghu N. Kacker, and Yu Lei

Software Test Attacks to Break Mobile and Embedded Devices
Jon Duncan Hagar

Software Designers in Action: A Human-Centric Look at Design Work
André van der Hoek and Marian Petre

CHAPMAN & HALL/CRC INNOVATIONS IN
SOFTWARE ENGINEERING AND SOFTWARE DEVELOPMENT

Software Designers in Action

A Human-Centric Look at Design Work

Edited by

André van der Hoek

University of California, Irvine

Irvine, California, USA

Marian Petre

The Open University

Milton Keynes, UK

CRC Press
Taylor & Francis Group
Boca Raton London New York

CRC Press is an imprint of the
Taylor & Francis Group an **informa** business
A CHAPMAN & HALL BOOK

CRC Press
Taylor & Francis Group
6000 Broken Sound Parkway NW, Suite 300
Boca Raton, FL 33487-2742

Printed on acid-free paper
Version Date: 20130607

International Standard Book Number-13: 978-1-4665-0109-6 (Hardback)

Library of Congress Cataloging-in-Publication Data

Software designers in action : a human-centric look at design work / edited by Marian Petre and Andre Van Der Hoek.
pages cm
Includes bibliographical references and index.
ISBN 978-1-4665-0109-6
1. Computer software--Development. 2. Computer programmers. I. Petre, Marian, 1959- editor of compilation. II. Hoek, Andr? van der, 1971- editor of compilation.

QA76.76.D47S6185 2014
005.1--dc23 2013014924

Visit the Taylor & Francis Web site at
http://www.taylorandfrancis.com

and the CRC Press Web site at
http://www.crcpress.com

Contents

Foreword

EVERYTHING AROUND US HAS BEEN DESIGNED. Anything that is not a simple, untouched piece of nature has been designed by someone. The quality of that design effort, therefore, profoundly affects our quality of life. The ability of designers to produce effective, efficient, imaginative, and stimulating designs is therefore important to all of us. This is clearly evident in the case of hardware—physical products, buildings, environments, etc. It is perhaps less evident, but often more important in the case of software that resides not only in our computers but also increasingly in those more familiar hardware artifacts. Software design not only has a complex internal functionality to satisfy but also a subtle external functionality in mediating the ways in which we interact with hardware.

Unlike the design of physical artifacts, software design does not have lengthy traditions and numerous exemplars to draw upon. So, in many ways, it has had to invent its own ways of designing. There are recommended procedures and methods for designing software; however, as we see in the cases that are studied in this book, professional software designers do not always follow such procedures and methods. In that respect, they are like most other professional and experienced designers who usually seem to work "intuitively." But, usually, it is much more a matter of experience rather than intuition. What is reported in this book, therefore, is what we can learn from studying experienced behavior in a design context. That learning, in turn, can inform design practice and design education.

This kind of study of real design activity was missing from design research for a long time. There was an early emphasis on prescribing how to design, based on very little evidence. More recently, there has been a growing field of study of design activity and design thinking, spread across engineering, architecture, and product design—the hardware domains. Now, software design is being studied in the same way, leading to a more informed and considered appreciation and understanding of designer' behaviors and skills.

Recently, the study of design activity has been dominated by the observation and analysis of the actions and conversations of small groups or teams of designers engaged in either real or simulated design tasks. These studies have tended to replace the earlier, more formal, laboratory exercises relying on the "think-aloud" protocols of single designers. The verbal (and nonverbal) exchanges of members of a team cooperating in a joint task provide more diverse but richer data indicative of the range of cognitive and other activities that are being undertaken during designing. The software design studies in this book are of this kind, using two-person teams engaged in the same experimental task. The method of conducting these studies was simply to record the team members working together at a whiteboard. As this book demonstrates, the richness of the captured data inspired a rich diversity of analyses.

The Irvine software design analyses were conducted by researchers coming not only from the software domain, but also from architectural, engineering, and industrial design domains as well as from cognitive science and other disciplines. The fact that such a variety of disciplines could contribute and collaborate together indicates that there are fundamental similarities within the design process that underlie the range of domains in which it is practiced.

Some of the similarities that have been observed through studies of design activity over recent decades include how experienced designers:

- Exercise the skills of gathering and structuring relevant problem information and data
- Develop a coherent view of those data by means of problem "framing"
- Adopt a solution-focused, rather than a problem-focused approach to the task
- Promote a coevolution of the problem and solution structures together
- Reflectively interact with a variety of representations of the problem and the solution
- Move fluently between different levels of detail, keeping options and lines of development open

Many of these observations reoccur in the studies reported here, which clearly situate software design research within the canon of design research. The studies in this book also introduce new analysis techniques and insights that provide significant extensions to the field. This makes the book an important contribution to design research and lays broader foundations for further research for many years to come.

Nigel Cross

Acknowledgments

We thank Sol Greenspan and Mary Lou Maher, who, as National Science Foundation program managers, provided us with a grant to organize the Studying Professional Software Design 2010 workshop. We also thank the Royal Society and the Wolfson Foundation, who provided support to Marian Petre through a Research Merit Award.

We thank Alex Baker, an inspiring collaborator, whose contributions to the vision, the video collection, the workshop, and the ensuing special issues were seminal.

University of California, Irvine, graduate students Emily Navarro, Nick Mangano, Nick Lopez, Gerald Bortis, and Mitch Dempsey all helped in organizing the event, arranging all aspects of logistics, including food, transportation, and other practicalia.

Much credit goes to all of the designers who agreed to be filmed and to the organizations for which they work. Without their generous gift of time, the workshop, the research, and this book would simply not have happened. We thank especially Jim Dibble, Ania Dilmaghani, and David Holloway, who enriched the workshop through their candid participation. They must have had a little taste of what it feels like to be a rock star, given how thoroughly they were "interrogated" by the researchers...

Randi Cohen of CRC Press deserves our praise and gratitude for her patience in guiding us through the production of this book.

Finally, we thank our families: Kim Keller, Anne Marie van der Hoek, Annika van der Hoek, Clive Baldwin, Grace Petre Eastty, Max Petre Eastty, Adam Baldwin, and Danny Baldwin. We spent considerable time away from them in order to complete this book. Their support throughout has been much appreciated.

Contributors

Aldeida Aleti
Information and Communication Technologies
Swinburne University of Technology
Melbourne, Australia

Rita Assoreira Almendra
Department of Art and Design
Technical University of Lisbon
Lisbon, Portugal

Alex Baker
Department of Informatics
University of California, Irvine
Irvine, California

Linden J. Ball
Department of Psychology
Lancaster University
Lancaster, United Kingdom

Rachel K.E. Bellamy
IBM T.J. Watson Research Center
Hawthorne, New York

Alan F. Blackwell
Computer Laboratory
University of Cambridge
Cambridge, United Kingdom

Jim Buckley
Department of Computer Science and Information Systems
University of Limerick
Limerick, Ireland

David Budgen
School of Engineering and Computing Sciences
University of Durham
Durham City, United Kingdom

Janet Burge
Computer Science and Software Engineering
Miami University
Oxford, Ohio

Bo T. Christensen
Department of Marketing
Copenhagen Business School
Copenhagen, Denmark

Henri Christiaans
Department of Ergonomics
Delft University of Technology
Delft, The Netherlands

Michael L. Collard
Department of Computer Science
The University of Akron
Akron, Ohio

J.J. Collins
Department of Computer Science and Information Systems
University of Limerick
Limerick, Ireland

Michael Desmond
IBM T.J. Watson Research Center
Hawthorne, New York

Natalia Dragan
Department of Computer and Information Science
Cleveland State University
Cleveland, Ohio

John Gero
Krasnow Institute for Advanced Study
Fairfax, Virginia

Mark D. Gross
School of Architecture
Carnegie Mellon University
Pittsburgh, Pennsylvania

André van der Hoek
Department of Informatics
University of California, Irvine
Irvine, California

Michael Jackson
Computing Department
The Open University
Milton Keynes, United Kingdom

Sian Joel-Edgar
Department of Management Studies
University of Exeter Business School
Exeter, United Kingdom

Bonnie E. John
IBM T.J. Watson Research Center
Hawthorne, New York

Jeff Wai Tak Kan
Fairfax, Virginia

Clayton Lewis
Department of Computer Science
University of Colorado, Boulder
Boulder, Colorado

Jonathan I. Maletic
Department of Computer Science
Kent State University
Kent, Ohio

Ben Matthews
School of Information Technology and Electrical Engineering
The University of Queensland
Brisbane, Australia

Janet McDonnell
Central Saint Martins College of Arts and Design
University of the Arts London
London, United Kingdom

Kumiyo Nakakoji
Software Research Associates, Inc.
Tokyo, Japan

Jeffrey V. Nickerson
Stevens Institute of Technology
Hoboken, New Jersey

Balder Onarheim
Department of Marketing
Copenhagen Business School
Copenhagen, Denmark

Harold Ossher
IBM T.J. Watson Research Center
Hawthorne, New York

Marian Petre
Centre for Research in Computing
The Open University
Milton Keynes, United Kingdom

Paul Rodgers
School of Design
Northumbria University
Newcastle upon Tyne, United Kingdom

John Rooksby
School of Computing Science
University of Glasgow
Glasgow, United Kingdom

Bonita Sharif
Department of Computer Science and Information Systems
Youngstown State University
Youngstown, Ohio

Mary Shaw
School of Computer Science
Carnegie Mellon University
Pittsburgh, Pennsylvania

Andrew Sutton
Department of Computer Science and Engineering
Texas A&M University
College Station, Texas

Antony Tang
Information and Communication Technologies
Swinburne University of Technology
Melbourne, Australia

Barbara Tversky
Department of Psychology
Stanford University
Stanford, California

and

Teachers College
Columbia University
New York, New York

Hans van Vliet
Department of Computer Science
Vrije Universiteit Amsterdam
Amsterdam, The Netherlands

Lixiu Yu
Stevens Institute of Technology
Hoboken, New Jersey

Yasuhiro Yamamoto
Tokyo Institute of Technology
Tokyo, Japan

Introduction

Over the past decades, the study of *design in practice* has become a major focus across many different disciplines. It is crucial to understand how professionals work, both with an eye toward developing innovations that help them work better, and with an eye toward documenting practices that help train new generations. Studying designers while they work has become a discipline of its own, with established expectations, practices, and venues for presenting results.

A particularly significant event took place in 1994, when Cross, Christiaans, and Dorst organized a workshop (Analysing Design Activity [1]) for which they invited participants from a number of different disciplines to study a set of videotapes of both an individual designer and a team of designers at work. This workshop was highly influential. First, in following up on an earlier workshop (Research in Design Thinking, 1992 [2]), it established an ongoing forum for a community of researchers interested in design research to come together and talk about their ideas. This community has since carried forward the workshops as a loosely coupled series of events with the umbrella name "Design Thinking Research Symposia," the latest of which (Articulating Design Thinking) took place in August 2012 at Northumbria University, UK. Second, the 1994 workshop produced a book that documented a host of different and often novel analysis techniques that were applied by the researchers. These techniques have since been refined and broadened, and continue to shape the performance of this kind of research today. Finally, the structure of the workshop was novel in sharing a common data set with a broad range of researchers, giving them a common reference point from which to compare their different perspectives.

This book documents the results of Studying Professional Software Design (SPSD) 2010, a workshop in the same vein as the Design Thinking Research Symposium series, though focused exclusively on software design. Until SPSD 2010, the study of professional software designers "in action" was relatively limited. A handful of isolated academic papers had appeared. Each provided valuable insights, but they did not aggregate readily. The work was too sparse to expose software design practice to the sort of scrutiny that has enabled the broader design research community to put other types of design practice into perspective.

The cause of this deficiency, seemingly, was that the software engineering research community had tended to focus its efforts on studying the *artifacts* of design. While this led to the availability of many different kinds of software design notations, together with the associated tools for specifying and analyzing the designs captured in these notations, the notations and tools were designed for expressiveness and for the *presumed* needs of software designers. However, studying design *products* is different from studying the situated design *practices* by which developers produce these products. Consequently, even though all software is designed in one way or another, how software design is performed today remains one of the least understood activities in which software developers engage.

Our aim was to "jump-start" the study of professional software designers and their work practices. Indeed, much is said about what software developers *should* do when they design, but little is known about what they *actually* do. We hope that this book lays a strong foundation for bringing the study of software design "up to par" with the study of design in other disciplines. Only then will we be able to learn from experts, document crucial practices, understand the relative strengths and weaknesses of different approaches and strategies, develop improved software design notations and tools, and, perhaps most important of all, better educate our students in the practice of software design.

HOW THIS BOOK CAME ABOUT

In 2008, Alex Baker—one of the SPSD 2010 coorganizers—was a graduate student at UC Irvine. A designer at heart, Alex wanted to study professional software designers. Having read the broad literature on design, including seminal works by Newell and Simon [3] and Schön [4], more recent perspectives by Cross [5] and Lawson [6], and foraying into the academic literature (e.g., Dekel and Herbsleb [7] and Cherubini et al. [8]), Alex proclaimed that he could never look at software engineering in the same way again. Instead, he considered software development a *design discipline*, one that had, to date, largely failed to recognize itself as such and one that lacked a proper understanding of what it meant to be driven by design from beginning to end.

Alex decided to focus his efforts on understanding *early software design*. This choice was motivated by the observation that much crucial design activity takes place in settings away from the desktop and the notations and tools that are available. Instead, it typically takes place in conversations with customers, planned meetings of the software architects, or impromptu design sessions shaping some aspect of the software to be produced. Two key characteristics differentiate early design: (1) its focus tends to be on important decisions, shaping the overall software application in what functionality it is to provide and how it is to do so; and (2) it tends to proceed through conversation, with auxiliary diagrams, sketches, and text drawn and written to support the conversation.

In this light, many questions arise. What representations do software designers use in early software design? What strategies do they employ in navigating the vast space of possible design choices? How do they collaborate and communicate to address a problem effectively? How do they accommodate context in their practice and thinking? What are some of the unique challenges pertaining to early software design compared to conceptual design in other disciplines?

These and other questions offer a rich and broad challenge, not just to the software engineering community but also to the design research community at large. Indeed, the idea of the SPSD 2010 was born out of exactly this realization: questions of this magnitude require a community that examines the practices exhibited by software designers from a multitude of different perspectives and disciplines. By organizing the workshop around the videos and transcripts that Alex had been amassing as part of his PhD research, we sought to begin this process.

With the help of a U.S. National Science Foundation grant, over 50 researchers from a broad range of disciplines were invited to participate in the workshop. Each researcher was asked to analyze one or more of the three videos we distributed, and to produce a white paper to share at the workshop, which was held 8–10 February 2010, at the UC Irvine campus in Irvine, CA. The response was overwhelmingly positive, and the workshop produced fascinating discussions, results, and collaborations.

This book contains selected output from the workshop. It includes 18 chapters, each representing a significant extension of the respective original workshop white paper. Introducing the chapters are four invited commentaries, each placing the chapters of its section into a broader design context. Together, these chapters and commentaries represent a comprehensive look at early software design, examining the designers' work and approach from a range of different perspectives, including various theoretical examinations of the nature of software design and the particular design problem posed, critical looks at the processes and practices that the designers followed, in-depth examinations of key supporting elements of design, and the role of human interaction.

NATURE OF SOFTWARE DESIGN

Before we introduce the videos that were the subject of the workshop, it is useful to examine the nature of software design, particularly in the context of other design disciplines. What can be learned from all the design research that has already been performed in other disciplines, especially since significant commonalities in the nature of design work and the practices exhibited by designers across disciplines have been found (e.g., Cross [5])? Could the observations made and the lessons learned transfer to software design? What adjustments might be needed?

In its basic form, software design concerns determining the abstraction structure of the intended software. However, the abstraction structures in software design tend to be deeper than those in other disciplines, whether an expressive domain such as dance, or an engineering domain such as electronics. The problem is that the *behavior* of the code must be designed as much, if not more so, than the structure of the code in terms of its overall components and connectors, or classes and associations. As a result, the final design of a program is its source code, as even just minor variations in the code can radically alter its behavior. Because of this, the distinction between design and manufacture is blurred, making software design a highly interwoven process, with design and implementation in a continuous and close relationship.

A confounding factor is that complexity in software is not inherently limited by the material in which software is created—programmatic instructions that can be made to do almost anything. It is not that other design disciplines are not complex, but that complexity in software design is bounded largely by our imagination and reasoning, rather than by properties of the medium: we can design as deep and multivariate a structure and behavior as we can imagine. Managing that complexity—and scale—takes the discussion back to abstraction and the need for rich notations to support design. Unlike other disciplines, however, software lacks stable notations. Indeed, the discipline is forever inventing new notations to match some perceived problem, yet, in practice, words and narrative continue to feature strongly in software design. Design notations tend to borrow heavily from mathematical and other formalisms, leading to a more limited expressive range than domains such as product design.

We also observe that, because we are all trained to understand change as an inherent part of software development, it has become built into the evaluation of software, and our understanding of "quality." Because software exists in a culture of patches, versions, and releases, flaws may be tolerable. Much software, indeed, is shipped with known flaws. Consequently, the priorities for software design include designing for change, and designing for understandability (by other designers and by developers). The notion that software is read by people as well as executed by computers is fundamental to the discipline, and further complicates the design process.

This discussion reiterates Brooks's classic observation that software is a different material to be designed, one in which essential difficulties arise that are unlike the kinds of challenges seen in other disciplines [9]. Brooks recognizes that software exhibits unique challenges in terms of its inherent complexity, need for conformity, constant changeability, and intangibility of form—challenges that drastically influence how software designers approach a design problem.

How, then, software designers deal with the unique matter of software is worthy of serious study, perspective, and comparison.

EXISTING WORK

The interest in better understanding how software is designed arose as early as the 1980s, when a few pioneers began to critically analyze both what it means to perform design as part of the software life cycle and how developers—especially experts—approached their design tasks. For several decades thereafter, however, no sustained engagement with the topic emerged. On the contrary, the topic has appeared only sporadically in the literature, and it is noticeably absent in high-profile disciplinary venues such as the International Conference on Software Engineering (ICSE) and the International Symposium on the Foundations of Software Engineering (FSE).

Nevertheless, the work that *has* appeared to date is starting to lay important foundations and to identify promising directions. The following examples, taken from our introduction to an *IEEE Software* special issue containing some results from the workshop [10], highlight several key results over the years (a fuller bibliography is available on the *IEEE Software* website http://doi.ieeecomputersociety.org/10.1109/MS.2011.155).

- Raymonde Guindon and colleagues were among the first to study in great detail how software designers work through a design problem, showing a relative absence of the exploration of alternatives [11].
- Bill Curtis and colleagues showed how the linear notion of design being a phase in the life cycle was a fallacy [12].

- Mauro Cherubini and colleagues highlighted just how prevalent and transient whiteboard software design is, serving a crucial role in the development process [8].
- Uri Dekel and Jim Herbsleb demonstrated that designers engaged in collaborative design used formal notations much less than expected, although the notations they did use resembled existing notations [7].
- Robin Jeffries and colleagues studied differences between experts and novices, observing that experts tend to work initially to understand the problem at hand and have better insight into how to decompose a problem into subproblems, for example, by choosing the appropriate subproblem to work on next [13].
- Marian Petre documented several strategies that designers use, often subconsciously, as part of their design repertoire and experience (for instance, by using provisionality in designs to leave room for future options) [14].
- Sabine Sonnentag found that high-performing professional software designers structured their design process using local planning and attention to feedback, whereas lower performers were more engaged in analyzing requirements and more distracted by irrelevant observations [15].
- Willemien Visser argued, with empirical support, that the organization of actual design activities, even by experts involved in routine tasks, is opportunistic—in part, because opportunistic design provides cognitive economy [16].
- Linden Ball and Thomas Ormerod, on the other hand, argued that opportunistic design behavior is actually a mix of breadth-first and depth-first solution development, and that expert behavior is informed by longer-term considerations of cost effectiveness [17].
- Carmen Zannier and colleagues proposed an empirically based model of design decision making in which the nature of the design problem determines the structure of the designer's decision-making processes. The more certain and familiar the design problem is, the less a designer considers options [18].

In recent years, the study of software design has undergone something of a resurgence. A number of publications have appeared on topics including tools [19–21], literature reviews [22,23], and empirical studies of design, sketching, and sketches [24–26]. This book adds to this emerging literature, and we hope that it amplifies the resurgence and helps to establish a community of researchers who together will push the agenda forward.

DESIGN PROMPT AND VIDEOS

Following the format of the 1994 Analysing Design Activity Workshop, particularly in providing all participants with a common data set consisting of videos and transcripts of software designers at work, we needed to decide on what kind of design activity we wanted to capture. While there is great appeal to capturing software design work "in the wild," as it is performed by professional software designers on a commercial project with real customers, demands, and deliverables, we ultimately decided against doing so for two reasons. First, it would disallow a comparative analysis across multiple designers, or teams of designers, working on the same project, thereby denying any chance of distilling differences in approaches and, more ambitiously, understanding how those differences might influence the design quality. Second, such design work carries with it an enormous context that is difficult to capture and communicate properly for the researchers analyzing the videos.

Hence, we decided to create an artificial design prompt. In drafting this prompt, we kept a number of goals in mind:

- *Challenge the designers to shape the overall product and its functionality as much as how it achieves that functionality.* The modern view of software design is not one of a phase in the software life cycle that translates a given set of requirements into a design document to guide implementation; rather, it acknowledges that "designerly thinking" takes place across the entire life cycle [27]. The design prompt should be faithful to this view, particularly because we wanted the workshop to focus on early design, which is when the interplay between what is desired and what is technically feasible is prevalent.
- *Provide a reasonably complex design problem that at the same time allows the designers to gain traction fairly quickly.* Complexity is the driver of design; without sufficient complexity in the design prompt, design-like behavior would be unlikely to appear. At the same time, we knew we would have access to the designers for only a limited amount of time. We thus needed a prompt that was not too trivial, nor too complex so as to inadvertently stifle meaningful progress.
- *Keep the design prompt accessible while avoiding domain bias.* Because we wanted to study multiple sets of designers working on the same problem, we needed to ensure that the design problem would be accessible to designers with a range of different expertise. At the same time, we did not want to induce domain bias in case one team was overly familiar with the domain of the prompt and another team was not. We required a prompt that was general in topic, yet unlikely to be part of the day-to-day work of the designers we were to recruit.

The result was a prompt for an educational traffic signal simulator to be used by an engineering professor in her class to teach her students about traffic signal timing. The prompt comprised two pages, motivating why the professor wanted this software and providing various relatively detailed requirements regarding its purpose and what she wanted her students to be able to do with it. The prompt was intentionally incomplete to leave room for interpretation, extrapolation, and innovation; that is, the designers were free to make assumptions about additional requirements, context, and the overall design approach to be followed. The full prompt is provided in Appendix I.

Once we had developed the prompt, we recruited professional software designers to participate in the experiment. Recruitment was performed through the UC Irvine alumni network, with a solicitation sent to all of the alumni of the Donald Bren School of Information and Computer Sciences. The solicitation explicitly asked for software designers who were considered expert designers in their company; that is, those whom the company would trust with key software design problems. We received a number of responses, which led to us visiting seven companies in total. Two of those companies provided us with access to two pairs of designers each, the other five companies one pair each, for a total of nine videotaped sessions.

Each design session was captured using two cameras, positioned to capture the board at large and the designers at work. A typical view is shown in Figure 1. Each pair of designers was given 1 hour and 50 minutes to work on their design. One team spent roughly 1 hour and then declared itself done, but all the other teams spent the full amount of the allocated time and in a few cases even an additional few minutes to tie up some loose ends. We set this time limit as we did not want to overly burden the designers who were volunteering their time and also because, in our experience, design meetings seldom run longer than 2 hours and, if they do, tend to be ineffective after that time.

After they had completed their work, each pair of designers was given some time to collect their thoughts, and then asked to briefly explain their design and overall approach, and, finally, each pair was interviewed at the end of the session.

The result was a set of nine videos, from which we selected three for distribution to the workshop participants, with a team each from Adobe, AmberPoint (which in the meantime has been acquired by Oracle), and Intuit. The videos were selected based upon two factors: the degree to which each team appeared to be working naturally and effectively, and the quality of the design and the design process that they followed in comparison to the other teams. Each pair of designers had worked with each other in the past, although the degree to which they had done so ranged from occasionally to nearly continuously.

The three videotapes, as well as transcriptions of them, were made available to the workshop participants roughly 4 months in advance of the workshop to allow the participants ample time for studying the videos and working through them in performing their analyses. The videos and transcripts are available on the workshop website: http://www.ics.uci.edu/design-workshop.

FIGURE 1 Typical camera view, taken from Adobe video.

WORKSHOP

The workshop featured 54 participants, mostly academics but also researchers at a number of companies and three of the designers who were featured in the videos. The researchers ranged from a broad variety of disciplines, the majority of them being in software engineering, with the others drawn from communities such as design studies, computer-supported cooperative work, human–computer interaction, cognitive science, and psychology.

The workshop spanned two-and-half days, and it was organized over 10 sessions (the full program is available on the workshop website: http://www.ics.uci.edu/design-workshop):

- Software design strategies (presentations)
- What did you not see? (break-out)
- Conceptualizing design (presentations)
- What should the communities present learn from each other? (break-out)
- Elements of software design (presentations)
- Implications for current tools and practices (break-out)
- Interaction (presentations)
- Education (break-out)
- Panel (discussants)
- Research agenda (break-out)

Each of the first 2 days concluded with a reflection session: three different discussants summarized what they observed, reflected upon what was discussed, and offered provoking thought challenges to the other participants. The six discussants (Fred Brooks, Mary Shaw, and Nigel Cross on day one, Clayton Lewis, Michael Jackson, and Barbara Tversky on day two) also served together as the final panel on the third day.

Overall, the workshop alternated between sets of 10-minute presentations by the participants and lively discussions in break-out groups of 6–10. Although the break-out sessions each had a theme, we took it as an indication of success that the discussions tended to go beyond the assigned topics. As a result, a large number of additional topics passed the revue, including cognition, representation, discourse, collaboration, requirements, problem analysis, assumptions, rationale, and design theory [28].

Two surprises emerged during the workshop. The first was that the three designers who were featured in the videos and attended the workshop ended up playing a very significant role. The three who were available to attend told us that they agreed to join out of curiosity, with a mild apprehension of having their every design move scrutinized by the researchers. As the workshop progressed, however, they became thoroughly engaged in the discussions, equal partners in developing an improved understanding of software design

FIGURE 2 Participant presenting with the wall of notes in the background. Cards on the left were already organized; those on the right were not.

practice. Indeed, because they had a unique perspective in being the designers whose work was being studied, their voices added significantly to the depth of the discussions and the agenda being pursued throughout the workshop.

The second surprise was the engagement with the Wall of notes (see Figure 2). We had provided all of the participants with index cards in order to keep track of their thoughts, commentaries, and proposals—as a means of documenting the discussion within the break-out groups. We invited the participants to share their notes by sticking them to a wall. Our previous experience with such a technique had had variable success, but at this workshop "the Wall" became a popular working resource and focus of conversation. Throughout the workshop, different participants contributed notes with their thoughts and observations to the Wall, while others spontaneously gathered for discussions around it, and some participants organized the content into groups of various kinds. We return to the Wall in the postscript of this book, because it provided the documentation for many of the observations that we make in that chapter.

THIS BOOK

On conclusion of the workshop, we set out to produce this book. To allow time for the participants to perform additional analysis, for the discussions of the workshop to influence the papers, and for others to perhaps analyze the videos and contribute, we issued a separate call for papers with a new deadline. Each paper we received in response to the call was peer-reviewed, with extensive comments provided to the authors. A subset of 18 of the revised papers was chosen for inclusion in the book. As a result of this process, each of the papers included represents a significantly updated and extended version of the corresponding white paper from the workshop.

The book is organized into four sections:

- *Design-in-theory.* In this first section, five chapters provide theoretical examinations of the nature of software design and the particular design problem posed. The chapters range from a detailed analysis of the essential design decisions that are embedded in the design problem, to an exposé of the full design space and how the solutions offered by the various design teams fit within this space but are actually quite different, to a discussion of how software design should be treated as a representational problem, to examining whether software design exhibits traits similar to other design disciplines through the lens of the function-behavior-system framework, to a meta-analysis that looks at the effects of the videos being based on an artificial prompt and setting rather than "in situ" design work.

- *Process of design.* The second section of the book discusses topics surrounding the process of software design; that is, the overarching approaches that the designers took in exploring the design space and articulating a solution. The chapters range from a comparison of software design to product design from the perspective of decision making, to a deconstruction and analysis of each of the sessions in terms of the high-level activities in which the designers engage, to a fine-grained look at the requirements and the design activities, to an examination of how ideas emerge and build upon one another, to a discussion of why current software engineering tools fail to support early design.
- *Elements of design.* The third section of the book focuses on several factors that cross-cut the design process. The chapters range from an analysis of the videos with respect to uncertainty and how it governs where the designers choose to focus throughout the design process, to the role of assumptions in early software design, to a close examination of the representations used by the designers through the cognitive dimensions of a notations framework, to how concerns play out and are considered throughout the process.
- *Human interaction.* In the final section of the book, four chapters examine the role of human interaction in the design process. The chapters range from a discussion of how designers must go "meta" and talk about design dimensions and evaluation criteria outside of making normal design progress, a look at disagreement and how it is accommodated and resolved, a new methodology for analyzing designers' conversations based on network analysis, and a very detailed examination of how designers interact with the design representations they use.

Each section of the book is introduced by an invited perspective. These perspectives, authored by Barbara Tversky, Jim Buckley and J. J. Collins, Mark D. Gross, and Alan Blackwell, not only reflect on each of the individual chapters that follow, but also place them in a much broader and often personal perspective on design at large.

Concluding the book is a postscript, which reflects on the chapters, the workshop, discussions, presentations, and the overall direction of the field. With the help of the Wall and other notes that were recorded during the workshop, it identifies a set of issues, each of which merits further research. One of those directions is especially important and was a persistent topic of discussion at the workshop: the collection of additional videos of designers in action on real software design problems. Despite the drawbacks and difficulties involved, the participants felt strongly that an attempt should be made to collect—and share—a meaningful collection of videos showing a series of design sessions over time. Design, after all, does not solely pertain to early design, but is practiced throughout software development. Having snapshots of design throughout the software life cycle would be a tremendous opportunity to advance the field.

We hope this book provides the inspiration for such an endeavor, as well as the many other endeavors that are necessary to truly understand, advance, support, and teach software design. We enjoyed organizing the workshop, participating in it, and putting this volume together. We hope the reader will appreciate and enjoy its many contributions as much as we do.

REFERENCES

1. N. Cross, H. Christiaans, and K. Dorst, *Analysing Design Activity*, Wiley, Chichester, 1996.
2. N. Cross, K. Dorst, and N. Roozenburg, *Research in Design Thinking*, Delft University Press, Delft, 1992.
3. A. Newell and H. Simon, *Human Problem Solving*, vol. 104, Prentice-Hall, Englewood Cliffs, NJ, 1972.
4. D. A. Schon, *The Reflective Practitioner: How Professionals Think in Action*, vol. 5126, Basic Books, New York, 1984.
5. N. Cross, *Designerly Ways of Knowing*, Birkhäuser, Basel, 2007.
6. B. Lawson, *How Designers Think: The Design Process Demystified*, Architectural Press, Boston, MA, 2006.
7. U. Dekel and J. D. Herbsleb, Notation and representation in collaborative object-oriented design: An observational study, *SIGPLAN Notices*, 42, 261–280, 2007.
8. M. Cherubini, G. Venolia, R. DeLine, and A. J. Ko, Let's go to the whiteboard: How and why software developers use drawings, in *Proceedings of the SIGCHI Conference on Human Factors in Computing Systems*, pp. 557–566, ACM Press, New York, 2007.
9. F. Brooks, *The Mythical Man-Month*, Addison-Wesley, Reading, MA, 1995.
10. A. Baker, A. van der Hoek, H. Ossher, and M. Petre, Guest editors' introduction: Studying professional software design, *IEEE Software*, 29, 28–33, 2012.
11. R. Guindon, Designing the design process: Exploiting opportunistic thoughts, *Human-Computer Interaction*, 5(2), 305–344, 1990.
12. B. Curtis, H. Krasner, and N. Iscoe, A field study of the software design process for large systems, *Communications of the ACM*, 31, 1268–1287, 1988.
13. R. Jeffries, A. A. Turner, P. G. Polson, and M. E. Atwood, The processes involved in designing software, in J. R. Anderson (ed.) *Cognitive Skills and Their Acquisition*, pp. 255, 283, Psychology Press, San Diego, CA, 1981.
14. M. Petre, Insights from expert software design practice, in *Proceedings of the 7th Joint Meeting of the European Software Engineering Conference and the ACM SIGSOFT Symposium on the Foundations of Software Engineering*, pp. 233–242, ACM Press, New York, 2009.
15. S. Sonnentag, Expertise in professional software design: A process study, *Journal of Applied Psychology*, 83, 703–715, 1998.
16. W. Visser, Designers' activities examined at three levels: Organization, strategies and problem-solving processes, *Knowledge-Based Systems*, 5, 92–104, 1992.
17. L. J. Ball and T. C. Ormerod, Structured and opportunistic processing in design: A critical discussion, *International Journal of Human-Computer Studies*, 43, 131–151, 1995.
18. C. Zannier, M. Chiasson, and F. Maurer, A model of design decision making based on empirical results of interviews with software designers, *Information and Software Technology*, 49, 637–653, 2007.

19. D. Wuest, N. Seyff, and M. Glinz, Flexible, lightweight requirements modeling with Flexisketch, in *Requirements Engineering Conference (RE), 2012 20th IEEE International*, pp. 323–324, IEEE, Piscataway, NJ, 2012.
20. N. Mangano and A. van der Hoek, The design and evaluation of a tool to support software designers at the whiteboard, *Automated Software Engineering*, 19(4), 381–421, 2012.
21. F. Geyer, J. Budzinski, and H. Reiterer, IdeaVis: A hybrid workspace and interactive visualization for paper-based collaborative sketching sessions, in *Proceedings of the 7th Nordic Conference on Human-Computer Interaction: Making Sense Through Design*, pp. 331–340, ACM Press, New York, 2012.
22. R. A. Pfister and M. J. Eppler, The benefits of sketching for knowledge management, *Journal of Knowledge Management*, 16, 372–382, 2012.
23. M. J. Eppler and R. Pfister, Sketching as a tool for knowledge management: An interdisciplinary literature review on its benefits, in S. Lindstaedt and M. Granitzer (eds) *Proceedings of the 11th International Conference on Knowledge Management and Knowledge Technologies*, pp. 1–6, ACM Press, New York, 2011.
24. J. Walny, S. Carpendale, N. Henry Riche, G. Venolia, and P. Fawcett, Visual thinking in action: Visualizations as used on whiteboards, *IEEE Transactions on Visualization and Computer Graphics*, 17, 2508–2517, 2011.
25. J. Walny, J. Haber, M. Dork, J. Sillito, and S. Carpendale, Follow that sketch: Lifecycles of diagrams and sketches in software development, in *Visualizing Software for Understanding and Analysis. IEEE International Workshop. 6th 2011. (VISSOFT 2011)*, pp. 1–8, IEEE, Piscataway, NJ, 2011.
26. H. Tang, Y. Lee, and J. Gero, Comparing collaborative co-located and distributed design processes in digital and traditional sketching environments: A protocol study using the function–behaviour–structure coding scheme, *Design Studies*, 32, 1–29, 2011.
27. R. N. Taylor and A. Van der Hoek, Software design and architecture the once and future focus of software engineering, in L. C. Briand (ed.) *Future of Software Engineering (Fose 2007)*, pp. 226–243, IEEE, Piscataway, NJ, 2007.
28. M. Petre, A. van der Hoek, and A. Baker, Studying professional software designers 2010: Introduction to the special issue, *Design Studies*, 31, 533–544, 2010.

I

Design in Theory

CHAPTER 1

Designing Designs, or Designs on Design

Barbara Tversky

Stanford University/Columbia Teachers College

CONTENTS

1.1 STUDYING SOFTWARE DESIGN *IN VIVO*

In 2009, computer scientists Andre van der Hoek, Alex Baker, and Marian Petre invited colleagues—some computer scientists, some not, some novices, some veterans—to participate in a collaborative research project. The starting point was three sets of videotaped design sessions in which small teams of experienced designers from three software companies were asked to design software to simulate traffic flow. The system was not meant to control traffic in the world, but rather to simulate controlling traffic in order to instruct students of software design about the complexities and subtleties of what might appear to be a simple everyday problem. Although the teams were experienced software designers, they were novices to this kind of design problem.

I was one of the outsiders invited to participate. However, three blocks prevented me from analyzing the videotapes. The first was the remnants of my rigid rigorous training in experimental psychology: in order to draw generalizable conclusions, research must have reasonably large samples and preferably manipulations and several different examples. The second block was technical. Trying to download and view the tapes delivered a severe concussion to my early model MacBook Air, resulting in permanent brain damage. The third block was resources. Even if the sacred principles of experimental psychology were foregone, coding the data would have required countless hours as well as reliability checks, and I had no one to help. I apologized, saying that I could not meet the terms of the workshop. To my good fortune, I was asked to attend anyway, and was delighted by the enormous enthusiasm of those who had undergone a shared arduous experience and had not only come out alive, but also with fresh insights that they were eager to share. Adding to the atmosphere was the surprise appearance of the designers, their own curiosity outweighing their apprehension at being examined under so many microscopes. Instead of analyzing the videotapes, I was asked to reflect on the analyses of others, specifically, those of Jackson, Lewis, Shaw, and Kan and Gero, as well as on the workshop in general. The latter first.

1.1.1 Some Overall Reflections

Given the complexity and layering of the design problem and the design process, it was inevitable that the perspectives of the design teams differed, and the perspectives of the researchers differed even more. What did their papers do? Many things. The task of the teams of authors was also a design task, to analyze these data, and the analyses produced were as varied as the teams of researchers. Not surprisingly, the designs and analyses derived in large part from the perspectives and experiences and expertise and theoretical predispositions and predilections of the teams of researchers. What was common, perhaps because these are researchers of design, was an attempt to use the particular to inform the general, and, in particular, to draw lessons for design practice.

Despite the enormous quantities of behaviors, some chose to analyze what was not there rather than what was there. At least one author (Matthews) called attention to unspoken assumptions. Assumptions are inherent in any discourse, wherever there is shared knowledge, and without shared knowledge, coherent discourse would be impossible. The greater the shared knowledge, the greater the unspoken assumptions, think of twin talk. Assumptions make communication more efficient, but they can lead to problems, for example, when the assumptions themselves are incorrect or not shared, despite beliefs to the contrary. One observer (Rooksby) noted many places where the teams showed awareness that they were participants in an experiment, and that this awareness affected their behavior, but perhaps only intermittently. At least one author (Petre) pointed out that design thinking is unobservable (anyway). The implication is that design thinking must be inferred from behavior, which cognitive psychologists possessively view as their territory.

Most chose to analyze the behavior that was there (rather than the behavior that was absent) in ways as varied as the backgrounds of the researchers. Some coded the data rigorously into predetermined theory-driven categories; others took a bottom-up impressionistic approach. Some focused on the task, others on the process, others on the interactions, and others on the modes of expression. Whether top-down or bottom-up, whether theory driven or bottom-up, the analyses were necessarily selective. Now to my more specific assignment.

1.2 ANALYZING AND REPRESENTING THE TASK

Jackson, Lewis, and Shaw focused on the task given to the designers, comparing that to the way that the designers analyzed the task. As they noted, the task had layers and layers of complexity, despite being a simplification of a real-world situation. Some examples: the layer of the structure of the roads; the layer of the operation of the signal controlling the flow of traffic; the layer of the traffic, its changing directions, pace, and quantity; the layers of the software that simulates all of these; and the layers of the user interface. All three noted failures of the design teams to adequately analyze the layers and the complexity of the task.

1.2.1 Importance of Analyzing Structure

Software and the systems that it embodies have structural components as well as operational components. Jackson noted the critical importance of thinking through the structure of the system to be simulated, put differently, of properly representing the structure. The structure of a system is the yin to the yang of the operation of the system. Often, the design desiderata are in terms of the operation of the system but the task entails designing structure, the, or a, structure that will yield the desired operation. Jackson's analysis revealed the many components of the structure of the problem as well as their operations and interrelationships, important features not always articulated by the design teams. He urged designers to take a far broader view of the design problem than even seems necessary in order to ensure that as many as possible of the important features be incorporated or at least considered in the design, a concern echoed by Lewis and Shaw.

1.2.2 Importance of Representing the Domain

Lewis observed that a simulation represents a domain, a representation that embodies the essential features of the domain but in simpler more accessible ways than the target domain. He lamented that rather than thoroughly analyzing and representing the target domain, the complex system of traffic, and the control of traffic, the designers delved straight into issues of implementation.

1.2.3 Importance of Considering the Design Space

As analyzed in detail by Shaw, representing a specific design task can be viewed as selecting a representation from a larger context of a space of possible designs. Considering a general design space can provide a general framework for a design problem, helping designers to keep the design goals in mind and to deliberate alternatives. Shaw noted that although the design teams reported that in their natural practice, they typically consider broader issues before focusing on a specific design, none of the design teams examined the general space explicitly; instead, they were quickly absorbed in the details of a specific design, often failing to consider alternative solutions and losing track of the design goals.

Narrowly or broadly, explicitly or implicitly, the designers needed to represent the design task. Representing the task entails abstracting the features and interactions that are essential. This is a key factor in design because the representation of the problem is a major determinant of its solution. The representation focuses the search for a solution. Jackson, Lewis, and Shaw all emphasized the importance of a considered and adequate representation of the problem. Each observed that the teams did not put efforts into adequately representing the task and pointed to difficulties in the designs and in the designing that they believe could have been avoided had the task been adequately represented. Jackson noted that each team assimilated the new problem to problems that were familiar to them, implying that their very expertise in a related domain prevented an adequate analysis of the new problem. Lewis went further, noting that "software designers not only may not have the necessary application domain knowledge, but also may not recognize the need for it." There are lessons here.

1.3 ANALYZING THE DESIGN PROCESS

Representing the task, even if it is done iteratively, is part of the design process. Some authors, such as Jackson, Lewis, and Shaw, analyzed the task to compare the ways that the designers analyzed and represented their task. Representation can be viewed as part of the structure of design. Others, such as Kan and Gero, sought to characterize the process of design. Characterizing the design process also entails representing a problem; in this case, selecting the categories that presumably best characterize the design process. Kan and Gero's categories—function, structure, behavior, and various subcategories of these—are meant to characterize the current focus of the designers' conversation. The categories are content-free, allowing comparison across vastly different design tasks. They also allow computing transition probabilities from one state to another and computing the frequencies of the categories across design sessions. Using this system, the authors found that, for the most part, the categories were fairly evenly distributed across the design process, except for a slight increase in problem representation at the beginning and a slight increase in evaluation at the end. On the one hand, the categorization scheme shows clearly that conversations in design were iterative, that aspects of function, behavior, and structure were considered and reconsidered throughout the design session, rather than in sequence as some prescriptive design advocates. On the other hand, the design sessions did progress and culminate in processes that were not captured by the coding scheme. Of course, there are many different ways to categorize design activities. It is possible that another coding scheme would highlight the course of the progression, and that such a coding scheme would help to illuminate when designing goes well and when it does not.

1.4 COGNITIVE TOOLS

1.4.1 Variety of Media

The modes of expression of the design teams and the design task were similar in complexity and layering. The designers worked at a whiteboard, and used words, prosody, gesture, diagrams, spatial position, and each other as they worked. These varied media served as design tools for the designers and data for the researchers. Just as designers need cognitive tools to externalize and sort their thoughts, researchers of design need cognitive tools to think about and analyze the data. One job of the tools is to reduce the data. Data reduction would be useless if it were random. Just as a telegram selects the words that carry the greatest meaning in a configuration that is likely to be informative, data reduction has to be systematic, to select the most meaningful portions and create the most revealing configurations. Reducing data entails taking a perspective on the data. In many cases, the perspectives were theory driven, categories and sometimes configurations that came from theoretical accounts of the problem, of design, of the tools. At one level, the design problem can be analyzed into its specific components, cars, roads, and signals. More generally, design can be viewed as a set of requirements and a set of means to fulfill them. Each of these perspectives provides a means for putting behaviors, especially language, into boxes, coding categories. The boxes of categories then need to be summarized and then configured, most typically by time or structure. The configured categories can be communicated most forcefully as diagrams, time lines, bar graphs, flowcharts, and networks. Thus, diagrams and sketches were used by the designers as well as by the researchers. To paraphrase one observer, designers create tools to create tools. The designers created diagrams on many levels, from the physical aspects of the problems to the controlling software. Like coding systems, diagrams reduce the information and structure it in ways that are meant to be productive for design.

1.4.2 Diagrams

Diagrams and sketches have numerous advantages over many other tools of thought, such as language, literal and metaphoric, gesture, and models (Tversky, 2011). They are quick to produce and easy to revise. They expand the mind by putting thought into the world, allowing the mind to act on objects in the world rather than objects in the mind and thereby augmenting limited-capacity memory and information processing. They are public and permanent, so they can be inspected, reinspected, and revised by a community. They extract, organize, and externalize thought so all can see, interpret, query, arrange, and rearrange. They use elements and relations in space to represent elements and relations that are literally or metaphorically spatial. They provide a spatial foundation for abstract inferences about distance, direction, similarity, category, hierarchy, height, size, color, shape, containment, symmetry, repetition, embedding, one-to-one correspondences, and more. They serve to check and ensure coherence, correctness, and completeness of design. They serve as a platform for inference and discovery. They facilitate collaboration by providing a shared platform for understanding and reasoning, as well as a shared product of the group, not of a single individual (Heiser et al., 2004). Designers have what has been called a conversation with their sketches (Schon, 1983), seeing new things in their own sketches, making unintended discoveries, and taking advantage of the ambiguity of sketches by reconfiguring them (Suwa et al., 2001; Tversky and Suwa, 2009; Tversky and Chou, 2010).

1.4.3 Gestures

Gestures serve some of the same ends as diagrams. Like diagrams, they externalize and express thought using space and actions in space to convey concepts and relations and actions that are literally or metaphorically spatial. Unlike diagrams, they are fleeting. But more than diagrams, they can convey actions, behavior, and function as well as structure. Problem solvers alone in a room gesture; their gestures correspond spatially to the problems that they are trying to solve. Remarkably, their gestures are instrumental to problem solving (Kessell and Tversky, 2006). Gestures serve those who observe them as well as those who make them. Specific gestures are key to conveying certain concepts. The structure of complex systems is relatively easy for learners to grasp in contrast to the action of complex systems, which is typically much more difficult for learners. In one experiment, two groups of participants heard the same verbal explanation of how an engine works. For one group, the explanation was accompanied by gestures that conveyed the actions of the engine; for the other group, the explanation was accompanied by gestures that conveyed the structure of the engine. Those who saw the gestures conveying the actions answered correctly more questions about the action of the engine and expressed more action in their diagrams of the system as well as in their verbal and gestural explanations of the system (Kang et al., 2012). Thus, gestures promote thinking both in those who make them and in those who view them. Like diagrams, they map thought directly to space, rather than arbitrarily to symbolic words. Indeed, several researchers pointed to the various roles that gestures served in the design interactions.

1.5 WHERE DO INNOVATIONS COME FROM?

1.5.1 Radical Breaks

Studying design is descriptive, and design is a fascinating process to watch. Yet, implicit in studying how designers design is the desire to improve how designers design, a prescriptive goal. Were there any lessons here? Much design is incremental, reasonably steady progress toward a set of goals that are increasingly refined as the design progresses. But there is also extraordinary design that is not incremental, what Jonathan Edelman calls a radical break: a seemingly sudden innovation that dramatically changes the design (Edelman, 2011). Edelman gave some 20 experienced teams an artifact to redesign. Most of the sessions were incremental, but there were radical breaks in a few. Radical breaks occurred when someone had an insight with a dramatically new idea. The single change in perspective led to a cascade of new ideas by all members of the team. The cascade of ideas was accompanied by new metaphors and by large gestures, many enacting the using of the artifact or its behavior.

How can radical breaks be encouraged? Perhaps a key is in the word "break." One pervasive and persistent bias of thought is the confirmation bias, the tendency to look for evidence that supports the hypothesis at hand (Wason, 1968). The same for design. Once we have an idea, we persist in finding support for that idea. Not only do we stop searching for other solutions, we may also fixate on the current solution, making it even harder to create alternatives. Could a contrarian strategy work better? For her dissertation, the artist and art educator Andrea Kantrowitz has been studying successful professional artists for whom drawing is a central part of their practice. She interviews them as they draw, as they make unintended discoveries in their own sketches. Many report a particularly productive strategy that they adopt: deliberately getting themselves in trouble in order to challenge themselves—in a safe place, breaking what they have done, violating old habits and routines to create something new, fresh, and innovative.

1.6 IN SUM

Design is everywhere, and often the same. A bit of the individual, a bit of the group; a bit of the requirements, a bit of the interpretations; a bit of the past, a bit of the future. The contributions here are triple portraits. They are of the design. They are of research on design. They are of the perspectives of the researchers. I offer this paraphrase of words attributed to Edwin Schlossberg, designer of experiences and information spaces: "The skill of designing [he was supposed to have said, 'writing'] is creating contexts in which other people can think."

ACKNOWLEDGMENTS

I am grateful for the opportunity to participate in the workshop, for the encouragement and support of Andre van den Hoek, Alex Baker, and Marian Petre, who so gracefully conducted the workshop and the aftermath, a far larger project than they could have imagined, and to the lively participants at the workshop for their stimulating and provocative talks, interchanges, and papers. I am indebted to Marian Petre for expert editing, and to Andre van den Hoek for expert guidance on this manuscript. I am also grateful for the support of National Science Foundation HHC 0905417, IIS-0725223, IIS-0855995, and REC 0440103.

REFERENCES

Edelman, J. (2011). Understanding radical breaks: Media and behavior in small teams engaged in redesign scenarios. PhD dissertation. Stanford University.

Heiser, J., Tversky, B., and Silverman, M. (2004). Sketches for and from collaboration. In J. S. Gero, B. Tversky, and T. Knight (Eds), *Visual and Spatial Reasoning in Design III*, pp. 69–78. Sydney: Key Centre for Design Research.

Kang, S., Tversky, B., and Black, J. B. (2012). From hands to minds: How gestures promote action understanding. In N. Miyake, D. Peebles, and R. P. Cooper (Eds), *Proceedings of the 34th Annual Conference of the Cognitive Science Society*, pp. 551–557. Austin, TX: Cognitive Science Society.

Kessell, A. M. and Tversky, B. (2006). Using gestures and diagrams to think and talk about insight problems. In R. Sun and N. Miyake (Eds), *Proceedings of the 28th Meeting of the Cognitive Science Society.* Mahwah, NJ: Lawrence Erlbaum Associates.

Schon, D. A. (1983). *The Reflective Practitioner.* New York: Harper Collins.

Suwa, M., Tversky, B., Gero, J., and Purcell, T. (2001). Regrouping parts of an external representation as a source of insight. *Proceedings of the 3rd International Conference on Cognitive Science*, pp. 692–696. Beijing: Press of University of Science and Technology of China.

Tversky, B. (2011). Visualizations of thought. *Topics in Cognitive Science*, 3, 499–535.

Tversky, B. and Chou, J. Y. (2010). Creativity: Depth and breadth. *Design Creativity*, 2011, 209–214.

Tversky, B. and Suwa, M. (2009). Thinking with sketches. In A. B. Markman and K. L. Wood (Eds), *Tools for Innovation*, pp. 75–84. Oxford: Oxford University Press.

Wason, P. C. (1968). Reasoning about a rule. *Quarterly Journal of Experimental Psychology*, 20, 273–281.

CHAPTER 2

Representing Structure in a Software System Design*

Michael Jackson

The Open University

CONTENTS

2.1 INTRODUCTION

Introducing a digital computer into a system brings an unprecedented level of behavioral complexity. Anyone who has ever written and tested a small program knows that software, even when apparently simple, can exhibit complex and surprising behaviors. To deal effectively with this complexity, the program text must be structured to allow the programmer to understand clearly how the program will behave in execution (Dijkstra 1968). That is, the program's behavior structure must be clearly represented by its textual structure.

In a system of realistic size, a further level of complexity is due to the combination and interaction of many features. For example, a modern mobile phone may provide the

* Reprinted from *Design Studies* 31 (6), Jackson, M., Representing structure in a software system design, 545–566, Copyright 2010, with permission from Elsevier.

functions of a telephone, a camera, a web browser, an e-mail client, a global positioning system (GPS) navigation aid, and several more. These features can interact in many ways—a photograph taken by the camera can be transmitted in a telephone call—and the resulting complexity demands its own clearly understood structuring if the features interactions are to be useful and reliable. The functional structures created and understood by the designer must be intelligibly related to the parallel structures perceived by the users of the system. It is failure in this user-interface structuring that explains the difficulty experienced by many users of digital television recorders, and the multitude of puzzled customers asking on Internet forums how to set the clock on their digitally controlled ovens.

Yet another demand for clear structure comes from the world outside the computer, where the problem to be solved by the system is located. For some systems, including critical and embedded systems, the essential purpose of the system is to monitor and control this *problem world*: the purpose of an avionics system is to control an airplane in flight, and the purpose of a cruise control system is to control the speed of a car. The designers of such a system must study, structure, analyze, and clearly depict the properties and behaviors of the problem world where the software execution will take its effect; they must respect and exploit those properties and behaviors in designing the software to produce the effects required.

Structure, then, is of paramount importance in software design: understanding, capturing, analyzing, creating, and representing structure furnish the central theme in this chapter. The chapter arises from the Studying Professional Software Design Workshop (SPSD 2010) sponsored by the U.S. National Science Foundation in February 2010. The workshop brought together a multidisciplinary group of participants, all with an interest in the theory and practice of design in its many manifestations. The workshop focused on previously distributed video and transcription records of design sessions, each of 1 or 2 hours, in which three teams of software professionals worked separately on a given system design problem. The introduction to the special issue of *Design Studies* in which this chapter appears gives a fuller account of the design problem itself and of the capture of the design teams' activities. The time allotted for the work was very short, and none of the teams could be said to have addressed the problem effectively. However, many interesting issues arose in the design sessions and were discussed in the workshop.

In this chapter, these issues are considered both at the level of software development generally and in the specific context of the given problem, focusing on the discovery, invention, and representation of structure in a few of its many dimensions and perspectives. The chapter progresses broadly from the task of understanding the initial requirement in Section 2.2 toward the design of the eventual software system in Section 2.8. Section 2.3 considers the designer's view of the reality to be simulated—in this case, a road traffic signaling scheme—and Section 2.4 discusses the simulation of this reality by objects within a computer program. Sections 2.5 through 2.7 explore some general structural concerns as they arose in the problem and its solution. Section 2.9 discusses some broader aspects of the design case study, and comments critically on the current software development practices and notations. Section 2.10 concludes the chapter by offering some more general observations relevant to design practice in other fields.

2.2 UNDERSTANDING THE REQUIREMENT

The problem, outlined in a two-page *design prompt*, is to design a system to simulate traffic flow in a road network. The sponsor or client is Professor E, a teacher of civil engineering, and the system's intended users are her civil engineering students. The system's purpose is to help the students to achieve some understanding of the effects on traffic flow of different schemes of traffic signal timings at road intersections. The design teams were required to focus on two main design aspects: the users' interactions with the system and the basic structure of the software program.

This traffic simulation problem is not a critical or embedded system: it is no part of the system's purpose to monitor or control traffic in the real world. Nevertheless, the system has a close relationship to the real world because it is required to simulate the behavior of real traffic under different regimes of traffic signaling at road intersections. An obviously important criterion of a successful design is that the simulation should be sufficiently faithful to the behavior of real traffic.

The core function of the system is to support the creation and execution of traffic simulations. Each executed simulation has the following basic elements:

- A *map*: a road layout containing several four-way intersections, all with traffic lights
- A *scheme*: a traffic signaling scheme defining the sequence and the duration of the light phases at each intersection
- A *load*: a traffic load on the system imposed by the statistical distributions over time of cars entering and leaving the layout

The student users will specify maps, schemes, and loads; they will use them to run simulations and observe their results on a visual display. They must be able to alter the scheme or the load and see how the change affects the traffic flow. By using the system in this way, the students are expected to obtain a broad intuitive understanding that the practical engineering of traffic signal timing is a challenging subject, and to gain at least a rough feeling for the variety of effects that the scheme can have. As Professor E points out, "This can be a very subtle matter: changing the timing at a single intersection by a couple of seconds can have far-reaching effects on the traffic in the surrounding areas. Many students find this topic quite abstract, and the system must allow them to gain understanding by experimenting with different schemes and seeing firsthand some of the patterns that govern the subject." The understanding they gain will prepare and motivate them for the systematic mathematical treatment in their civil engineering course on the subject of traffic management.

Understanding the requirement—that is, the problem that the software is intended to solve—is a crucial task in developing software (Jackson 2001). Kokotovich (2008) cites the investigation by Matthias (1993) of novice and expert designers in several fields. Matthias found that expert designers treat problem analysis as a fundamental early design activity, but novices typically neglect it. Major misunderstandings of the system's requirement were detectable both in some of the recorded design sessions and in some of the subsequent workshop discussions.

One design team was—temporarily—misled by the mention in the design prompt of an "existing software package that provides relevant mathematical functionality such as statistical distributions, random number generators, and queuing theory." They assumed that the purpose of the software package mentioned was to support a *mathematical analysis* of the traffic flow. The system would take a specified map, scheme, and load, and derive a system of equations; by solving these equations, the system would determine performance measures such as waiting times at intersections, average speeds over each road segment, the expected number of cars held up at each intersection, and so on. These results would be displayed on a computer screen; when the user changes the map, scheme, or load, the effect of these changes on the performance measures would be recalculated and made visible on the screen.

In fact, the design prompt did not call for such a mathematical analysis: instead, it called for a *simulation* of the traffic flow. The system would contain program objects corresponding to the parts of the real-world traffic and its environment: vehicles, roads, and signals. The properties and individual behaviors of these program objects would be specified by the user in the map, scheme, and load. The execution of the system would cause these program objects to behave and interact in ways closely analogous to their real-world counterparts. The resulting traffic flows would then emerge from the interactions of the program objects just as real-world traffic flows emerge from the individual behaviors and properties of the vehicles, roads, and signals. As the simulation progresses, the flows would be displayed as a pictorial representation of cars moving along roads, the display changing over time just as real traffic flows do.

A related misunderstanding of the requirement was the assumption that the simulation should produce not only a visual representation of the light settings and the resulting traffic flow, but also numerical outputs such as average journey times. In fact, the design prompt document says nothing about numerical outputs, but emphasizes visual representation throughout.

In one workshop discussion, it was argued that the purpose of the system was to help the student users to find the optimal signaling scheme for a given map and a given load. This, too, was a misconception. The purpose of the software system, clearly described in the design prompt, was to help the students to develop an informal intuitive understanding by running and observing clearly visualized simulations. Finding optimal schemes would be the goal of the formal mathematical treatment that the students would encounter later, in their academic course. The system would prepare the students for this formal mathematical treatment by introducing them to the problem and stimulating their interest by animating some practical examples of the complexities it can exhibit.

2.3 ROAD LAYOUT AND TRAFFIC IN THE REAL WORLD

Simulations are necessarily simplifications, and the design prompt directs the system designer to consider only a simplified area of a real-world road layout. Many commonly found features are eliminated. There are no one-way roads and no T-junctions; all intersections are four-way and all are controlled by traffic lights; there are no overpasses and no giratories. The part of the road layout to be considered is bounded; it is quite small, containing only a few—"six, if not more"—intersections.

Even with these layout constraints there are variations that the system must be able to handle, and some that the designer may be free to ignore. In considering the design choices here, it is important to bear in mind the purpose of the system, and to consult Professor E whenever the consequences of a design choice are obscure. For example, the simplest road arrangement that satisfies the layout constraints is a rectangular grid; for many reasons the system design will be easier if more complex layouts are ignored. However, as Figure 2.1 shows, the layout constraints given in the design prompt are also satisfied by a rectangular grid with added diagonal roads. Whether this complication must be handled depends on a question of traffic management theory. All cycles in the rectangular grid are quadrilaterals, while the complex layout also has triangular cycles. Do these triangular cycles give rise to unique characteristic variations in the traffic flow effects that the students are expected to learn about from the system? This is a question that Professor E must answer for the designer.

Another design choice concerns the way cars enter and leave the traffic flows in the layout. Since there are no one-way roads, cars can enter and leave at any point at which a road crosses the map boundary. If the map contains parking lots, it is also possible for cars to enter or leave the traffic flow at any point at which a parking lot is connected to a road. This possibility seems to introduce a large complication. The system will need to keep track of the number of cars in each parking lot, and the points at which cars enter the traffic flow on leaving the lot will constitute intersections of a new and different kind, requiring special and potentially complex treatment. This additional complication seems to add little or nothing to the value of the system for its proclaimed purpose, and all the teams wisely ignored the possibility.

The real-world traffic signal schemes that must be considered are less constrained. One or more lanes at each intersection may be equipped with sensors: the traffic signal scheme can then take account of the presence or absence of vehicles in a lane, avoiding prioritization of directions in which there is no traffic. The design prompt does not stipulate whether a right turn on red is permitted. It states that the system must "accommodate left-hand turns protected by left-hand green arrow lights," but it is unclear whether unprotected left turns are permitted anywhere. It is also unclear whether "lane" has its usual real-world meaning in which a wide road may have several lanes in each direction, or instead denotes one of the two possible travel directions along a stretch of road. A "variety of sequences and timing schemes" should be accommodated, but the variety is left to the designer to choose.

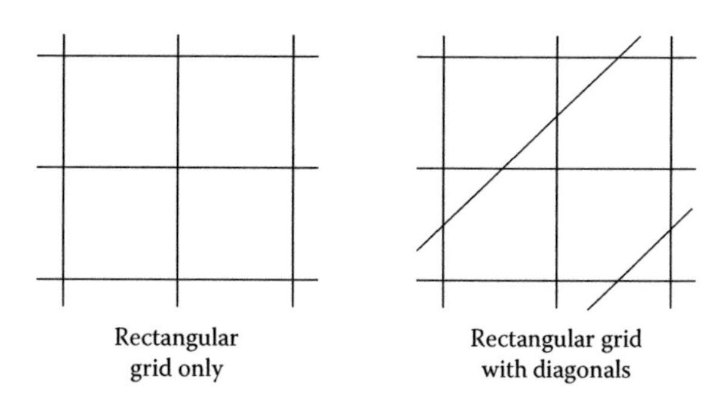

FIGURE 2.1 Possible road layouts.

The design prompt does not mention large trucks or emergency vehicles, and individual vehicle behavior is implicitly limited to the normal lawful behavior of passenger cars. The design prompt says nothing about speed limits, and the designer may perhaps assume that a uniform speed limit applies everywhere. Cars do not break down or collide; their drivers always obey the traffic signals and never block an intersection. No road segments or lanes are closed for repairs.

2.4 SIMULATION OF REAL-WORLD TRAFFIC BY PROGRAMMING OBJECTS

The physical elements of a real-world traffic system and their relationships are shown in Figure 2.2. The diagram shows only the physical elements of the real-world traffic, represented by the symbols, and the interfaces at which they interact, represented by the connecting lines. It does not represent such abstract things as signaling schemes or the drivers' planned journeys through the layout.

Vehicle sensors and signal units are permanently located (c and d) at positions in the road layout. The vehicle drivers see and obey (e) the traffic signals; they move into road space only if it is not occupied (f) by another car; and their movements affect (g) the states of any vehicle sensor they encounter. The sequencing and timing of the signal lights are controlled (a) by one or more signal controllers, which read and take account of (b) the states of any relevant vehicle sensors. It is the behavior of such real-world traffic systems that is to be simulated by an executable computer program.

All the design teams assumed—realistically, in today's software development culture—that the executable simulation was to be programmed in an object-oriented language such as Java. So a significant part of the design task is to decide what programming objects should participate in the simulation and how they should interact. In effect, the properties and behaviors of these objects when the simulation is executed will constitute the designer's chosen model of the properties and behaviors of the simulated reality.

Starting from the depiction in Figure 2.2, it seems initially that the necessary objects are: one for each car; one for each signal unit; one for each sensor; one for the road layout; and one for each signal controller. However, while it is reasonably clear that one object is

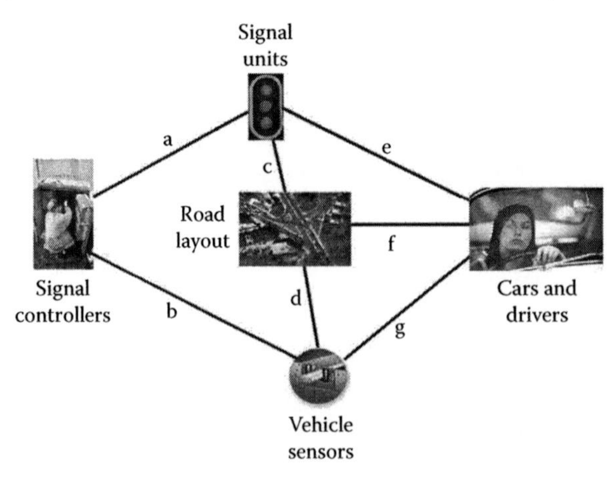

FIGURE 2.2 Real-world traffic elements and interactions.

needed to represent each car and one for each sensor, it is not yet clear how many individual objects will be needed to represent the road layout, the signal units, and the signal controllers.

The road layout is made up of roads that cross each other at intersections, so we may initially expect to have one object for each road and one for each intersection. These objects must be associated so that each intersection object indicates which roads cross and each road indicates the sequence of intersections encountered along the road. The signal units might be represented in several different ways. For example, one object for each lamp, one for each set of lamps physically contained in the same casing, or one object for all the casings located at one intersection. The signal controllers too might be represented in several different ways, ranging from one controller for the whole layout to one for each set of lamps or even for each lamp.

We will return later to these more detailed concerns about the numbers of objects of the various classes. Broader concerns demand attention first. They reflect potent sources of confusion and difficulty that were dramatically obvious in some of the recorded design sessions.

2.5 INITIATIVE AND CAUSALITY

The five element classes shown in Figure 2.2 interact to give the overall behavior of the real-world traffic. Where does the initiative for this behavior come from? Which elements cause the behavior, directly or indirectly, and how?

Initiative in the real world comes from only two sources: the signal controllers and the car drivers. First, the signal controllers. They take the initiative, according to their designed control regime and, perhaps, the states of the vehicle sensors, in changing the visible states of the lamps in the signal units. Even if there were no cars on the roads, the signals would, in general, go through their sequences as commanded by the controllers. Second, the car drivers. Each driver is engaged in a purposeful journey from one place to another, driving their car in order to progress along a chosen route at chosen speeds. Progress is constrained by traffic laws that limit the car's permissible road positions and speeds, by the need to avoid a collision with other cars, and by obedience to the signals.

There are no other sources of initiative. The road layout, signal units, and sensors all change their states as a result of the initiatives taken by the signal controllers and the drivers, but they are entirely passive: they take no initiative on their own account. A part of the road may or may not be currently occupied by a car, but the road itself can neither import nor expel a car. A sensor may be on or off, but it can neither permit nor block a car's movement. Up to an acceptable approximation, the executed simulation must reflect this behavior faithfully. It will be harder to achieve a sufficient degree of fidelity if the simulation design deviates gratuitously from the behavior of the real-world traffic.

Setting aside the internal object structures of the signal controllers, the signal units, and the road layout, Figure 2.3 shows an initial software structure for the simulation. The solid rectangles and the solid lines connecting them directly represent the software objects and interactions that model the real-world traffic elements shown in Figure 2.2. However, these five object classes are not enough to run the simulation, and additional objects are needed.

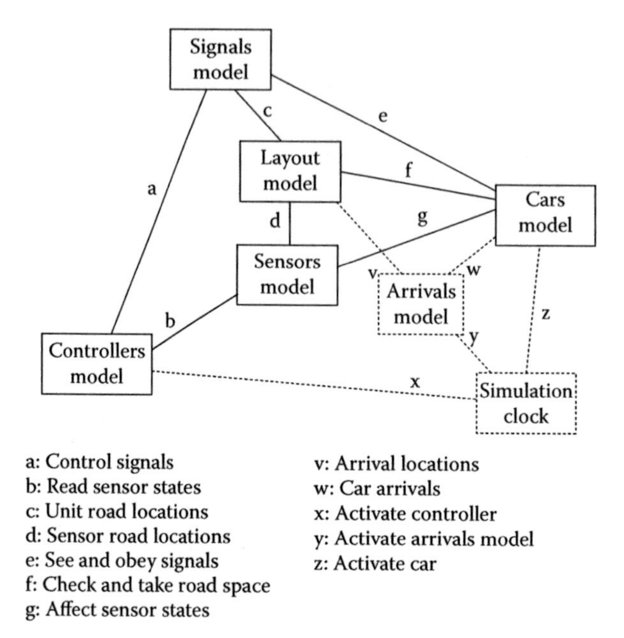

FIGURE 2.3 Synchronous simulation elements and interactions.

The *arrivals model* object, represented by the smaller dotted rectangle, will certainly be needed. Because the road layout is bounded, it is necessary to consider the arrivals and departures of cars: cars arriving from outside the boundary will appear to the simulation to be newly created, and on departure they will appear to have been destroyed. The software function of creating a car object cannot be a function of the car itself, so it is necessary to introduce one or more objects to create cars. These objects are represented by the arrivals model object in Figure 2.3. Like the signals model, arrivals are multiple objects, one located (v) at each entry point to the layout and creating (w) new cars model objects there, according to statistical distributions representing the frequency of arrivals at that point.

The *simulation clock* object, represented by the larger dotted rectangle, may or may not be needed, depending on a technical design choice. Given the arrivals objects, the simulation could be implemented by a *single-threaded* or a *multithreaded* design. In a multithreaded design, each independently active object has its own *thread of control*: that is, it behaves as if it were running on an independent processor, concurrently with all other objects, and synchronizing its behavior with the other objects only when it interacts with them. The simulation depicted in Figure 2.3 could then be implemented by the five solid-line objects with the sole addition of the arrivals object. Independent control threads would be provided for each signal controller model, each arrivals model, and each newly created car. This would be an *asynchronous* design for running the simulation. The passage of time—for example, to distribute arrivals over time, to delimit signal phases, or to maintain a car's desired constant speed along a road segment—would be represented by *wait* statements in the programmed behaviors of the arrivals objects, signal controllers, and cars, respectively.

The concurrent behaviors in a multithreaded design offer many opportunities for subtle programming errors that are hard to avoid, diagnose, and eradicate: their effects are difficult to reproduce in program testing. There may also be efficiency penalties in execution. For these reasons, a single-threaded *synchronous* design may be preferred, in which only one object can be executing at a time. The single-threaded design uses the simulation clock object represented by the larger dotted rectangle. This clock object would be the only independently active object in the simulation. It would emit (x, y, and z) periodic ticks to activate the controller objects, arrivals objects, and car objects, each one progressing in its simulated lifetime as far as it can progress in the short time associated with the tick. One controller, for example, might count down one tick while it waits for the next signal change to become due, while another makes a signal change that became due immediately after the preceding tick. One arrivals object might make a randomly determined choice to create a new car, while another makes a randomly determined choice to do nothing. One car might continue to wait at a stop signal while another makes a unit of progress along a road.

The key point in such a design is that the simulation clock does nothing except mark the progress of time in units of one tick, for each tick briefly activating each arrival, car, and controller object. It would be a fundamental confusion, therefore, to think of the simulation clock as a kind of master controller, like a global policeman determining the behavior of every individual part of the simulation. What each controller, arrival, or car object does in each tick should be determined purely by its own current state, its defined behavior, and its interactions with neighboring objects. The simulation clock is merely a technical software implementation mechanism to represent the passage of global time while saving the designer from the well-known programming pitfalls of asynchrony and concurrency in a multithreaded system.

2.6 CAUSALITY AND DETERMINATION

The most notable confusion about initiative and causality that was displayed in the recorded design sessions centered on the movement of cars in the simulation. In one session, a designer suggested that when a signal changes to green, the intersection "asks its roads to feed certain numbers of left-turn, right-turn and straight-ahead cars." This surprising suggestion almost exactly reverses the direction of causality in the real-world traffic: it makes the intersection and the road into active objects, while the cars become passive objects, moved by the road's action, which, in turn, is caused by the intersection.

A related confusion about the movement of cars concerns the path to be taken by each car when it encounters an intersection. This confusion permeated many of the recorded design sessions, and also appeared in discussion sessions in some workshop groups. The confused idea was to regard each incoming road at each intersection as having the property that a certain proportion of the cars arriving at the intersection would turn left, a certain proportion would turn right, and the remainder would continue straight ahead. The confusion here is quite subtle. It is, of course, true that empirical observations of the traffic in a real-world road system will allow these proportions to be measured and recorded. Traffic engineers do make such measurements and use them to determine efficient signal timings. For example, if most northbound and southbound vehicles at a particular intersection continue straight ahead or turn right, it would obviously be inefficient to provide a lengthy phase for protected left turns. The confusion lies in treating these proportions as a *determining cause* of the vehicles' behaviors, when in fact they are an *emergent effect* of the routes that the drivers are following. Treating the effect as if it were a cause will give rise to senseless journeys by fragmenting each journey into a series of unrelated choices at successive intersections: in effect, each journey becomes a random walk through the road layout. If real-world traffic patterns are to be simulated with a reasonable degree of fidelity, this confusion must be avoided.

A deeper philosophical point is exposed here about object-oriented programming. It was not by chance that the first usable object-oriented programming language (Dahl et al. 1970, 1972) was named Simula: it was a language designed for the programming of discrete behavioral simulations. Objects in Simula possessed defined behaviors, independent but interacting, and the goal of the simulation was to reveal the properties that emerged from their interaction. To an important extent, the original purpose of such a language is perverted when it is used in design to proceed in the reverse direction—starting from the given required global behaviors and properties of a system to derive the interacting behaviors and properties of the objects from which these global properties are required to emerge. In such a design process, the system is decomposed into objects with independent behaviors and properties: but it is then hard to ensure that the global results that emerge will be what are required.

For designing the traffic simulation considered here, an object-oriented programming language is an excellent choice: the signal controllers and the cars have given behaviors, and the system's explicit purpose is to reveal the emergent properties of the traffic flow.

2.7 DESCRIPTION SPAN

The problem of describing the behavior of individual cars reveals another fundamental and ubiquitous structural concern. In designing, describing, or analyzing any subject of interest, it is necessary to decide on the *span* of the description. The span of a description (Jackson 1995) is the extent to which it describes its subject matter, the extent being defined in time, in space, or in the size of a subset. For example, a map of Europe has a smaller span than a map of the world; a history of the Second World War has a smaller span than a history of the twentieth century; a line in a text has a smaller span than a paragraph; a day has a smaller span than a year; and a study of automobiles has a smaller span than a study of road vehicles. The penalty for choosing too small a span for the purpose in hand is that the subject matter of interest is then described only by a number of fragments, and obtaining an understanding of the whole from these fragments is difficult and error-prone. In software development, many unnecessary difficulties arise from the choice of an insufficient description span.

The initial focus of two of the three teams provided an early illustration of the consequences of an ill-chosen span: they began their discussion of the problem by considering just one intersection. This probably seemed natural because it is primarily at intersections that traffic control is exercised, and certainly there are aspects of the problem for which one intersection is the appropriate span for a discussion and a description. However, after considering one intersection, both teams eventually turned their attention to the question: How can departures from one intersection be related to arrivals at a neighboring intersection? The focus on individual intersections now became a serious disadvantage: the rules were clearly going to become very complicated. It took a little time before the teams realized that they must consider the question in the larger span of a whole road or even the whole layout.

On applying the notion of span to describe the behavior of an individual car as it encounters successive intersections, it becomes immediately clear that the appropriate span is the car's complete purposeful journey across the bounded layout. Nothing else can adequately represent the behavior of real cars and drivers. (Of course, in the real world there are exceptions such as learner drivers practising left turns, joyriding teenagers, and mechanics road testing a repaired car: the purpose of their journeys is not to reach a destination. Such exceptions can be ignored in the simulation. They may be assumed to be relatively rare, and since these drivers have no desired destination, their efficient and convenient movement is not a matter of significant concern to traffic engineers.) The turning behavior of a car should therefore be determined by the route assigned to it when it was created by the arrivals model—this route, like the car's creation itself, is drawn from some statistical distribution defined by the users. Such a design corresponds well to the reality that traffic loads commonly result from such purposeful journeys as commuting between home and work, children's school runs, shopping expeditions, and visits to entertainment centers.

Another example of a span concern is the design of the object structure for the signal controllers. In an earlier section, it was suggested that the signal controllers might be represented in several different ways, ranging from one controller for the whole layout to one for each set of lamps or even for each lamp. The choice must be governed by the span of control to be exercised. Evidently, all the signals at one intersection must be brought under unified control. Without such unified control, it would be very hard to ensure, for example, that the signals are never simultaneously green for conflicting flows, and the software to control the signals at a single intersection would become a complex distributed system in its own right. Similarly, if the signals at the intersections along one main road are to be synchronized to present a steady uninterrupted sequence of green signals to cars travelling just below the maximum permitted speed, then some signal control must span the whole road. Signal control objects at different spans will be necessary, interacting in a structure of multiple objects corresponding in total to a single signal control for the whole layout.

Regardless of the synchronization of signals at multiple intersections, the representation of signal states and sequences at a single intersection raises its own issue of a description span. As the design prompt stipulates, "combinations of individual signals that would result in crashes should not be allowed." Intuitively, this means that only certain combinations of traffic flows can be permitted at any one time, and the allowed signal settings must correspond to these combinations only. For example, "left-turning northbound traffic may not be permitted in combination with straight-ahead southbound traffic" or "east-west traffic may not be permitted in combination with north-south traffic." Each permissible combination can be captured by enumerating the associated states of individual signals, and one of the design teams explored representations of signaling sequences in exactly these terms, using lines of colored dots or stripes to represent the states of individual signals.

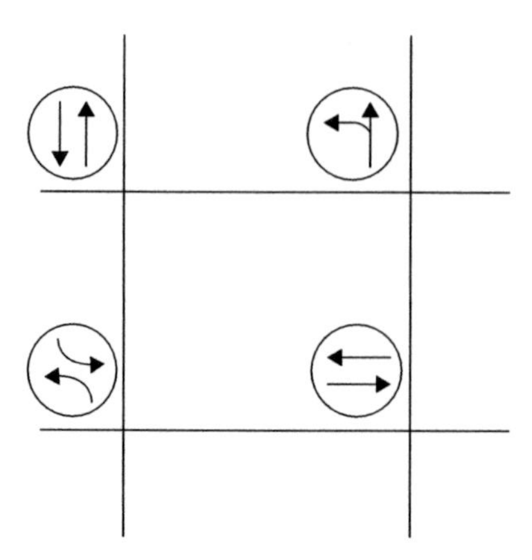

FIGURE 2.4 Representing signal states at intersections.

A description of a larger span for each combination at each intersection could be expressed in terms of permitted traffic flows rather than of individual signals, leaving the individual lamp settings in each case to be implied by the permitted flows. This span leads to a more compact, direct, and convenient representation by symbols in a style of the kind shown in Figure 2.4. Their meanings are immediately obvious, and the complete set of permissible symbols—not all of which are shown in Figure 2.4—directly represents the set of permissible combinations of signal states. The symbols can be used to good effect in design to represent signal sequencing and timing at each intersection, as shown in a state machine diagram in Figure 2.5.

Further symbols can represent transitional phases in which all lights are red, and no flow is allowed. State machines of this kind could play a useful role in the users' specification of signaling schemes. The symbols could also be used in the visualization of the traffic light states, as suggested by Figure 2.4. Although the users' own driving experience has accustomed them to recognizing configurations of colored lights as they approach them along the road at ground level, the suggested symbols are more immediately understandable when seen, as it were, from the bird's eye view, which is more appropriate to users of the system watching a simulation on a screen. Each symbol directly captures in a single glyph all the currently permitted and forbidden traffic flows at the intersection.

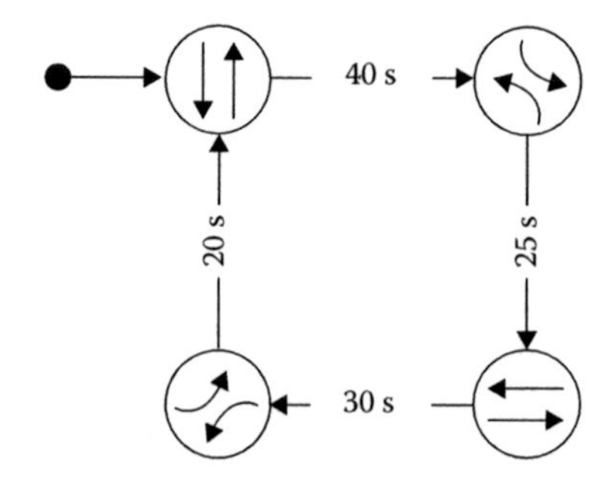

FIGURE 2.5 Representing signal sequencing and timing.

As a general rule, a designer who is in doubt should choose a larger rather than a smaller description span. The consequence of choosing an unnecessarily large span is small: the subject matter naturally fragments itself into mutually independent parts among which no relevant relationships or interactions are found. The consequence of an inappropriately small span is large: it is gratuitous complexity with all its attendant difficulties.

2.8 FORMING SIMULATIONS

The student users must define, run, and observe many simulations if they are to acquire the desired intuitive understanding of traffic control. To stimulate this understanding, how should these simulations be formed? What variations should be supported? How should the various simulations be related to each other? As briefly discussed in Section 2.2, a single simulation run requires a road layout map, a signaling scheme, and a traffic load. The users must be able to save maps, schemes, and loads for later use in repeated or varied simulation runs, and recombine them to observe the effects of different combinations.

One way of structuring the basic elements of each simulation into parts that can be usefully combined is shown in Figure 2.6:

- The *simulation infrastructure* part is common to all simulations executed by the system, and consists only of the clock object.
- The layout *map* part consists of the layout model and the sensors objects.
- The signaling *scheme* part consists of the controllers and the signals objects.
- The traffic *load* part consists of the arrivals and the cars objects.

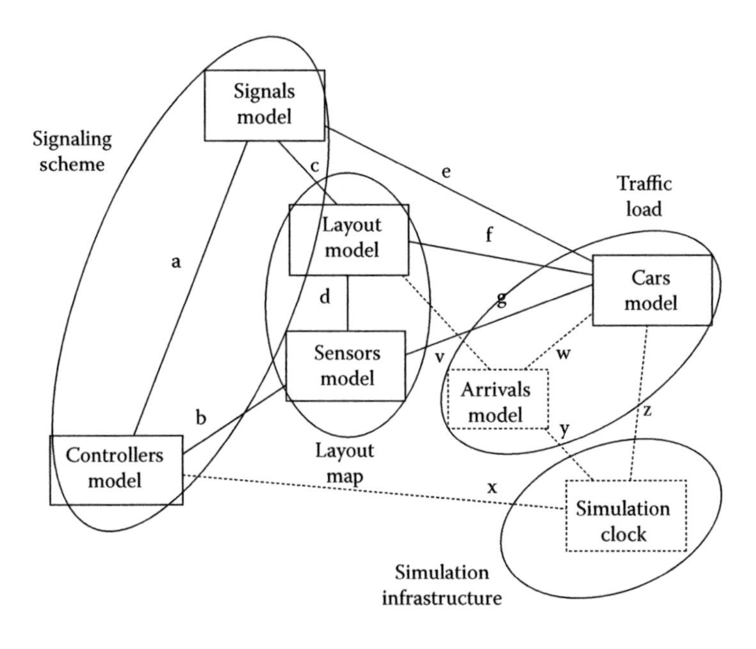

FIGURE 2.6 Simulation parts for combinations.

The objects belonging to the traffic load part are positioned (v, f) at specific locations in the layout map; the signaling scheme objects, too, are positioned (a, b) at specific locations. Each load and scheme can therefore be combined only with maps that provide exactly those locations.

Careful software design will allow the relationships (e) and (g) to be ignored for purposes of combining simulation parts. Intuitively, the signals that a real-world driver can see and a simulated driver must obey (e) are determined by the driver's location (f) in the layout. In the simulation, therefore, the cars model objects need not communicate directly with the signals model objects. This freedom will make it easier to exploit the treatment of the signaling states proposed in Section 2.7 and illustrated in Figure 2.4. In the same way, the cars model objects need not communicate directly with the sensors model objects in the simulation. A traffic load part is therefore compatible with any map providing the positions required by the relationships (a, b, v, f), regardless of the presence or absence of sensors at those layout positions.

The ease with which simulation parts can be defined and combined to form simulations will play a large part in the convenience and attraction of the system. As one of the teams expressed it, users should be able "to create something without fussing with it too much." There are many design concerns to be examined here, several of them focused on the dynamics of the user interaction at more than one level. For example, defining a signal control scheme will be easier if the system provides what one team called "cyclable defaults." Default local schemes are defined, each with the span of one intersection, and these can then be reproduced at as many additional intersections as desired, the user switching between different defaults for different groups of selected intersections. At a larger span, simulation parts can be conveniently defined by modification in the context of a particular simulation. Having run a simulation and observed its effects, the user can modify the scheme or load, defining and storing a new scheme or load for immediate use in the next simulation. In this way, both the definition of the parts and the comparison of their combined effects can be easy and convenient.

2.9 SOFTWARE SYSTEM DESIGN

Discrete complexity should be a major determinant of design decisions for software systems. The designer must address the inevitable complexity arising from nonnegotiable requirements, and must also aim to minimize additional system complexity by avoiding it wherever reasonably possible. The traffic simulation problem demands decisions where avoidance of complexity may be crucial.

One example is the definition of routes for individual cars. A possible way of defining a route that starts at a given entry point to the map is simply to specify its exit point. The simplest route has its entry and exit at opposite ends of the same road, and no turns are necessary. Slightly less simple is a route that leaves by a road perpendicular to its entry. At least one turn is necessary; if there are intersections at which a left turn is altogether prohibited, then more turns may be necessary. Even less simple is a route that leaves by a road parallel to its entry, allowing more than one choice of turns for crossing from the entry to the exit road. Several design questions arise. Is the car's route fully predetermined when the car is created on arrival, or does the cars model object execute an algorithm to choose appropriate intersections as turning points? In either case, how can routes be easily worked out if the map is not a simple rectangular grid? In a map in which some left turns are altogether prohibited, may there not be some pairs of entry and exit points between which there is no possible route? These issues must be considered—and addressed or circumvented—to avoid a growth of complexity that could produce an unreliable system, break the project's resource budget, or both.

An important example of circumventing difficult issues concerns the meaning of "lane" in the map. In real-world traffic the meaning is clear. Wide roads have two or more lanes in each direction. At intersections, cars are often funneled into lanes dedicated to straight-ahead travel or to left or right turns. The signals at an intersection may have a phase of the kind represented by the lower left symbol in Figure 2.4, in which the left-turning traffic lanes in opposite directions proceed simultaneously while traffic in all other lanes must wait. A driver intending to negotiate such an intersection must be careful to arrive in the appropriate lane. Changing lanes between intersections is an unavoidable maneuver for many journeys, and is often hard to achieve in dense traffic: it may involve an interaction between the driver changing lanes and another who courteously gives way. This real-world behavior is hard to formalize and therefore hard to simulate.

The design question, then, is whether this aspect of traffic movement can be neglected, and, if not, how it is to be simulated. One possible option is to circumvent the difficulty altogether, by supposing each road to have only one lane in each direction. Each lane at an intersection necessarily contains both those cars intending to turn left and those intending to turn right or proceed straight ahead. To prevent cars with different turning intentions from blocking each other, signal phases of the kind represented at the lower left in Figure 2.4 must never occur at any intersection: left turns are permitted only in phases of the kind represented by the symbol in the top right of Figure 2.4. This design choice ensures that straight-ahead travel is never permitted when left turns are forbidden, and vice versa. The result is that in every signaling phase at an intersection, either every vehicle in a lane can proceed or none can proceed. Whether this option is an acceptable solution to the difficulty depends on how far it damages the fidelity of the simulation. Professor E must be consulted once again.

Software system design activity is subject to characteristic distortions of technique. One distortion arises from the widespread tendency to focus attention on the software artifact itself, most often narrowly conceived as an object-oriented computer program. The problem world and the users are then relegated to a secondary position in which they attract no explicit attention but are considered—if at all—only implicitly and indirectly. This distortion was conspicuous in some of the recorded design sessions. It was especially notable that the teams rarely, if ever, discussed the behavior of real-world traffic: for the most part, if any of the designers thought about it explicitly they did so only dimly through the distorting lens of programming objects.

To a large extent, this failure to pay sufficient attention to the problem world is engendered and encouraged by a confusion between things in the problem world and the programming objects that serve to model them. In theory, the software developer should consider these two subjects separately, and should also consider explicitly the respects in which each has properties and behaviors that are not reflected by the other. This is a counsel of perfection: in practice, most developers consider only the programming objects directly. The unified modeling language (UML) notations commonly used by most software developers were derived from notations for fragmented descriptions of object-oriented program code. In many respects, they are ill-suited to describe the problem world, but most software developers have few others in their toolbox.

The most important single factor in design is relevant previous experience. The three design teams were drawn from companies whose products are very different from the kind of system the teams were asked to design. The inadequacies in their design work were attributable in part to the very short allotted time for the design work, and in part to the artificial conditions of the design sessions; but the teams' evident unfamiliarity with this kind of simulation played a very large part indeed. In the term used by Vincenti (1993), they were engaged in *radical design*:

> In radical design, how the device should be arranged or even how it works is largely unknown. The designer has never seen such a device before and has no presumption of success. The problem is to design something that will function well enough to warrant further development.

Much—too much—software design is radical design in this sense. One team recognized their situation quite explicitly. They saw their design as a prototype at best. Later versions would have improved features motivated by a better understanding of the requirements to be acquired from users' experience of the system and from continuing discussion with the customer, Professor E. This recognition is more profound than it may first appear. Dependable success in designing software systems can only be a product of specialized experience and the attendant evolution of a discipline of *normal design*. In normal design, in Vincenti's words:

> ... the engineer knows at the outset how the device in question works, what are its customary features, and that, if properly designed along such lines, it has a good likelihood of accomplishing the desired task.

In software development, as in the established engineering disciplines, the existence of firmly established normal design is the essential prerequisite of dependable success. Many software developers hanker after the excitement of daring innovation when their customer would prefer the comforts of reliable competence based on experience.

2.10 SOME GENERAL OBSERVATIONS

A realistic software system has more complexity than any other kind of design artifact. The complexity arises from the essential purpose of the software: to automate the multiple system features, and their interactions with each other and with the heterogeneous parts of the problem world. An established normal design for artifacts of a particular class embodies a design community's evolved knowledge of the structures by which these complex interactions can be brought under intellectual control.

Although the highest degrees of discrete complexity are uniquely associated with software systems, the complexity of interactions more generally seems to be a common theme in many design fields. Certainly, this kind of complexity in architecture was recognized by Alexander (1979):

> It is possible to make buildings by stringing together patterns, in a rather loose way. A building made like this is an assembly of patterns. It is not dense. It is not profound. But it is also possible to put patterns together in such a way that many patterns overlap in the same physical space: the building is very dense; it has many meanings captured in a small space; and through this density it becomes profound.

Alexander writes here of the aesthetic qualities of density and profundity, but this kind of composition—"many meanings captured in a small space"—is also of the greatest practical importance in engineering. The introduction of the tubeless tire was a major advance in the design of road vehicles. Previously, a tire consisted of an outer cover providing the interface with the road surface and an inner inflatable tube to absorb shocks; the tubeless tire combined the two in one component providing both functions. Another example from automobile engineering is the unitary body, acting both as the framework to which the engine, transmission, driving controls, suspension, and running gear are attached, and as the coachwork providing seating and weather protection for the driver and passengers.

Even in quite simple design tasks, this need to combine different functions in one unified design—though not necessarily in one physical component—plays a major part. A teapot must be easily filled; it must keep the tea hot, and hold it without leaking; it must stand stably on a flat horizontal surface; it must be easy to pick up by the handle and must pour without dripping; the handle must not become too hot to hold comfortably; the teapot must be easy to carry and easy to clean; and it must have an attractive appearance. Each function considered in isolation is easily achieved. The virtue of a classic teapot design—that is, a normal design—is to combine all of these functions in a way that reliably serves the needs of a significant class of teapot users.

In any design task there must be some combination of radical and normal design. An entirely normal design demands no design work whatsoever: it is merely an exact copy of an existing design. Realistic design work can never be completely normal: even a modest difference in the scale of an artifact can reveal otherwise unsuspected functional factors. In typography, the 8-point version of a typeface is not simply a copy of the 12-point version reduced in size, and the teapot designer should not expect the 0.75 L version to be an exact copy of the 1 L teapot. At the other end of the spectrum that runs from radical to normal, there can be no entirely radical design. A designer must bring some relevant experience and knowledge to the task: this is necessarily experience and knowledge of existing designs of similar artifacts. An essential aspect of tackling any design problem and its difficulties is to relate it to similar problems already known.

The traffic simulation problem was of a kind unfamiliar to all three design teams. All three quite rightly tried immediately to assimilate the problem to something they already knew. Two teams took the view that the problem was an instance of the model-view-controller

(MVC) software pattern; the third identified the problem as "like a drawing program." Unfortunately, these hasty classifications were inadequate; but in every case they were accepted uncritically and never explicitly questioned. By locking themselves into these first assumptions, the teams determined the direction, perspective, and content of most of their subsequent work. Interestingly, although the MVC software pattern is used in drawing programs, the focus of the third team's work was very different from the first two. Perhaps motivated by the word "drawing," the third team focused on the user interface for specifying the road map and the signaling scheme, while the first two teams focused on the control of execution during the simulation. Prior knowledge and experience were immediately translated into preconceptions about the design problem and its solution. No team devoted a balanced amount of time and effort to the two design aspects called for by the design prompt: the user interface and the software structure.

This readiness to fix preconceptions is strengthened by a designer's focus on the artifact that seems to be the obviously direct product of the design work. In software design this means focusing on the software itself—the program texts and the program execution within the computer—at the expense of the problem world and the customer's requirements. In design more generally, it means focusing on the object being directly designed rather than on the experience and affordances that it will provide to its users in the environment for which it is intended. The result is likely to be very unsatisfactory. Norman (1988) presents many classic cases, including the ubiquitous designs for doors that offer no clue as to whether they should be opened by pushing, pulling, or sliding.

The true product of design is not the designed artifact itself: the true product is the change the artifact brings about in the world, and especially the experience it eventually provides to its users. A fundamental factor for design success is therefore a good understanding of what is expected from this experience. Conveying this understanding is a crucial purpose of a statement of requirements, but it may be difficult to convey this understanding by a written document alone. The design prompt document for the traffic simulation system was skillfully written; yet every team misunderstood some important aspect of the system's intended purpose. Unsurprisingly, the team that focused on the user interface reached a better understanding than the two teams that focused on the internal mechanisms for executing the simulation. Focusing on the user interface led them naturally to think about how the users would experience the system, and to consider how to make that experience more convenient.

A good understanding of the system's purpose should inform the design choices, especially those concerned with modeling the road layout and traffic behavior. The designer must balance costs and benefits. A more elaborate model of the road layout and traffic behavior would allow simulations to correspond more closely to the complexities of real-world traffic. So each complication—diagonal roads, vehicle breakdowns, parking lots, complex car routings, and a realistic treatment of left-turn lanes—brings a benefit. But it also brings many costs. Designing and programming the software will be more expensive. The program text will be larger, and the additional complexities may damage the system by increasing the likelihood of execution failures. More elaborate simulations will also demand more elaborate specification of the maps, schemes, and loads, making the system harder to learn and more cumbersome to use. The system's clearly intended purpose is to convey an intuitive understanding: as the design prompt says, "This program is not meant to be an exact, scientific simulation." Most of the additional complications, then, offer little or no benefit to set against their costs. Adapting the words of Antoine de Saint-Exupery, the design of this system will be complete, not when there is nothing left to add, but when there is nothing left to take away.

ACKNOWLEDGMENTS

The workshop, Studying Professional Software Design (SPSD 2010), was partially supported by NSF grant CCF-0845840; this support included the travel costs of the author of this chapter. The three design teams, from Adobe, AmberPoint, and Intuit, provided the indispensable initial material for the workshop by working on the design problem under intense time pressure and in the glare of a video camera. They generously permitted the records of their efforts to be transcribed, and viewed and analyzed by the workshop participants before the workshop began. This chapter has benefited from comments on an earlier version by the three reviewers and by the editors of the special issue of *Design Studies.*

DISCLAIMER

This is the author's version of a work that was accepted for publication in *Design Studies.* Changes resulting from the publishing process, such as peer review, editing, corrections, structural formatting, and other quality control mechanisms, may not be reflected in this document. Changes may have been made to this work since it was submitted for publication. A definitive version was subsequently published in *Design Studies* DOI 10.1016/j.destud.2010.09.002.

REFERENCES

Alexander, C. (1979) *The Timeless Way of Building*; Oxford University Press, New York.

Dahl, O-J, Dijkstra, E.W., and Hoare, C.A.R. (1972) *Structured Programming*; Academic Press, London.

Dahl, O-J, Myhrhaug, B., and Nygaard, K. (1970) The Simula67 common base language; Technical Report, Oslo, Norway.

Dijkstra, E.W. (1968) Go To statement considered harmful. Letter to the editor. *Communications of ACM*, 11 (3), 147–148.

Jackson, M. (1995) *Software Requirements & Specifications: A Lexicon of Practice, Principles, and Prejudices*; Addison-Wesley, New York.

Jackson, M. (2001) *Problem Frames: Analysing and Structuring Software Development Problems*; Addison-Wesley, Boston, MA.

Kokotovich, V. (2008) Problem analysis and thinking tools. *Design Studies*, 29 (1), 49–69.

Mathias, J.R. (1993) A study of the problem solving strategies used by expert and novice designers: An empirical study of non-hierarchical mind mapping; PhD thesis, University of Aston, Birmingham.

Norman, D.A. (1988) *The Psychology of Everyday Things*; Basic Books, New York.

SPSD (2010) International Workshop Studying Professional Software Design, February 8–10, University of California, Irvine, CA; http://www.ics.uci.edu/design-workshop.

Vincenti, W.G. (1993) *What Engineers Know and How They Know It: Analytical Studies from Aeronautical History*; The Johns Hopkins University Press, Baltimore, MD.

CHAPTER 3

Role of Design Spaces in Guiding a Software Design*

Mary Shaw

Carnegie Mellon University

CONTENTS

A CENTRAL TASK IN DESIGN is deciding what artifact will best satisfy the client's needs; this might involve creating a new artifact or choosing from among existing alternatives. A design space identifies and organizes the decisions to be made about the artifact, together with the alternatives for those decisions, thereby providing guidance for refining the design or a framework for comparing alternative designs.

3.1 DESIGN SPACES

The *design space* in which a designer seeks to solve a problem is the set of decisions to be made about the designed artifact together with the alternative choices for these decisions.

* This chapter is a major extension and revision of "The Role of Design Spaces," which appeared in *IEEE Software* in January, 2012 [14].

A *representation of a design space* is one of the explicit textual or graphical forms in which a particular design space—or one of its subspaces—may be rendered.

Intuitively, a design space is a discrete Cartesian space in which the dimensions correspond to the design decisions, the values on each dimension are the choices for the corresponding decision, and points in the space correspond to complete design solutions. A design method provides a strategy for locating good solution points or regions of the space that are likely to be rich in good solutions.

In practice, most interesting design spaces are too rich to represent in their entirety, so representations of a design space display dimensions corresponding to the properties of principal interest. Also in practice, the design dimensions are not independent, so choosing some alternative for one decision may preclude alternatives for other decisions or make them irrelevant. For example, if displaying a value is optional, then all decisions about the display format are irrelevant if the value is not displayed; if multiple values are to be displayed as a graph, then usually all should use the same units. As a result, it is convenient to represent portions of the design space as trees, despite the disadvantage of implying an order in which decisions should be made.

Often—and in most practical problems at scale—the design space is not completely known in advance. In these cases, the elaboration of the space proceeds hand in hand with the design process. Simon [1] addresses the difference between the two cases in Chapter 5, where he treats the task of selection from a fixed space as enumeration and optimization, and he treats the task of searching an unknown or open-ended space as search and satisficing. These cases align (very roughly) with routine and innovative design.

In this interpretation, the design space is very concrete. This stands in contrast to a common, much vaguer, usage in which "design space" refers loosely to the domain knowledge about the problem or perhaps to all decisions in a design activity, be they about the problem analysis, the designed artifact, or the process of producing the design. Indeed, I have encountered designers in other domains who resist this interpretation because it seems to constrain their design explorations with a rigid a priori, inflexible structure. To the contrary, I see it as a way to organize decisions, to guard against neglecting to consider alternatives, and as a base on which to add new considerations as they arise in the design.

The representation of a design space for a particular task is usually a slice of the complete design space that captures the important properties required of the artifact. By organizing the design decisions, a representation of a design space helps the designer to systematically consider the relevant alternatives. This supports the design of a new artifact (the *design* task), including the identification of interactions and trade-offs among decisions. It supports the selection of an artifact from among those already available (the *selection* task). It also provides a way to compare similar products (the *comparison* task), by highlighting the differences between designs and by allowing systematic matching to the needs of the problem at hand. Naturally, a good representation for a particular problem should emphasize the decisions with the greatest impact on the desired properties of the solution.

Design spaces can inoculate the designer against the temptation to use the first practical alternative that comes to mind. For example, in studying software architectures, I repeatedly saw a tendency for developers to use a familiar system structure (possibly one supported by a familiar tool) instead of analyzing the problem to select an appropriate structure. Evidently, software developers were often oblivious even to the existence of alternatives—that is, they were *defaulting* into familiar structures instead of *designing* suitable ones.

Design spaces have been used in computer science to organize knowledge about families of designs or systems since at least 1971 [2]. They have been used, among other things, to describe computer architecture [2,3], user input devices [4], user interface (UI) implementation structures [5], software architectural styles [6], distributed sensors [7], and typeface design [8]. The exploration of design spaces to find suitable designs, often by searching, provides a model of designer action in design studies [9]. In the early 1990s, Xerox used design spaces to capture design rationale in human–computer interaction (HCI) [10,11]; their representation of the design space was developed incrementally as a coproduct of design, and it captured not only questions (dimensions) and options (alternatives) but also criteria for making a decision.

3.2 REPRESENTING DESIGN SPACES

Representations for design spaces can be textual or graphical, and they can organize the information in different ways. I illustrate these options with a small design space for personal information sharing. This is only a small slice of the entire design space, selected to illustrate several different representations.

The three dimensions of this small example, each with two possible values, are

- *Activation:* Is the communication driven by the sender pushing the information to the reader or by the reader pulling the communication?
- *Privacy:* Is the communication private to a small set of known parties, or is it public?
- *Authorship:* Is the information authored by a single person or by an open-ended group?

Figure 3.1 depicts this example as a three-dimensional space with one sample artifact at each point in the space. Since the space is completely populated, this representation is well suited to a selection task—that is, choosing a point in the space by choosing values on the three dimensions. For example, if the design requires private broadcasting from a single sender, the appropriate point in the space is <Sender push, Private, Solo author>. For example, an e-mail is pushed by the sender to the mailbox of the reader, it is authored solely by the sender (any quoted material is selected by the sender), and it is private to the sender and the named recipients.

Of course, the points in this space may be occupied by more than one application. For example, instant messaging also lies at the <Sender push, Private, Solo> point. Selection among multiple items at a point depends on other dimensions (e.g., delivery latency or persistence). It is also possible, indeed common, for some points to be unoccupied. This might happen because the combinations of choices for the unpopulated point do not make sense, or it might indicate an opportunity for new products.

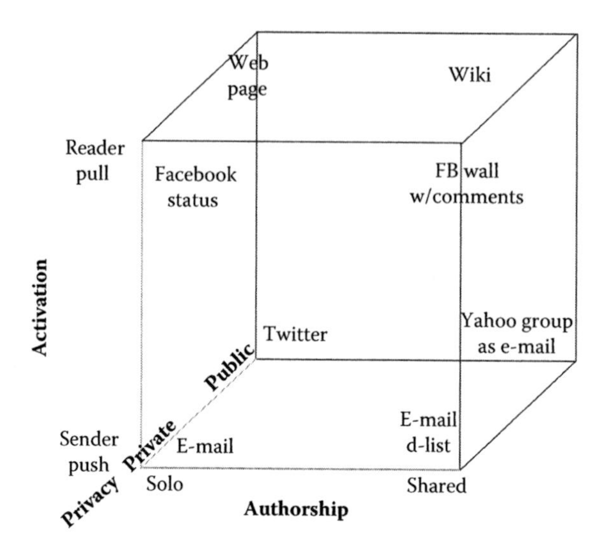

FIGURE 3.1 Design space for information sharing.

Sketching multidimensional spaces graphically clearly does not scale well to more than three dimensions. This same small space can be represented in other ways. Figure 3.2 uses a tabular form with rows corresponding to the points in the space and columns corresponding to the dimensions; it has the shortcoming that the points on each dimension are represented only implicitly, in the values in the body of the table. The design space can be represented in this format only to the extent that it is populated with a full range of examples. This representation is well suited to the comparison task, because the available options correspond to rows, and the columns facilitate a comparison.

This example can also be represented with an emphasis on the dimensions and their values. Figure 3.3 shows one such form. Following Brooks [12], the tree has two kinds of branches: choice and substructure. Choice branches, flagged with "##," are the actual design decisions; usually one option should be chosen (though a system with multiple operating modes might support multiple options). Substructure branches (not flagged) group independent decisions about the design; usually all of these should be explored.

Instance	Activation	Privacy	Authorship
Web page	Reader pull	Public	Solo
Wiki	Reader pull	Public	Shared
Facebook status	Reader pull	Private	Solo
Facebook wall w/comments	Reader pull	Private	Shared
Twitter	Sender push	Public	Solo
Yahoo group as e-mail	Sender push	Public	Shared
E-mail	Sender push	Private	Solo
E-mail d-list	Sender push	Private	Shared

FIGURE 3.2 Instance-oriented representation of the design space of Figure 3.1.

```
Activation
|   ## sender push      [e-mail]
|   ## reader pull      [wiki] [web]
|   ## interactive
Privacy
|   ## private          [e-mail]
|   ## known group
|   ## login controlled
|   ## public           [wiki] [web]
Authorship (edit or append rights)
|   ## solo             [e-mail] [web]
|   ## shared           [wiki]
```

FIGURE 3.3 Dimension-oriented representation of the design space of Figure 3.1.

This representation has the advantage of showing alternatives without relying on examples (three new possibilities are added here), and it handles hierarchical descriptions well. It is thus well suited to the design task, which requires decisions that correspond to traversing the tree. Its disadvantage is that a point in the space is represented diffusely by tagging all relevant values. This is illustrated here by placing e-mail, wiki, and (static) web page in this representation; the figure would get very busy if all the instances were included. This representation can, however, describe unoccupied points. This example has only one hierarchical level, but the format admits of a deeper structure, and indeed the traffic signal simulation space is much richer.

Naturally, if other properties are of interest, a different design space would be appropriate to the problem. For example, the representations in Figures 3.1 through 3.3 address the way that information flows between users. If the properties of interest were about the representation and the storage of content, the additional dimensions of interest might be <Persistence, Locus of State, Latency, Distribution Span, Content Type>. This would lead to an extended design space such as that shown in Figure 3.4. Although this representation contains more dimensions and therefore covers more design choices, it does not exhaust the possibilities, so it is still a slice of the total design space.

3.3 A DESIGN SPACE FOR TRAFFIC SIGNAL SIMULATION

To relate design spaces more closely to practice, I turn to a problem of a more realistic size, drawn from the National Science Foundation (NSF)-sponsored workshop Studying Professional Software Design [13]. In preparation for this workshop, three two-person teams were taped as they worked for 1–2 hours on a design problem. The tapes and transcripts of these sessions were analyzed by a multidisciplinary group of design researchers, who met for 3 days to discuss the sessions.

The design task was a simulator of traffic flow in a street network, to be used by civil engineering students to appreciate the subtlety of traffic light timing. The full text of the task statement (the "prompt") is given in Appendix I.

This chapter expands and revises the author's *IEEE Software* overview of design spaces [14], principally by adding extensive commentary on the transcripts that were the subject of that workshop. This commentary presents a representation of the design space implicit

```
Activation
|   ## sender push
|   ## reader pull
|   ## interactive
|   ## sender-push & reader-pull
Privacy
|   ## private
|   ## known group
|   ## login controlled
|   ## public within application
|   ## public
Authorship (edit or append rights)
|   ## originator
|   ## originator plus comments
|   ## all present
|   ## anyone
Persistence
|   ## none
|   ## hours to days
|   ## months
|   ## archival
Locus of state
|   ## none
|   ## per user
|   ## server
|   ## cloud
Latency
|   ## interactive
|   ## seconds
|   ## minutes
|   ## hours to days
Distribution Span
|   ## K known recipients
|   ## N anonymous recipients
Content Type
|   ## short text
|   ## text
|   ## text + attachments
|   ## structured text
|   ## images
|   ## videos
|   ## web page
```

FIGURE 3.4 Representation of the expanded design space of Figure 3.3.

in the transcripts and discusses the ways that considering the design space helped, failed to help, and might have helped the designers.

Figure 3.5 presents a representation of the design space implicit in the transcripts. None of the teams explicitly considered a design space; therefore, to develop the content of Figure 3.5, I studied the videos and transcripts of the design sessions to identify the principal conceptual

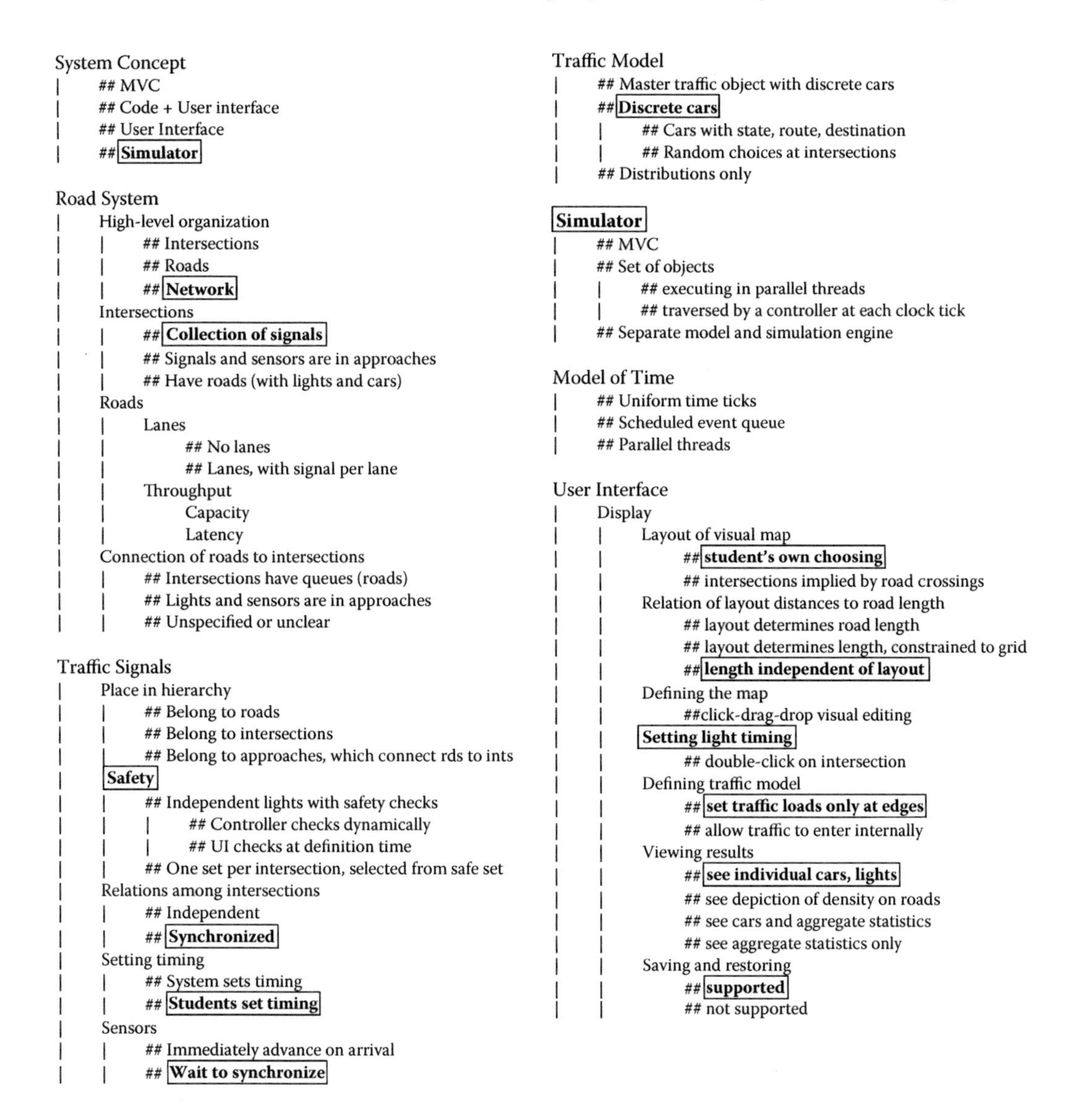

System Concept
MVC
Code + User interface
User Interface
Simulator
Road System
High-level organization
Intersections
Roads
Network
Intersections
Collection of signals
Signals and sensors are in approaches
Have roads (with lights and cars)
Roads
Lanes
No lanes
Lanes, with signal per lane
Throughput
Capacity
Latency
Connection of roads to intersections
Intersections have queues (roads)
Lights and sensors are in approaches
Unspecified or unclear
Traffic Signals
Place in hierarchy
Belong to roads
Belong to intersections
Belong to approaches, which connect rds to ints
Safety
Independent lights with safety checks
Controller checks dynamically
UI checks at definition time
One set per intersection, selected from safe set
Relations among intersections
Independent
Synchronized
Setting timing
System sets timing
Students set timing
Sensors
Immediately advance on arrival
Wait to synchronize
Traffic Model
Master traffic object with discrete cars
Discrete cars
Cars with state, route, destination
Random choices at intersections
Distributions only
Simulator
MVC
Set of objects
executing in parallel threads
traversed by a controller at each clock tick
Separate model and simulation engine
Model of Time
Uniform time ticks
Scheduled event queue
Parallel threads
User Interface
Display
Layout of visual map
student's own choosing
intersections implied by road crossings
Relation of layout distances to road length
layout determines road length
layout determines length, constrained to grid
length independent of layout
Defining the map
##click-drag-drop visual editing
Setting light timing
double-click on intersection
Defining traffic model
set traffic loads only at edges
allow traffic to enter internally
Viewing results
see individual cars, lights
see depiction of density on roads
see cars and aggregate statistics
see aggregate statistics only
Saving and restoring
supported
not supported

FIGURE 3.5 Representation of the design space for traffic signal simulation. Decisions implied by the prompt are in boxed boldface.

entities that the teams included in their designs, along with any alternatives they considered. In some cases, the teams did not explicitly mention alternatives, and I inferred the conceptual entity from the team's direct selection of an implementation element.

Trying to be generally faithful to the (often informal, implicit) structure that emerged from the design discussions, I identified a set of principal dimensions to organize the alternatives and annotated these with details from the discussions. I also reviewed the task prompt, noting choices that were implied by the prompt itself. In some cases, I added fairly obvious alternatives that did not otherwise appear in the transcripts.

In addition, I subsequently examined the demonstration version of a commercial traffic simulation tool [15]. This tool has clearly received more design and development effort than the workshop exercise, and it is a professional tool rather than a classroom simulator. Nevertheless, it is informative to see where this tool lies in the design space.

The resulting representation of the design space is incomplete in two important ways. First, the alternatives identified in the representation do not exhaust the possibilities. It is easy to imagine other alternatives for many of the decisions. Indeed, the commercial traffic simulation tool [15] presents many such possibilities. Nevertheless, the representation of Figure 3.5 provides a uniform framework for comparing the designs studied by the workshop and for reflecting on ways that considering the design space explicitly might have helped the designers.

Second, this representation captures only the larger-grained and (apparently) most significant design decisions, and it does not cover the full scope of the task. Omitted, for example, are the characterization of the traffic entering and leaving at the edges of the map, the handling of left turns, and analytics. Nevertheless, the representation is evidently detailed enough to draw out interesting observations and distinctions; adding another level of detail would increase its complexity without the assurance of commensurate benefit.

Finally, this design space is expressed in terms of the solution space, not of the problem space. Thus "road" means the abstraction of an implementation construct (which might have representations of lanes, capacities, and so on) that is intended to capture the aspects of real-world roads that are relevant to the simulation. It is important to bear this in mind, because the teams' discussions usually did not distinguish clearly between the problem space and the solution space.

Figure 3.5 shows a representation of the design space implied by the prompt and the transcripts. It has seven principal dimensions, each of which has further substructures:

- System Concept
- Road System
- Traffic Signals
- Traffic Model
- Simulator
- Model of Time
- User Interface

The elaboration of each of these dimensions is hierarchical. This provides a uniform basis for comparing the designs studied by the workshop.

The prompt clearly implies certain design decisions. For example, it says "Students must be able to describe the behavior of the traffic lights at each of the intersections," which quite clearly indicates that students should set and vary the timing of the signals. These implied decisions are flagged with **bold** boxed text in Figures 3.5 and 3.6.

The three teams were from Adobe, Intuit, and AmberPoint. Figure 3.6 overlays on Figure 3.5 the decisions apparently made by these three teams, denoted AD, IN, and MB, respectively.

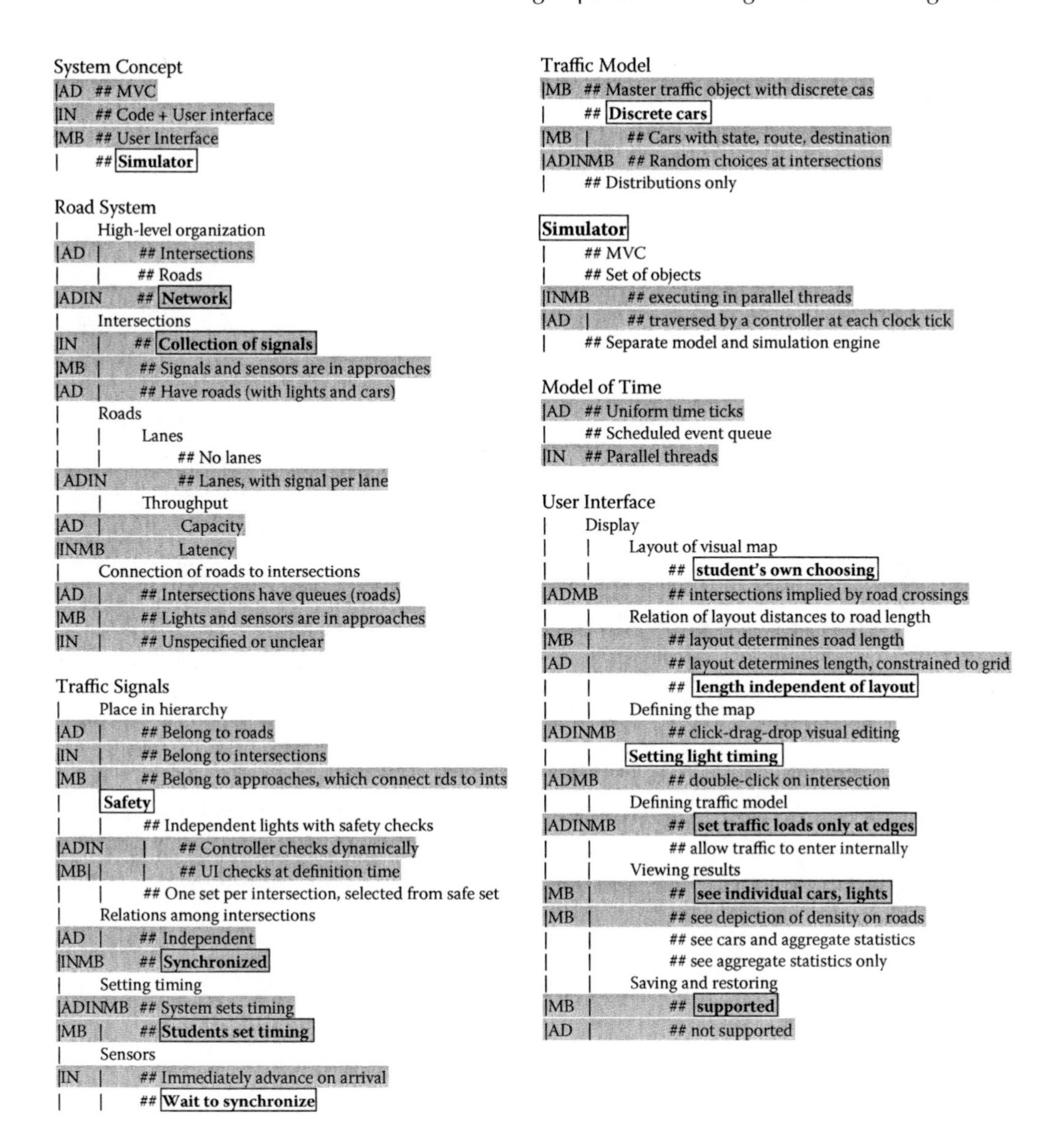

System Concept
|AD ## MVC
|IN ## Code + User interface
|MB ## User Interface
| ## Simulator
Road System
| High-level organization
|AD | ## Intersections
| | ## Roads
|ADIN ## Network
| Intersections
|IN | ## Collection of signals
|MB | ## Signals and sensors are in approaches
|AD | ## Have roads (with lights and cars)
| Roads
| | Lanes
| | ## No lanes
| ADIN ## Lanes, with signal per lane
| | Throughput
|AD | Capacity
|INMB Latency
| Connection of roads to intersections
|AD | ## Intersections have queues (roads)
|MB | ## Lights and sensors are in approaches
|IN | ## Unspecified or unclear
Traffic Signals
| Place in hierarchy
|AD | ## Belong to roads
|IN | ## Belong to intersections
|MB | ## Belong to approaches, which connect rds to ints
| Safety
| | ## Independent lights with safety checks
|ADIN | ## Controller checks dynamically
|MB| | | ## UI checks at definition time
| | ## One set per intersection, selected from safe set
| Relations among intersections
|AD | ## Independent
|INMB ## Synchronized
| Setting timing
|ADINMB ## System sets timing
|MB | ## Students set timing
| Sensors
|IN | ## Immediately advance on arrival
| | ## Wait to synchronize
Traffic Model
|MB ## Master traffic object with discrete cas
| ## Discrete cars
|MB | ## Cars with state, route, destination
|ADINMB ## Random choices at intersections
| ## Distributions only
Simulator
| ## MVC
| ## Set of objects
|INMB ## executing in parallel threads
|AD | ## traversed by a controller at each clock tick
| ## Separate model and simulation engine
Model of Time
|AD ## Uniform time ticks
| ## Scheduled event queue
|IN ## Parallel threads
User Interface
| Display
| | Layout of visual map
| | ## student's own choosing
|ADMB ## intersections implied by road crossings
| | Relation of layout distances to road length
|MB | ## layout determines road length
|AD | ## layout determines length, constrained to grid
| | ## length independent of layout
| | Defining the map
|ADINMB ## click-drag-drop visual editing
| | Setting light timing
|ADMB ## double-click on intersection
| | Defining traffic model
|ADINMB ## set traffic loads only at edges
| | ## allow traffic to enter internally
| | Viewing results
|MB | ## see individual cars, lights
|MB | ## see depiction of density on roads
| | ## see cars and aggregate statistics
| | ## see aggregate statistics only
| | Saving and restoring
|MB | ## supported
|AD | ## not supported

FIGURE 3.6 Composite representation of the three teams' design decisions for traffic signal simulation. Decisions implied by the prompt are in boxed boldface. Decisions for the commercial tool are highlighted.

The diversity among the three designs is striking. For comparison, the decisions for the commercial tool are indicated in Figure 3.6 with highlighted backgrounds.

3.4 ANALYSIS OF TRAFFIC SIGNAL SIMULATION DESIGN EXERCISE

3.4.1 General Observations

The three teams were working on a problem outside their direct expertise, and they had only 2 hours to produce a design that was sufficiently complete to hand off to a programmer.

Locating the teams' design decisions in the design space, along with the apparent implications of the prompt, calls attention to several aspects of the design exercise.

First, although in discussions at the workshop the team members indicated that their normal process would involve high-level planning, in this exercise all three teams dove directly into the details. This design task was posed as an exercise to be done in a very short time, and it may have appeared deceptively simple. Perhaps as a consequence, the teams began by identifying objects or other implementation constructs. Or maybe the current passion for objects distracted them from the problem structure.

As time passed, the teams often lost track of the requirements of the prompt. Most notably, the Adobe (AD) and Intuit (IN) teams lost track of the principal purpose of the simulation, which was for students to try out different signal timings. Only at 49:41 minutes into the session did Team AD realize that they should not be hard coding the timings. The AmberPoint team (MB) had an extended discussion with specific light timing at 29:00 to 37:00, but it was in the context of how the visualization would work for students, not how the students would set the timings.

Also surprising was the small number of instances in which multiple design alternatives were explicitly considered and one was selected as being more appropriate—that is, there was little explicit consideration of the design space. The most extended discussion (if it was a discussion) of a pair of alternatives was in Team MB's design. At 41:00 to 47:00, one partner argued that the traffic should consist of distinct cars, each with an intended route and the other partner argued that distributions of turn frequencies at intersections were sufficient. The distinction came up again at 1:02:xx, 1:04:xx, 1:22:xx–1:23:xx, and 1:26:xx–1:28:xx. In the end, though, I could not identify an explicit decision.

Teams AD and IN used an object-oriented approach. Team AD used visitor and traverser patterns in describing the simulation, but neither used inheritance; for example, to add left turns and sensors to basic definitions. I wonder also whether the strong emphasis on hierarchy in object-oriented programming led to forcing a hierarchical relation between roads and intersections (as distinguished, for example, from saying that the map model was a graph in which the nodes represented intersections and the arcs represented roads).

3.4.2 Omissions

In addition to the design discussions that did take place, it is interesting to note some discussions that did not take place, though it would seem natural for them to happen.

Each of the transcripts begins with a period of silence in which the team apparently reads the prompt. Only Team IN followed this reading with a discussion of the problem structure from 5:36 ("looks like basically two pieces the interaction and the code for …") to 8:17 ("for this type of problem a good first attack is a data analysis"). Even Team IN, though, discussed this in terms of the software that would solve the problem. No team began with a discussion of the elements of the problem, their interactions, and the requirements for a solution.

In an ordinary design setting, I would expect a team that is not already expert in the problem to ask what is already known about the problem. This might be the domain knowledge, or it might be implementations of similar problems that could serve as exemplars.

This task presumably relies on common knowledge of traffic signal timing, a phenomenon that almost any driver has experienced. But, as the prompt indicates, experiencing the phenomenon is not the same as understanding how it really works. No team mentioned asking for outside expertise, perhaps because of the artificial setting of the task.

The safety requirement, "combinations of individual signals that would result in crashes should not be allowed," was pushed down into the details-yet-to-be-settled by all teams. Team AD assigned it to the controller at 9:24. Team IN recognized the problem at 15:38 and again at 53:32. They noted that it looks like a state machine at 16:32, but pondered how to share information across all signals at 16:42. Team MB delegated the checking to the UI, saying (at 35:24) that the system should not allow a student to manually set a light timing that could cause a collision. A design alternative that seems obvious would be to address the safety requirement by treating all the lights at an intersection as a single object rather than as a collection of independent lights. That single object could then have a predefined set of permissible configurations and some constraints on order (e.g., "sequence for each light must be {green-yellow-red}*"). After all, when one direction turns green, the other direction turns red at the same time. Students could then choose constrained sequences of these configurations and assign times for each configuration. This follows the UI principle that it is easier to select from a given list than to generate values from scratch.

There was also little attention to the synchronization of timing between intersections, even though this synchronization was in many ways the central issue of the simulation. Although this functionality got some passing recognition from Team IN, only Team MB recognized the need for setting relative timing between intersections, when they considered ways to set this up efficiently at 56:33.

In each of these cases, it is easy to imagine how a design process that relies on the design space might help.

3.4.3 Architectural Considerations: System Concept and Simulator Dimensions

The *system concept* dimension corresponds to the choice of the overall system architecture* and thereby provides the structure for the rest of the design, especially the choice of strategy for the *simulator* and, to a certain extent, the associated *model of time* for simulation. Perhaps because of the short time frame imposed by the workshop, the teams spent very little time explicitly discussing overall organization. Each of the three teams ended up with a different system concept, none of which was explicitly a simulator. Unsurprisingly, their treatment of simulation and time also differed. In each case, the team identified the top-level organization almost automatically, without considering and evaluating alternatives; it appears from the transcripts that the overall system concept was chosen implicitly, more by default than by deliberate design.

The first sentence of the prompt sets the task as "designing a traffic flow simulation program," and the remainder of the prompt describes both the phenomenon to be simulated and the ways that the students will set up and observe the simulation. Team MB focused on

* By "architecture" I mean the high-level concepts that guide the system organization, not the selection of data structures or the class structure of an object-oriented system. Thus "we'll use MVC" is architectural, but "a road is a queue" is not.

the simulation aspect of the problem, though they did not say much about the simulation engine. Recognizing the character of a simulator in this fashion might have helped Team AD separate "control" (i.e., running a simulation on a specific case) from the model-view-controller (MVC) (which would mediate between the UI and the data being defined by the student and perhaps between the simulation results and the display). Viewing the system as a simulator might have led Teams AD and IN to consider alternatives to a massively parallel execution of objects. It would have almost certainly helped Team MB separate the UI featuring a drawing tool from the model that was being created with the drawing tool.

Despite the lack of explicit discussion in the transcripts, the architectures—or at least the high-level organization—of these systems and the way they emerged shaped the teams' discussions.

3.4.3.1 Adobe

Team AD couched their discussion in terms of objects. They began by selecting data structures for intersections, roads, and the "cop," which was the main object to advance the state of the system via timer events. A high-level network object allowed users to add roads, from which intersections were inferred. Intersections contained information about how traffic flows through intersections, together with the states of lights. Roads were referenced by intersections, and they contained queues of cars and capacities. The system was driven by timer events; it worked by using the visitor pattern to visit each element at each clock tick to update the state. Interaction was through a window depicting the map, with the map features clickable for editing. The MVC pattern was mentioned, but instead of using the MVC pattern to organize the design activity, they periodically tried to decide whether an entity (the cop or the clock) was a model or a controller.

3.4.3.2 Intuit

Team IN saw a data-driven problem in the prompt, so they focused on the data items. Despite mentioning the MVC, they centered the design on a main map, on which the user added intersections connected to roads. The roads themselves were queues with sources of traffic. This main system map was the overall controller, and intersections were also active objects that queried roads for traffic and enforced safety rules on lights. The simulation would run as a collection of threads for these various parts. Their UI was a drawing package with drag-and-drop functionality.

3.4.3.3 AmberPoint

Team MB focused on the visual aspect of the problem, and their discussion considered things that the students must do. They started by building the map with roads and intersections and making a way to save the map. Given this map, they created the traffic patterns, perhaps varying by the time of day. Then they set the signal timings and the treatment of left turns and sensors; they recognized that students should be able to click on an intersection and fill in details, and they also need automated support for replicating the same pattern to different intersections. They provided ways to compose the parts, save the simulation, and run the simulation using the map to show how the traffic is flowing.

3.5 REFLECTIONS ON DESIGN EXERCISE

There was very little explicit discussion of architecture, and what I could identify seemed to be mostly architecture-by-default rather than architecture-by-design. Architectural or other high-level design provides a structure that reduces confusion in the detailed design. The transcripts show considerable retracing of steps. Even in their final reports, although these would seem to present an opportunity for sketching the architecture—or some high-level view—to set the context, the summaries were remarkably short on overarching vision. It is interesting to consider whether a few minutes devoted to architecture or high-level design at the outset would have clarified the detailed design discussions and, if so, what factors affect whether the architectural analysis is done.

In each case, the selection of the top-level organization was almost automatic. The teams did not consider and evaluate alternatives. They all recognized that there is a UI part and an "everything else" part. Teams IN and AD associated this with an MVC architecture and separated the UI for later consideration. Team IN did not appear to use the MVC architecture to organize the model and controller. Certainly, there will be a UI, but many architectures involve a UI. Indeed, considering the UI led teams to both the MVC and the "drawing tool" as the organization. So, it appears from the transcripts that the overall system concept was chosen casually or uncritically, more as a default than a deliberate decision.

As indicated in Figure 3.6, the design space includes at least four alternatives for an overall system concept. All teams chose different global organizations, and—curiously—none chose a simulator explicitly.

Both Teams AD and IN mentioned the MVC in discussion, but both used the MVC pattern informally and neither mentioned the MVC in the summary presentations. Team IN did not appear to use the MVC architecture to organize the model and controller. Team AD did refer to the model and controller; however, as noted above, instead of using the model–controller distinction to organize the design activity, they tried to map elements of their design into the MVC template. In classic MVC, the controller is chiefly a dynamic mediator between the user's actions (through the UI) and the domain logic in the model. Here, especially for Team AD, the controller handled the details of running the simulation. When the simulation is running, the user is not (or at least not necessarily) offering input to the system. So, incorporating the simulation logic in the controller makes sense for the common informal meaning of "controller," but it is not a good match for the controller of the MVC pattern.

For several design decisions, different teams chose different alternatives, but in each case the team simply adopted the alternative rather than considering the choices. One such example is the model of time. Team AD decided (at 11:25–12:50, without considering alternatives) on a master clock that synchronized the simulation with regular "ticks." This is one of the standard models for representing time in a simulation, the other being an event queue that sets off events in simulated time order; a third alternative is to design with parallel threads, which leaves the underlying time model to the implementation. The synchronous regular-tick time model is the better match for a visualization that should proceed smoothly (that is, with a uniform scale factor relative to real time), but that was not mentioned in the discussion. Team IN speculated about a clock at 38:12, but at 38:33 they described the simulation control as firing off a car thread for each input, and by 49:30 they had a massively parallel simulation. The teams did explicitly address the model for time, but they did not make deliberate choices; had a representation of the design space been available, they might have either made different decisions or developed confidence in the decisions they did make.

Team MB, on the other hand, began by considering the UI task, deciding immediately that the system would be a drawing tool. They began designing the interaction with the UI, but soon—tacitly—shifted the focus of the discussion to the simulation.

Would it have made a difference if the teams had asked early on what kind of system they were building and what architecture would be appropriate for that system? That is, would it have made a difference if they had explicitly recognized that the architecture was being chosen from a set of alternatives? It is hard to know, and it is hard to know whether the failure to do so was related to the somewhat artificial experimental setting.

The prompt calls for "designing a traffic flow simulation program." It asks the designers to address two major issues: "the interaction that the students will have with the system" and "the important design decisions that form the foundation of the implementation." For the latter, the prompt asks the designers to "focus on the important design decisions that form the foundation of the implementation." The framework of Figure 3.5 helps us to see whether these mandates are satisfied.

A simulator is a well-known type of software system, with a history that goes back many decades. Identifying a system as a simulator leads to recognizing—and separating in the design—four concerns:

- The model of the phenomenon to be simulated
- The means of setting up a specific case to simulate
- The simulation engine itself
- The current state of a simulation (including a way to report results)

Although the teams talked about simulation, I was surprised that none discussed how to structure the software for a simulator. Doing so could have clarified the system structure, helped to select an appropriate model for time, and reminded the teams about parts of the design that were neglected.

3.6 A TRAFFIC SIMULATION PRODUCT

Traffic signal simulation is not a novel problem. Commercial simulators support professional traffic engineers. One such product, from Trafficware, provides a free demonstration version of their simulation tool [15]. This tool accepts quite detailed descriptions of intersections, roads, vehicles, traffic light timings, and other information, and it simulates traffic loads at the individual car level with detailed graphic displays and data collection. This is, of course, a sophisticated tool in which considerable product development effort has been invested and that embodies a great deal of the domain knowledge. Nevertheless, it is informative to place this tool in the design space of the classroom version. After the workshop, I examined this demonstration version. Even without access to the code, many of the design decisions were evident. The system can be used in a variety of ways; different modes lie at different points in the design space. The decisions exhibited by the tool are indicated in Figure 3.6 with highlighted backgrounds.

3.7 IMPLICATIONS FOR PRACTICE

Capturing design knowledge—decisions and alternatives—in the form of a design space provides a framework for systematically considering design alternatives (the *design* task), for selecting among existing artifacts (the *selection* task), and for comparing designs (the *comparison* task). In addition, it helps the designer recognize interactions and trade-offs among decisions. In each of the examples discussed here, it is easy to imagine how a design process that incorporated the design space might have helped.

Using a design space to guide design involves adopting or creating an initial approximation, then refining it during design. Ideally, the domain knowledge for common domains will be captured in canonical design spaces for those domains; these could at least provide a starting point for a problem-specific design space. Even if no design spaces are available at the outset, the discipline of creating a partial representation should sensitize the designers to the existence of alternatives, bring out interactions among issues, and help to organize the design discussion. For a design exercise as time constrained as the exercise studied here, it might not be worth doing much more than laying out the natural choices systematically.

Thinking in terms of design spaces should sensitize the designer to the existence of alternatives. It might have reduced the number of omissions such as the ones described here. It would help identify the type of system that would be appropriate for the problem. Incorporating the use of design spaces in normal practice should lead the designer to ask whether a design space has already been developed for this or a similar problem, to avoid a problem analysis from scratch and to exploit domain expertise encoded in the design space. The design space could also help identify the right level of abstraction and granularity for design elements.

ACKNOWLEDGMENTS

The workshop on Studying Professional Software Design, including travel support for the participants, was partially supported by NSF grant CCF-0845840. The workshop would not have been possible without the willingness of the professional designers from Adobe, Intuit, and AmberPoint to work on the design task and to allow the recording of that work to be examined in public. This chapter is a major extension and revision of "The role of design spaces," which appeared in *IEEE Software* in January 2012 [14].

REFERENCES

1. Herbert A. Simon. *The Sciences of the Artificial.* MIT Press, Cambridge, MA, 3rd edition, 1996.
2. C. Gordon Bell and A. Newell. *Computer Structures: Readings and Examples.* McGraw-Hill, New York, 1971.

3. Dezsö Sima, The design space of register renaming techniques. *IEEE Micro* 20 (5), 2000: 70–83; doi:10.1109/40.877952.
4. Stuart K. Card, Jock D. Mackinlay, and George G. Robertson. A morphological analysis of the design space of input devices. *ACM Transactions on Information Systems* 9 (2), 1991: 99–122.
5. Thomas G. Lane. User interface software structures. PhD thesis, Carnegie Mellon University, May 1990.
6. Mary Shaw and Paul Clements. A field guide to boxology: Preliminary classification of architectural styles for software systems. *Proceedings of the 21st International Computer Software and Applications Conference*, IEEE Computer Society, Washington, DC, 1997, pp. 6–13.
7. Kay Römer and Friedemann Mattern. The design space of wireless sensor networks. *IEEE Wireless Communications* 11 (6), 2004: 54–61.
8. Kombinat-Typefounders. Design space for typeface design. http://www.typedu.org/lessons/article/designspace.
9. Robert F. Woodbury and Andrew L. Burrow. Whither design space? *Artificial Intelligence for Engineering Design, Analysis, and Manufacturing* 20 (2), 2006: 63–82.
10. Allan MacLean, Richard M. Young, Victoria Bellotti, and Thomas Moran. Questions, options, and criteria: Elements of design space analysis. *Human-Computer Interaction* 6, 1991: 201–250.
11. Xerox Research Center Europe. Design Space Analysis Project. http://www.xrce.xerox.com/Research-Development/Historical-projects/Design-Space-Analysis.
12. Frederick P. Brooks, Jr. *The Design of Design*. Pearson Education, Boston, MA, 2010.
13. International Workshop "Studying Professional Software Design," February 8–10, 2010. http://www.ics.uci.edu/design-workshop.
14. Mary Shaw. The role of design spaces. *IEEE Software* 29 (1), 2012: 46–50; doi: 10.1109/MS.2011.121.
15. Trafficware. SynchroGreen Adaptive Traffic Control System. http://www.trafficware.com/.

CHAPTER 4

Designing Programs as Representations

Clayton Lewis

University of Colorado, Boulder

CONTENTS

4.1 PROGRAMS AS REPRESENTATIONS: QUESTIONS ABOUT DESIGN

Like many software systems, the traffic simulator that is the subject of this design study is part of a *representational system*. In this chapter, I will define what a representational system is, and bring out some special requirements that must be met in designing an effective representation within such a system. These requirements suggest some key questions about the design process that we can explore in the records of the designer discussion captured in the design study.

In a representational system, questions in a *target domain* of interest, in this case the functioning of road networks with various arrangements of traffic signals and timing, and various traffic patterns, are mapped to questions in a *representation domain*, in this case, questions about the behavior of the traffic simulator program. For a representational system to work, when questions are mapped from the target domain to the representation domain, the same answers have to be obtained, as shown schematically in Figure 4.1.

Further, it has to be *easier*, in some sense, to answer the questions in the representation domain than in the target domain: that is the point of using a representation. One can easily see the value of the traffic simulator, looking at it in this way. It would surely be easier to experiment with the simulator, trying out different scenarios, than it would be to use real roads, cars, drivers, and traffic signals. Additionally, one can easily see that the behavior of the simulator has to correspond to that of real traffic: if it does not, we will not learn the right lessons from our experiments with the simulator.

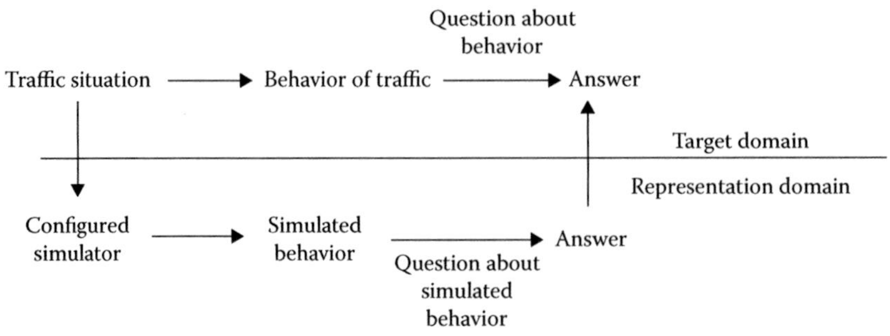

Question about behavior
Traffic situation → Behavior of traffic → Answer
Target domain
Representation domain
Configured simulator → Simulated behavior → Answer
Question about simulated behavior

FIGURE 4.1 Representation of traffic and behavior.

What are the "traffic situations" referred to in Figure 4.1? For some uses of traffic simulators, these would be real situations, involving actual roads, traffic signals, and so on. But in the design study here, the simulator is intended to be used by students, not to represent *real* traffic situations, but to represent *hypothetical* traffic situations that they want to learn from. The distinction between real and hypothetical situations complicates the picture somewhat, because there has to be some connection between hypothetical traffic situations and real ones, for the learning to have any value. But this connection is not shown in Figure 4.1. The Appendix has a more careful description of this representational system, as well as a more complete treatment of the theory of representation that provides a background for our discussion. In the body of the chapter, we will gloss over these complications.

Continuing to work with the scheme in Figure 4.1, the key relationship between the domains shown is that when a question is asked in the representation domain, using the simulator, the correct answer has to be given in the target domain. If this is not the case, the simulator is not faithful to reality, and will not be useful. But this relationship is hardly likely to hold by chance. Rather, the structure and behavior of the simulator have to be crafted to reflect the structure and behavior of the corresponding entities in the target domain; in this case, traffic situations and their behaviors. A key question thus arises: *How do software designers ensure that the systems they create have structure and behavior that correspond appropriately to entities in the target domain?*

When answering questions in a computational domain, such as the representation domain in Figure 4.1, it is often the case that a good deal of the required work can be performed automatically. Of course, that is one of the main reasons why computational representations are so useful, and so widely used. The traffic simulator illustrates this: the intent is that the simulator, once configured so as to represent some situation of interest, will automatically generate the appropriate behavior.

But it nearly always happens that some work has to be performed by *people*, the users of the system. In our case, the users of the simulator will need to make some judgment of the behavior that the simulator generates, so as to understand the behavior and learn from it.

This situation can be thought of as involving a further representational layer, in which some aspects of the structure and behavior in a computational domain are mapped into a domain of *human perception*, populated by structures and processes that can be perceived by people. The widespread use of *visualizations* in connection with computational systems is a response to this need, though many other approaches are possible, and sometimes used, such as auditory displays.

Figure 4.2 elaborates on Figure 4.1 to show this added layer. For the system as a whole to work, when users make judgments in the perceptual domain, they have to get the correct answers. Further, the perceptual judgments required have to be *easy*: if the judgments are difficult and error-prone, the overall system will be unsatisfactory. Hence, another key design question arises: *How do software designers create perceptual representations that allow users to answer the questions of importance to them easily and accurately?*

Nearly all computational systems involve another facet of representation, involved in *control* or *configuration* by users. Commonly, the structure and operation of the system are not fixed, but can be modified in some way by users. The logic of representations in this situation is more complex than in the situations already considered, in that not just questions but also *operations* have to be mapped between domains. Thus, the operation of changing the behavior of a program in some way is mapped onto an operation, or action, that a user can perform in a domain of *human perception and action*, such as pressing a key or making a particular selection on a menu. For this mapping to be effective, it has to be possible for the user to *perform* the needed action, but, usually of more concern, it has to be possible for the user to *select* the appropriate action from a space of possibilities, using *perceptible cues*. For this to work, the cues, as interpreted by the user, have to have a structure that corresponds to the structure of the operations in the computational domain that the user needs to control.

These rather abstract considerations can be made more concrete in connection with our design problem. The users of the simulator have to be able to set up the simulator so as to represent some situation of interest. In doing this, there must be actions that they can perform, such as clicking on buttons or selecting from menus or drawing a map of a street grid, that do the right things. Further, there have to be intelligible cues, such as button labels or menu items, that guide users in setting up the simulator in an appropriate way.

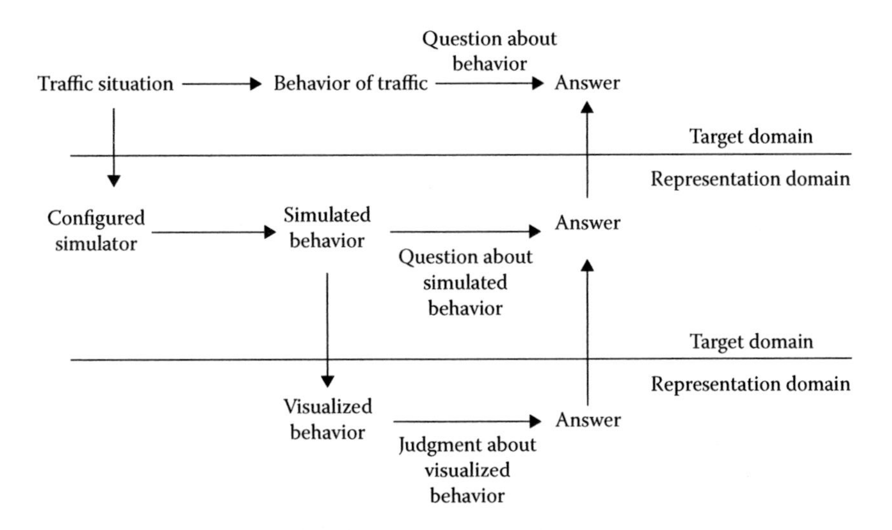

FIGURE 4.2 Visualization of behavior.

These cues have to connect to users' goals, that is, to the hypothetical traffic situations that users want to explore. But they also have to be linked, via the associated actions, to the way the simulator actually works, so that users can set up the simulator in the right way for exploring the situations they are interested in. That is, the requirement is that cues are provided that (a) correspond to the focal aspects of hypothetical traffic situations, that is to users' *goals*, and (b) correspond to *actions*, such as pressing keys, clicking buttons, and drawing, that the user can carry out, in such a way that (c) when the actions are performed, the simulator will be set up to correspond to the hypothetical traffic situation.

These considerations expose a third question about design: *How do designers devise cues that correspond appropriately to aspects of users' goals,* ***and*** *to appropriate application settings?*

4.2 EXPLORING THE QUESTIONS IN THE DESIGN STUDY

The recorded design sessions in the study offer direct evidence about all three of the design questions just raised. The design teams whose work was recorded will be labeled D, M, and N, for Adobe, AmberPoint, and Intuit, respectively. Portions of the recordings will be referred to by time stamps, giving hours:minutes:seconds.

As with all analyses of these study data, it is important to acknowledge the artificiality of the design situation, including the short duration of the sessions, and the fact that the designers had no access to outside informational resources. We can, nevertheless, find suggestions in what the teams do and do not consider in the sessions.

Do the software designers in the study ensure that the systems they create have structure and behavior that correspond appropriately to entities in the target domain?

The target domain in the design study is traffic, including roads, intersections, traffic signals, cars, and their interrelationships and behaviors. Because it is a domain of great practical importance, much has been learned about it. For example, the manual by Koonce et al. (2008) is a guide of 265 pages, devoted entirely to traffic signal timing. This is a complex domain, and so it is not easy to capture its behavior in a representational system.

There are a number of factors that are important in traffic simulation whose roles are far from obvious. One is *driver response time*: when a traffic signal changes, drivers do not perceive the change and respond immediately, and an accurate simulation of the traffic flow at an intersection has to take the delay in responding into account (and designers of actual traffic signal timings *do* take it into account).

Another relevant factor is the timing problem called "yellow trap" (see http://midimagic.sgc-hosting.com/lagtrap.htm; http://projects.kittelson.com/pplt/LearnAbout/Learn3.htm). A yellow trap occurs when the signals for traffic moving in opposite directions through an intersection are offset in time, so that the signal for the traffic moving (say) north changes to red while the signal for the traffic moving south remains green. Northbound drivers may enter the intersection to make a left turn, but find no gap in the traffic, and so wait in the intersection. When their signal changes to yellow, they may assume that the oncoming signal is also yellow, and attempt to complete their turn. But if the opposing signal is, in fact, still green, a collision is likely, as both the left-turning and oncoming drivers think they have the right of way.

Note that it is not a viable solution to the yellow trap problem simply to lock the signals for opposing flows together. This locking would rule out common signaling arrangements in which the traffic moving (say) north has both a circular green (for proceeding north) and a green arrow (for turning left), while southbound traffic sees red signals, and then southbound traffic gets circular green and a green arrow while northbound traffic sees red signals.

Another factor is the role of *gaps* in traffic. Both the right turns on a red signal and unprotected left turns require the driver to find an acceptable gap in a traffic stream to execute the turn safely. An accurate simulation has to capture this aspect of driver behavior, because, in fact, different signal timing regimes may provide gaps at different rates, affecting the traffic flow.

Unfortunately, none of the design teams are aware of these factors. Strikingly, two of the teams, D and A, never express any concern about needing more understanding of the target domain than they feel they have, or can reason through unaided.

Team M do recognize limitations in their understanding of the target domain at two junctures. At 22:46.7, they say, "I mean, we might have to code whatever [the client's] understanding is of traffic light theory into it, right?" At 1:13:25.8, they say that it would be good to consult the client about how to specify signal timing. Note that these questions do not address the specific aspects of the domain just discussed, but at least the team recognize that there are important things about the domain that they do not know.

Do the software designers in the study create perceptual representations that allow users to answer the questions of importance to them easily and accurately?

As laid out in the prompt for the designers (with italics added here for emphasis), users of the simulation need to "be able to *observe* any problems with their map's timing scheme, alter it, and *see* the results of their changes on the traffic patterns." Users should be able "to learn from practice, by *seeing* first-hand some of the patterns that govern the subject." The users' ability to "observe" or "see" changes and results depends on how the results of the simulation are presented to users.

Teams D and M devote some attention to this issue, while Team N, in their brief session, do not consider it at all (apart from a bare mention at 42:02.3: "Because I could see they'd want to click the system and then sit there and watch it run and say, oh yeah cars are piling up here, okay.").

Team D discuss the issue showing "analytics" at 48:20.4, 1:13:16.9, 1:14:19.6, and 1:34:11.5. They also consider visual presentations at 1:13:16.9, 1:14:19.6, 1:24:35.4, 1:25:34, and 1:26:00.0. Their discussion includes showing data for cars and for intersections, as well as some aggregate measures, "[48:20:4] … you have an analytics piece looking in and assessing questions like, okay how congested is it? What's the average wait time for an average car going through our city streets."

Toward the end of the session, starting at 1:31:01.8, the team consider how to "evaluate success" from the simulation. This consideration comes *after*, not before, considerable discussion of "analytics." That is, the team leave until late in the session the fundamental issues of what questions are, or should be, important to users, and how the system can help users answer these questions easily.

They propose that users could evaluate the success of a scenario by making a graph of the data from different simulation runs, and examining the slope of wait time against "capacity":

> 1:33:37.7: Right, so you want the wait time to go flat with increasing capacity. Then, in being able to determine this graph, because that then really shows the success of the network overall.

(It is apparent from the discussion that what is meant is actually *load*, not "capacity"; as they see it, a successful timing scenario is revealed by small increases in wait time as load, traffic introduced into the network, is increased.)

Two things are interesting about this discussion, with respect to the representational issues in the design. First, having developed a way in which users might evaluate simulation results by drawing this graph, the designers do *not* consider arranging for the simulation tool to actually create the graph. Rather, they are content that the simulation simply produces data from which the graph could, in principle at least, be created.

Second, the designers do not consider whether the proposed graph actually captures what users should know about the performance of the simulation. They just assume that it does.

In contrast, Team M recognize that there is a gap in their knowledge in this key area:

> 55:26.6: I think we have to talk to Professor E again, because she didn't specify the most important aspect of it, which is how do you know that your lighting system is working to its best potential? I can draw the map, I can specify all these parameters and now I run it, how do I know this is good?

They return to this crucial gap later, as part of a discussion of how data from the simulation might be displayed (like Team D, they consider data at the level of intersections, and the idea of showing individual cars moving through the road grid):

> 1:24:30: I don't know yet what to do with all that data but it seems like somehow that data has to produce some meaningful feedback to the user and say that traffic is going well.

The contrast with the other teams is notable. Team M are not willing to assume that they can just make up reasonable performance indicators.

Do the designers in the study devise cues that correspond appropriately to aspects of users' goals, and to appropriate application settings?

For one aspect of the target domain, the street grid, all of the teams use a familiar representation, maps. The designers presume that users can readily connect features of a hypothetical traffic situation to a map, and that the map can be associated with appropriate simulator settings by the software.

Another key aspect of the simulation that users must control is traffic signal timing. As specified in the problem, this is a complex business, taking in not only relatively simple matters such as the length of the period of illumination of a given signal face, but also quite subtle issues such as ruling out unsafe operating cycles (such as the yellow trap described earlier), and, especially complex, defining the effect of traffic detectors. How do the designers approach these issues?

As discussed earlier, Team M express a wish to learn from the client how to represent signal timing. As they expect, traffic engineers do, in fact, have a specialized representation for signal timing, called the *ring-and-barrier* diagram. Figure 4.3 shows an example.

The boxes in the diagram represent *phases*, with the arrows showing the traffic flows (and pedestrian movements) permitted during the phases. The two rows of boxes represent *rings*; the phases in each ring are entered consecutively, with a loop back from the right edge to the left edge. The vertical gray lines are *barriers*, representing synchronization points for the rings. Despite their vertical alignment, Phases 01 and 05, on the left do not have to be the same length. But the signals cannot move from Phase 02 to Phase 03 in the upper ring unless Phase 06 has also finished, so that the lower ring moves synchronously to Phase 07, when the upper ring moves to Phase 03.

None of the design teams use this representation. Team D say very little about how users would control the timing of lights. They discuss timing as a consequence of "rules" executed by a "traffic cop" controller object in their simulation:

> 20:36:2: Yeah and also, so that's one thing, and also like how long these signals stay green. Is that also something the cop looks at?
>
> Well, that could be, maybe the cop …
>
> That's part of the rule. There's certain rules, the rules encapsulate maybe the length of the time that the light, maybe the rate at which the cars come, I don't know.

The team do not address how these rules would be specified, or how users would use them to control signal timing. This omission is striking, since signal timing is the key independent variable that users are supposed to be able to manipulate, according to the problem statement.

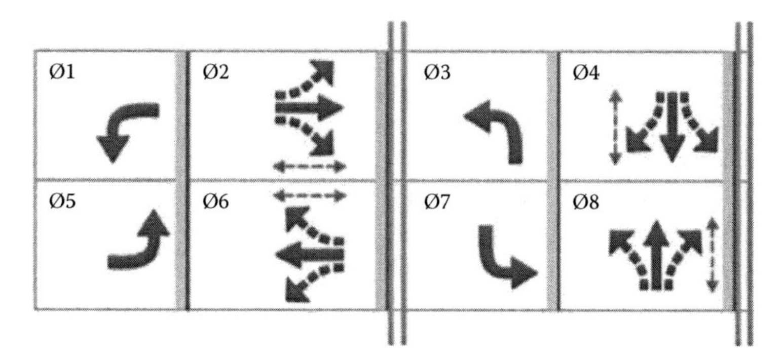

FIGURE 4.3 A ring-and-barrier diagram showing protected permissive phasing. (From Koonce, P., Rodegerdts, L., Lee, K., Quayle, S., Beaird, S., Braud, C., Bonneson, J., Tarnoff, P., and Urbanik, T., *Traffic Signal Timing Manual*, Federal Highway Administration, Washington, DC, 2008, available online at http://www.signaltiming.com.)

Team N also say very little about control of the signal timing, leaving the matter until late in their 60-minute session:

> 51:32:3: So which time—is this time for each one of these? Or is this the time between state changes? Because there are two different times. There's the time between a red light turning green—or is that just the time between all state changes? Are we saying those are equal?

After a brief discussion of the potential existence of separate times for red, yellow, and green signals, the team move into a consideration of the impact of traffic sensors, and they do not return to the matter of how signal timing would actually be controlled.

Team M are the only team to make a serious effort to describe how users would control signal timing. They propose a system of color-coded time lines:

> 37:11:4: Really what you kind of want is a timeline. I think it would help visualize it. Kind of like, now I've got these little 5 second increments out of 120. So then over here, you can say, because I think with those numbers I had it was really hard to see where was the overlap between them, but if I could do—I could choose to either do north south, east west, or I could even split them into four different ones if I wanted to. And then, I can say from 0 to 50 on the timeline, is going to be green, and then it's going to be yellow for like 15 seconds, and then red for the remainder, and then it's easy for me to layout on here well I definitely want it red for this whole period, and definitely red while it's yellow, and then there should be a little bit of an overlap.

They go on to consider the desirability of controlling the time lines either by drawing, or by entering numbers, using an appropriate widget:

> 41:13:3: Oh yes, but you'd also have to see the numbers somewhere. So like either way you set the number and the thermometer changes or you drag the thermometer and the number changes.

Having come this far in the design, the team are aware of limitations in their concept, and, as mentioned earlier, they recognize, unlike the other teams, that it would be good to understand more about how signal timing could be controlled:

> 1:13:25.8: Yeah, so lets assume you chose pairs so there's some way to say north, south, east, west. And then we still have to do the detail design for how you specify these numbers but we'll get to that through prototyping. But it does seem like it would be good to be able to visualize the overlap of the red and green. But the actual mechanism that we're going to use for setting it is a little unclear, and we probably need to talk to Professor E a little more about that.

The relationship between traffic sensors and signal behavior, expressly included in the scope of the problem, was not addressed adequately by any of the design teams. This is related to the teams' attempts to work out a representation of signal timing on their own, in that a key feature of the ring-and-barrier diagram is to structure the effects of sensor inputs. Signal controllers can respond to the detection of the presence or absence of traffic at selected locations by extending phases, or terminating them early, but only within the constraints specified by the ring-and-barrier diagram. There is very little chance that the design teams could just happen on a representation that captures these requirements.

4.3 DISCUSSION

Some of the designers neglected the structure of the target domain in design problems. As we have seen, only one of the design teams expressed any wish to understand more about the target domain of traffic. There is a passage in the problem statement that might seem to justify this indifference:

> The program is not meant to be an exact, scientific simulation, but aims to simply illustrate the basic effect that traffic signal timing has on traffic.

Could it be argued that this tells designers that their common sense appreciation of traffic is an adequate basis for design? Not really: only if designers can be confident that they know what "the basic effect that traffic signal timing has on traffic" actually is could they proceed. The problem statement goes on:

> She [the client] anticipates that this will allow her students to learn from practice, by seeing first-hand some of the patterns that govern the subject.

Here again, an appropriate design has to capture "some of the patterns that govern the subject." Without an understanding of these patterns, design really cannot go forward successfully. One of the teams recognized this problem; the others did not.

Some of the designers privileged questions of implementation over questions of representation.

If the designers were not attending to the structure of the target domain, what were they doing? First, it is not the case that they were not attending to *any* of the structures of the target domain. All of the teams recognized the value of maps in capturing some of this structure, namely, the topology and (at least sometimes) the geometry of a road grid. Some design attention did go into this matter, for all three teams, even though key features of real road grids such as the length of the turning lanes were not dealt with. Two of the teams did give some attention to subsidiary representational questions of visualization and control, as discussed earlier.

Having said that, two of the teams, D and N, gave their attention to aspects of their designs that were not representational at all: the structure of the program itself. Thus, the integrity of the program, apart from its correspondence to the target domain, was a matter of concern for these designers.

One cannot quarrel with the decision to devote some attention to this. A program that captures the structure of the target domain, but cannot be implemented correctly, or feasibly maintained, would not be useful. But a "correctly" implemented and eminently maintainable system that does not represent the target domain is not useful, either.

Further, as Agre (1994) has pointed out, applying a computationally economical apparatus to problems, in a way that neglects the actual structure of the problems, leads to systematic distortion. Real problems are replaced by their computational shadows; for example, workers in an organization find that sensible behaviors that are not captured by the computational systems they work with are forbidden. There is a striking example of this phenomenon in Holland (1994), in which a hugely expensive machine tool is held on a shipping dock for lack of a single small part. The part cannot be supplied because the production control software that has been imposed on the operation thinks that the part is in inventory, when it is not.

There are examples in the study of the computational cart being put before the representational horse. Team D are wrestling at 1:25 with how to manage the speed of cars, given that they have previously decided to use *queues* to represent stretches of road. There is no reasonable way to handle speed this way, because only the order of the cars, and not their positions and velocities, is captured, and indeed cars with different speeds would scramble the queue, as faster cars overtake slower ones. But the appeal of the queue, a quintessentially computational notion with which the designers are very comfortable, masks its inappropriateness as a representation for this problem. Team N are also committed to queues as a key device, for managing vehicles at intersections, and while they note the issue of "travel time" in two parts of their discussion (27:06.5 and 28:08), they do not propose a means of dealing with it. Using queues to represent stretches of road makes it difficult, if not impossible, to manage travel time.

Team N also weigh how to accommodate *lanes* in their computational view, starting at 11:02. Though lanes are crucial to traffic modeling, for example, in determining how many left-turning cars can accumulate before interfering with the through traffic at an intersection, the team resist including them:

> 11:45.3: "Lanes seem like a complexity we don't want to get into."
>
> 12:01.1: "I don't think we need to make assumption of lanes, period."

They return to this matter later, after realizing that left-turn lanes are important. But they are still intent on "simplifying" their treatment:

> 21:35.6: "Does it say anything about the number of lanes? Because we can make the assumption of just two lanes everywhere. Make life slightly easier."
>
> 26:20.8: "… so what does the left-hand turn really mean though, it's just another—you're just basically turning from intersection to intersection no matter what we do at this point, right? And so instead of going from intersection A to B, it'd be from A to C. In the actual modeling of it. That's why I'm trying to think, the lanes aren't necessarily so important as which intersection does it get you to next?"

Trying to "simplify" the simulator, for computational reasons, is crippling when it results in omitting key features of the target domain.

In the approach of the other team, M, computational considerations do not seem to shape their analysis. But that does not mean that they approach the problem as one of capturing the structure of the target domain. Rather, they concentrate on developing a *user view* of the application. While they do note that more understanding of the domain would be useful, as mentioned earlier, they focus on what users will do and see. Like computational integrity, this is certainly important, but it cannot compensate for an inadequate representation. Just as an adequate computational structure cannot be created without regard for the structure of the target domain, so an adequate user view cannot be created without regard for that structure.

Where do we go from here?

As long ago as 1988, Curtis et al. cited "the thin spread of application domain knowledge" as a key factor in the failure of many software projects. As Curtis et al. noted, "Software developers had to learn and integrate knowledge about diverse areas such as ... the structure of the data to be processed and how it reflected the structure of objects and processes in the application domain [p. 1271]," but too often they failed to do so. In the terms used here, lack of domain knowledge translates into the failure of software systems to capture the structure of the target domain adequately, so that using the software fails to produce the correct answers.

There is only a little evidence in the transcripts that the designer participants in this study were sensitive to this key problem. As we have seen, only one of the teams expressed any need to understand more about the application domain for which they were designing.

The representational perspective on software suggests that it is important to see a piece of software not just as a thing in itself, whose crucial virtues are internal (e.g., performance, testability, and so on), but also as a part of a larger structure. The virtues of software have to include the structural correspondences in which it participates. Because these correspondences link the software to the application domain, software designers have a professional responsibility to understand that domain adequately.

Technical developments are underway that may help to establish and support this perspective in software engineering practice. Model-driven engineering (Schmidt, 2006) and intentional software (Simonyi et al., 2006) both aim to bring explicit models of application domains into software engineering practice. Perhaps we can look forward to a time when software designers recognize the importance of application domain structure, and routinely seek out and use proven representations that capture key aspects of this structure.

REFERENCES

Agre, P.E. 1994. Surveillance and capture: Two models of privacy. *The Information Society*, **10**(2), 101–127.

Curtis, B., Krasner, H., and Iscoe, N. 1988. A field study of the software design process for large systems. *Communications of the ACM*, **31**(11), 1268–1287.

Holland, M. 1994. *When the Machine Stopped: A Cautionary Tale from Industrial America*. Boston, MA: Harvard Business School Press.

Koonce, P., Rodegerdts, L., Lee, K., Quayle, S., Beaird, S., Braud, C., Bonneson, J., Tarnoff, P., and Urbanik, T. 2008. *Traffic Signal Timing Manual*. Washington, DC: Federal Highway Administration. Available online at http://www.signaltiming.com.

Krantz, D., Luce, D., Suppes, P., and Tversky, A. 2007. *Foundations of Measurement.* Mineola, NY: Dover.
Mackinlay, J.D. 1986. Automating the design of graphical presentations of relational information. *ACM Transactions on Graphics*, **5**(2), 110–141.
Mackinlay, J. and Genesereth, M. 1985. Expressiveness and language choice. *Data & Knowledge Engineering*, **1**(1), 17–29.
Schmidt, D.C. 2006. Model-driven engineering. *IEEE Computer*, **39**(2), 41–47.
Simonyi, C., Christerson, M., and Clifford, S. 2006. Intentional software. *ACM SIGPLAN Notices*, **41**(10), 451–464. See also http://www.intentsoft.com/technology/overview.html.

APPENDIX

Theory of Representations

What is a representational system? There is some theory of representation that we can draw on, which is best developed in the case of measurement systems, a special case. For the theory of measurement, see the three-volume *Foundations of Measurement* by Krantz et al. (2007); for the broader theory of representations, see Mackinlay and Genesereth (1985) and Mackinlay (1986).

A representational system consists of a target domain, in which there is something we want to accomplish, and a representation domain, with mappings connecting them. The point of representation is that we map work in the target domain into work in the representation domain, where it can be done more easily or faster or better in some other way. Then the results are mapped back to the target domain, where we need them.

For any representational system to work, the answer we get when we map our question into the representation domain, and then map the answer back to the target domain, has to be the same as the answer we would have obtained if we had done the work to answer the original question in the target domain.

Many representational systems involve *approximations*, situations in which answers obtained in the representation domain do not map exactly to the answers that would have been obtained by working directly in the target domain. Here, the usefulness of the representation hinges on how important the discrepancies are, in practical terms.

Measurement systems are representational systems in which the representation domain is a fairly simple mathematical structure, such as numbers. But in many representational systems, the representation domain is not a simple mathematical structure, or a mathematical structure of any kind. For example, for a simple bar chart the representation domain is not a mathematical structure, it is marks on a piece of paper or a pattern of colored dots on a computer screen. In many representational systems, including the simulators considered in this chapter, the representation domain is neither a mathematical domain nor physical marks, but a computational structure.

While the representational system sketched in Figure 4.1 has just two related domains, representational systems are characteristically *layered.* That is, as suggested in Figure A4.1, the target domain is mapped to representation domain 1, but then representation domain 1 serves as the target domain for representation domain 2, which is further mapped to representation domain 3, and so on. Each mapping delivers value by mapping questions from one domain into another domain in which they can be answered more easily. For example, many questions about physical systems are answered by mapping them first

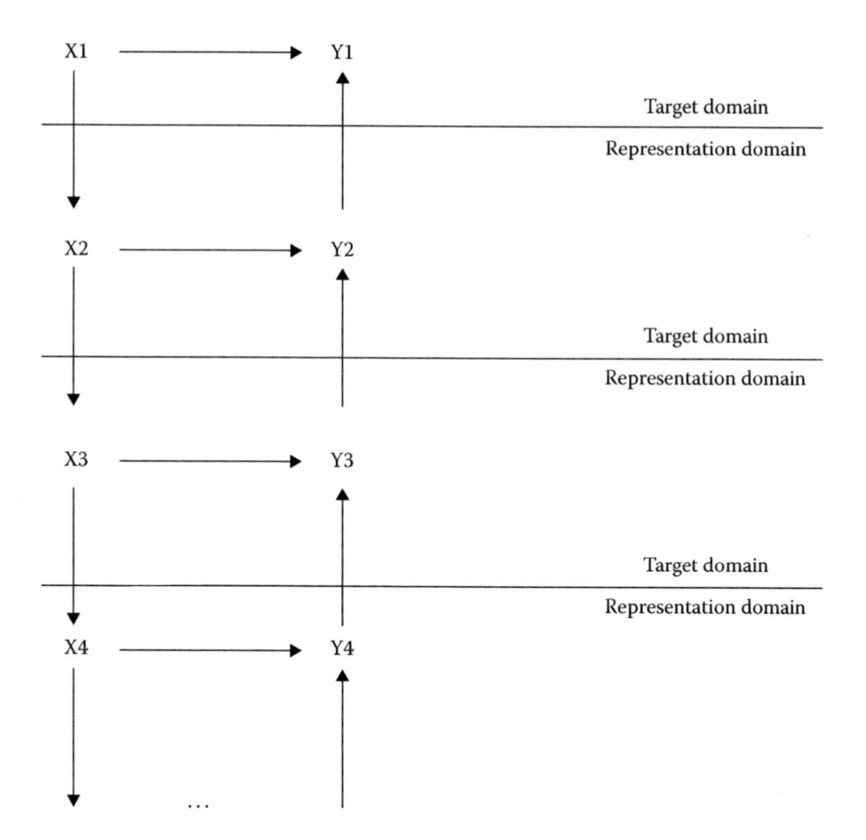

FIGURE A4.1 Layered representational systems.

into a mathematical representation, capturing the key relationships, and the mathematical representation is then mapped to a computational one, allowing answers to questions about the original physical system to be answered by automatic computation, which is fast and cheap.

In the traffic simulation application, layering arises in a different way. As mentioned in the body of the chapter, the student users of the simulator are not expected to deal with any real traffic simulations. Rather, they will be studying *hypothetical* traffic situations, and using the simulator to work out the behavior to be expected in these imaginary situations.

Figure A4.2 shows the layering involved in this application. Note that the relationship between real and hypothetical situations, and their behaviors, shown in the figure, must hold if the behavior of hypothetical situations is to be constrained in a useful way.

Expressiveness and Effectiveness

Mackinlay (1986) introduced terms for two key aspects of representational systems, in the context of data visualization. "Expressiveness" is structural adequacy: does the structure of the representation domain accurately mirror the structure of the target domain? Thus, our first design question, *How do software designers ensure that the systems they create have structure and behavior that correspond appropriately to entities in the target domain?* could be restated, *How do software designers ensure the* ***expressiveness*** *of the systems they create?*

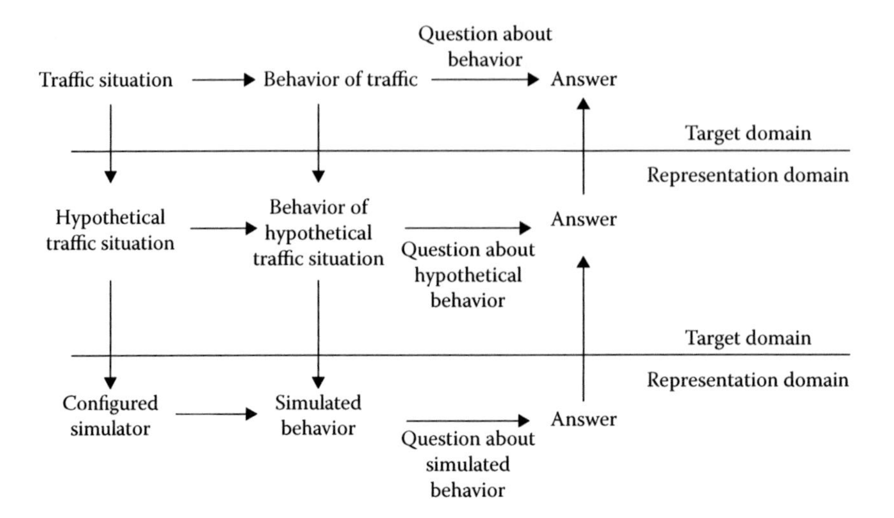

FIGURE A4.2 Representation of hypothetical traffic situations.

Mackinlay distinguishes expressiveness from *effectiveness* as the extent to which the human perceptual operations required to use a visualization can be easily and accurately carried out. Thus, our second design question, *How do software designers create perceptual representations that allow users to answer the questions of importance to them easily and accurately?* could be restated, *How do software designers ensure the* ***effectiveness*** *of the systems they create?*

Target-Oriented and Implementation-Oriented Control

The aspects of the operation of a computational system that users may need to control can be separated into two groups: *target* oriented and *implementation* oriented. Target-oriented aspects are ones that reflect the characteristics of the target domain, in this case imagined roads, traffic signals, traffic, and so on. For example, the user needs to be able to configure the simulation in such a way that it represents a situation in which the timing of certain signals is managed in some desired manner. An implementation aspect is one that modifies something about how the simulation runs, but not the situation it represents. For example, the time step used in a simulation is an implementation aspect.

Appropriate representations in the human perception and action domain would be expected to differ between the target and the implementation aspects of control. If useful representations of the target domain are already available, these would be candidates for representing the target-oriented aspects of control, but would not be expected to help with the implementation aspects.

The implementation aspects can be expected to pose a challenge in many situations, for two reasons. First, users often have some understanding of the target domain, and none of the implementation, and this makes providing target-oriented cues easier. Second, even if users have little or no understanding of the target domain, by definition they are interested in it, and they may be willing to invest in understanding the representations of it. Understanding implementation considerations is only of secondary interest.

None of the design teams in the study addressed implementation-oriented controls, though these would undoubtedly be needed in a real system.

CHAPTER 5

Studying Software Design Cognition

Jeff Wai Tak Kan

Independent Design Researcher

John Gero

Krasnow Institute for Advanced Study

CONTENTS

5.1 INTRODUCTION

What do software designers do when they design? While there have been many studies of software designers, few if any have focused on their design cognition—the cognitive activities while they design. How does a team of software designers expend their cognitive effort during a design session? This chapter addresses this question by presenting a case study of a team of software designers where their cognitive effort is measured and analyzed.

Over the past three decades, protocol analysis has become one of the most widely used methods to study human design activities and cognitive design processes [1,2]. However, unlike many other research domains, there is a lack of agreement in both the terminology and the research methodology. In protocol studies of designers, there is no uniformity in the segmentation and coding of design protocols. The papers in DTRS2 [1] and DTRS7 [3] demonstrate the range of methods and coding schemes applied to studying the same raw data. This chapter uses a generic coding scheme to study a team of professional software designers. By using this generic method, the results from this study can compared with those from other studies that use the generic scheme.

5.2 FUNCTION–BEHAVIOR–STRUCTURE DESIGN ONTOLOGY

In order to establish a common ground to study design activities, we propose to use an ontology as an overarching principle to guide the protocol study. Gruber [4] defines ontology, apart from its use in philosophy, as an explicit specification of a conceptualization. The knowledge of a domain is represented in a declarative formalism in terms of a set of objects and their relationships. An ontology may be thought of as the framework for the knowledge in a field.

A design ontology is described by defining the representational terms and their relationships. Gero [5] viewed designing as a purposeful act with the goal to improve the human condition. Gero [5] stated: "The meta-goal of design is to transform requirements, more generally termed functions which embody the expectations of the purposes of the resulting artifact, into design descriptions. The result of the activity of designing is a design description."

The coding of the protocols will be based on a general design ontology, namely, the function–behavior–structure (FBS) ontology [5] as a principled coding scheme (as opposed to an ad hoc one) (Figure 5.1).

The FBS design ontology [5], as a formal model, represents designing in terms of three fundamental, nonoverlapping classes of concepts called design issues: function or purpose (F), behavior (B), and structure (S); along with two external classes: design descriptions (D) and requirements (R). In this view, the goal of designing is to transform a set of functions into a set of design descriptions. The function of a designed object is defined as its teleology; the behavior of that object is either expected (Be) or derived (Bs) from the structure (S), that is, the components of an object and their relationships. A design description cannot be transformed directly from the functions, which undergo a series of processes among the FBS variables. Figure 5.1 shows the eight FBS design processes in relation to the function (F), behavior (B), and structure (S) state variables. The formulation (R → F, F → Be) is the transformation of the design intentions from requirements, expressed in terms of functions, into expected behavior. Synthesis (Be → S) is the transformation of the expected behavior (Be) into a solution structure. Analysis (S → Bs) is the derivation of "actual" behavior from the synthesized structure (S). Evaluation (Bs ↔ Be) is the comparison of the actual behavior (Bs) with the expected behavior (Be) to decide whether the solution structure (S) is to be accepted. Documentation (S → D) is the production of the design description.

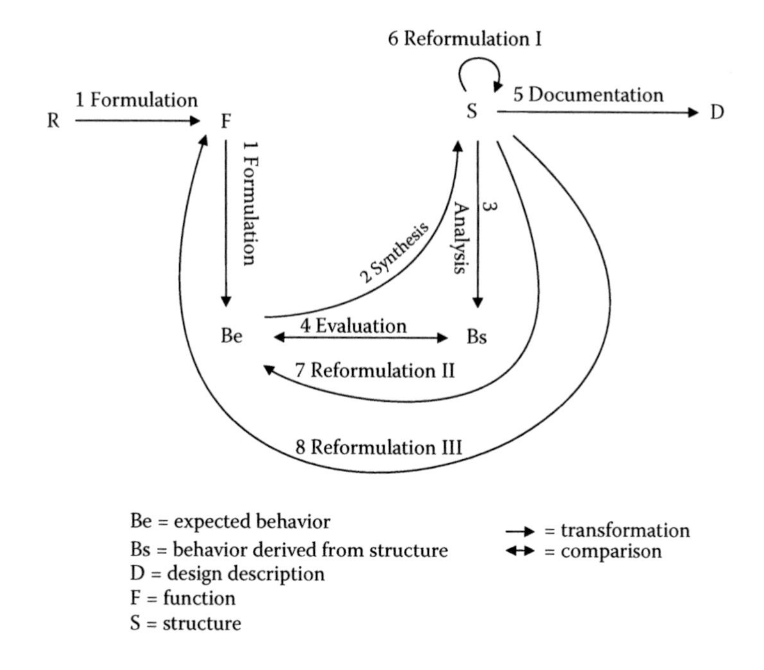

FIGURE 5.1 The FBS ontology of design processes.

Traditional models of designing iterate the analysis-synthesis-evaluation processes until a satisfactory design is produced. The aim of introducing the three types of reformulations is to expand the design state space so as to capture the innovative and creative aspects of designing. These have not been well articulated in most models because they have not been adequately understood.

Reformulation type I ($S \rightarrow S'$) addresses changes in the design state space in terms of the structure variables or the ranges of values for them. Reformulation type II ($S \rightarrow Be'$) addresses changes in design state space in terms of the behavior variables. A review of the synthesized structure may lead to the addition of expected behavior variables. Reformulation type III ($S \rightarrow F'$) addresses changes in the design state space in terms of the function variables.

5.2.1 Software Designing and FBS

The basis of the FBS view is that all design tasks, irrespective of their domain, can be described using a generic framework such as the FBS framework, as any design task will have functions or purposes (F) to fulfill for which the object is being designed, and the resulting behaviors (Be and Bs) and structure (S). Therefore, one significant advantage that such an approach has over those not based on an ontology is that the FBS approach can be used to represent designing from various domains under a common ontological structure.

In designing physical objects, the manifestation of the structure (S) variables will usually be some physical reification. For example, to design a portable shelter, the function (F) of the shelter can be formulated to an expected behavior (Be) of a space with a simple erection method that provides protection. This may be synthesized into an "A-frame" structure (S). With this structure, one can analyze its behavior (Bs), for example, headroom, and its structural stability. After evaluating these behaviors, the designer may accept or reject this structure (S).

However, in the design of software, the structure (S) will not have any physicality. In a typical object-oriented programming paradigm, the objects or patterns of objects and their relationships are the structure. The software designers formulate the expected behaviors of objects from the functions of the software they design. With these expected behaviors, they can synthesize the objects or the relationships of the codes of those objects. With these objects, they can derive their behaviors by running that part of the program, or formally or mentally simulating their behaviors. For example, in this protocol, which is the result of designing a traffic signal simulator (based on the design requirements outlined in http://www.ics.uci.edu/design-workshop/files/UCI_Design_Workshop_Prompt.pdf), the input, roads, and intersections are objects that were considered as part of the structure (S). During designing there were discussions concerning the expected behavior, such as the drag-and-drop environment of the input, and the signal timing of the intersections. An example of the behavior derived from the structure is the time for cars to travel between intersections.

It is useful to think of the concepts of independent and dependent variables in a software design formalism to provide an analogy to the relationships between structure (S) and behaviors (Be and Bs). An independent variable is one that may be varied freely or autonomously by a designer. A dependent variable is one that derives from the interaction that occurs between the independent variables, such that its behavior derives from the changes that are effected through the independent variables. In our coding of the protocol of software designers, structure (S) comprises those variables that the designers manipulate independently (or conceive that the users of the program may wish to do). In this protocol, examples include the number of lanes, cars, and intersection geometry. Behavior comprises those properties that derive from these, for example, the number of cars that can clear through the intersection in a given amount of time.

5.3 OBSERVATIONS OF THE SESSION

In this section, we present some basic qualitative and quantitative observations of the Intuit protocol session in the National Sciences Foundation workshop on Studying Professional Software Design (http://www.ics.uci.edu/design-workshop/index.html). Since we are concerned with showing the applicability and utility of using a generic method to study software design cognition, the specifics of the design task are not of particular interest and are not repeated here. They can be found at http://www.ics.uci.edu/design-workshop/files/UCI_Design_Workshop_Prompt.pdf. Similarly, the protocol videos and the transcripts of the utterances in the protocol videos are not presented here. They can be found at http://www.ics.uci.edu/design-workshop/videos.html.

5.3.1 Qualitative

We refer to the two participants as Male 1 and Male 2. Male 1 seemed to be more senior and took the leadership role; he started by asking Male 2 to give a summary and how he saw the program structure could be broken down. Male 1 did most, if not all, of the drawing and writing on the whiteboard. They spent more than 5 minutes at the beginning to protocol the design brief individually without verbalizing. Then, Male 1 suggested to separate the user interface (UI) from the underlying data structure and to focus on the types of data; a few entities/variables were proposed—“signals,” “roads,” “cars,” etc. Afterward, they examined if it was necessary to have the “lanes” as an object, the expected behaviors of the software were discussed, and the scenarios of traffic at intersections were simulated on the whiteboard, especially with the left-turning situation. With the “signal” object, the expected behavior of the “rules” to control the signals was reintroduced by Male 2; he suggested very early on that in the program structure there are rules applied to each intersection. “Intersection” as an entity and its control were discussed. With these, they established cases with the discussion of “distance,” “speed,” and “travel time”; an object-oriented software structure was proposed at around 30 minutes. “Use cases” was proposed and physical structure analogies, such as “protected left,” “add car,” “road becomes a queue,” “a map contains intersections,” and “meta-controller” were deliberated. Manual simulations were carried out throughout the session to evaluate and design the required program. Near the end of the session, the structure of the program—different from the physical structure—was put forward.

5.3.2 Quantitative

The turn taking by the two designers was very even; Male 1 had 125 utterances and Male 2 had 123 utterances. The word counts of Male 1 and Male 2 were 2777 and 2894, respectively. Figure 5.2 shows the word count of the two individuals in 5-minute intervals without breaking up the utterances. It shows the variation of the number of words of both

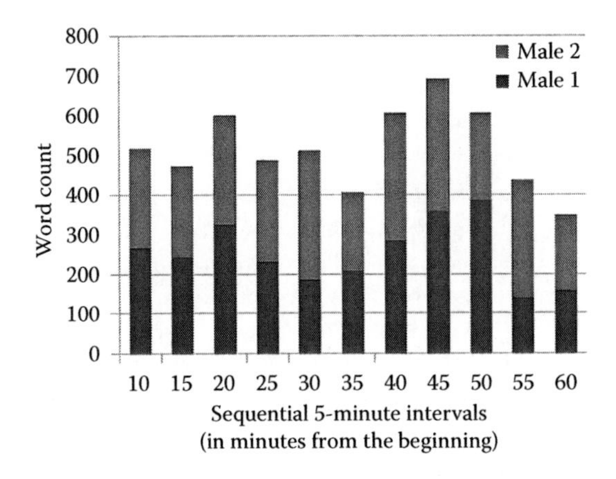

FIGURE 5.2 Word count in sequential 5-minute intervals over the protocol for Male 1 and Male 2.

designers across the session. After around 45 minutes, there was a constant drop in the word counts. At the 50 and 55 minute marks, there was a reverse of dominance.

5.4 FBS DESIGN ISSUES SEGMENTING AND CODING SCHEME

The generic coding scheme applied here consists of function (F), expected behavior (Be), behavior derived from structure (Bs), structure (S), documentation (D), and requirement (R) codes. The protocols are segmented strictly according to these six categories. Those utterances that do not fall into these categories are coded as others (O); these include jokes, social communication, and management. Table 5.1 shows some examples of the segmentation and coding.

This framework does not assume any domain of designing nor does it assume any particular number of participants being studied. It only abstracts and describes the cognitive focus of designing in terms of the FBS design issues of a segment. With this generic coding scheme, although simple, we aim to capture some fundamental aspects of designing which will form a foundation for further analysis. Unlike other coding systems, there is a nexus between segments and codes using this coding scheme: one segment per code and one code per segment. This forms the basis of segmentation. This is likely to produce a higher degree of uniformity and robustness than other methods since the same design issues are always used.

The segmentation/coding involved two independent coders who segmented/coded separately (in different geographical locations with no intercoder influence) and who then carried out an arbitration session to produce the final arbitrated segmented/coded protocol. The agreement between the two coders was 82.8% and the agreement between the coders and the final arbitrated results was 90.4%. The high percentages of agreement provide additional support for the claim that the FBS ontology-based coding methodology provides a consistent and coherent way to view a protocol by different people. In the absence of such

TABLE 5.1 Examples of Protocol and Codes

Segment	Name	Transcript	Code
1	Interviewer	Feels like school again	O
2	Male 1	Well, I want to start by hearing your summary of this	O
3	Male 2	Gotcha, well. Looks like basically two pieces:	R
…	…	…	…
27	Male 1	you know I think about this as a software application.	F
28	Male 1	Looks totally like you want to pull out some kind of patterns and some kind of patterns and [inaudible] controller type of thing.	Be
29	Male 1	And so if we extract the UI piece first,	S
30	Male 1	and then let's focus on kind of the underlying stuff	S
31	Male 1	in order to support you know kind of the traffic flow.	Be
…	…	…	…
38	Male 1	So focus on the data pieces for this particular thing.	S
39	Male 1	Writes: "Data"	D

a unifying, guiding ontological framework, different coders can produce widely differing segmentation schemes, making any kind of quantitative analysis not robust and, more significantly, the results from individually generated coding schemes cannot be compared unless the codes can be directly mapped onto each other.

5.5 RESULTS AND ANALYSIS

5.5.1 Descriptive Statistics

The segmented protocol contains 640 segments, 603 of which are design issue codes, those that do not relate to design issues are coded as "O" and they are not counted further in this analysis. Figure 5.3 shows the distribution of the FBS codes. Male 2 referred to the requirements more than Male 1, but overall the "R" coded segments were sparse (17 in total). Only three function (F) issues were raised (Male 1:2, Male 2:1). Male 1 did most of the documentation, where by "documentation" we mean the externalization of thoughts by drawings, symbols, and words. The structure (S) and expected behavior (Be) issues dominated this session. Under the FBS ontology, the interaction of S and Be can be either a synthesis process (Be → S) or a reformulation process (S → Be′). A Markov chain will be used in the next subsection to analyze this kind of interaction.

The design session was divided into three equal-sized thirds based on the number of segments, to provide the basis for a more detailed analysis. Figure 5.4 shows the change of distributions of the FBS design issues across the three thirds. Changes can be observed qualitatively in this figure and statistical analyses can be carried out to confirm the qualitative results.

Requirement issues were raised mostly in the first third of the session with Male 2 contributing double the number of requirement issues. In the second and third parts of the session, each of the participants only raised one requirement issue. In this design session, function issues were rare and only occurred in the first third of the session—no function issue was observed in the second and third thirds of the design session.

Expected behavior issues represented one-third of the cognitive effort of these software designers. Male 1 contributed more behavior issues than Male 2 in the first third of the session. Their roles reversed as the design session progressed. This is an indication of a change

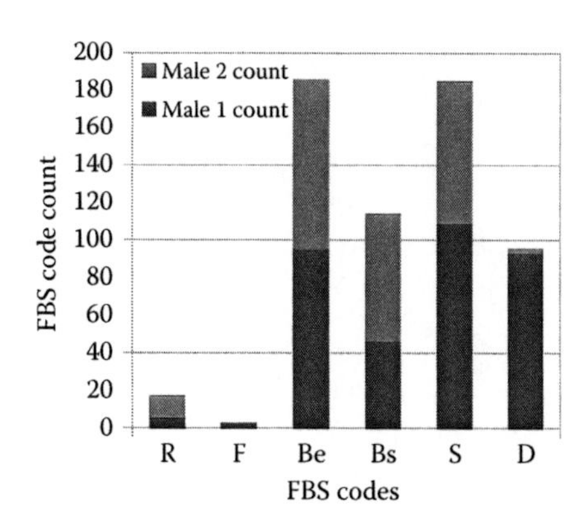

FIGURE 5.3 The distribution of design issues via FBS codes for Male 1 and Male 2.

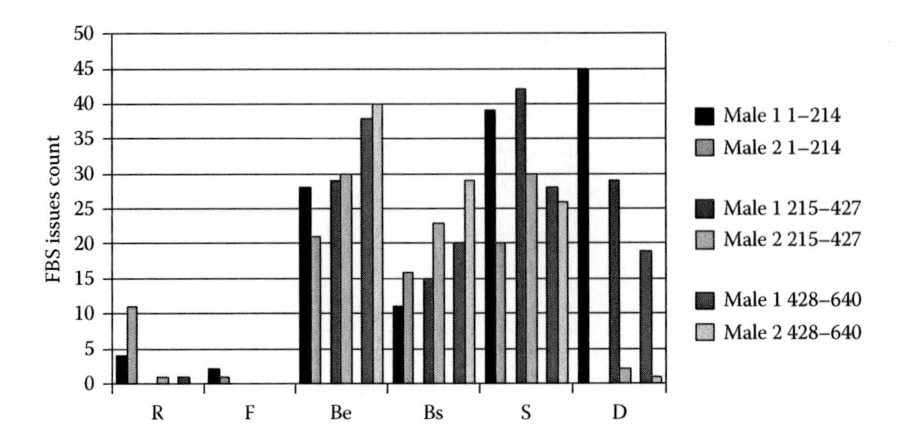

FIGURE 5.4 The design issue, via FBS code, distributions for the beginning, middle, and end of the session for Male 1 and Male 2.

in the relationship between the two team members over time. There was a steady increase in the focus on behavior from structure as the design progressed. This change in cognitive focus is in line with theories about software design.

The number of structure issues peaked in the middle of the session with Male 1 being more dominant. Like other design domains, structure is the dominant design issue in software design.

Unlike domains involving the design of physical artifacts, the action of documentation decreased as the session progressed. Male 1 did most of the documentation, while Male 2 did not have any documentation action in the first third of the session. We can also observe that Male 1 focused more on structure while Male 2 focused more on behavior. This division of contribution within a team is an indication of either background experience or role playing and requires further analysis to disentangle.

5.5.2 Sequential Analysis

Traditionally, design protocols have been analyzed by static statistical methods that are often based on the assumption that each segment is an independent event. Markov chains can be used to model or describe the probability of one event leading to another. Kan [6] proposed using a Markov analysis to examine the sequence of FBS states. Kan and Gero [7] used a Markov analysis to compare multiple design sessions. Markov chains have also been used to analyze writer's manuscripts and to generate dummy text [8]; to rank web pages by Google [9]; and to capture music compositions and synthesize scores based on analyses [10], to name some of its applications.

A probability matrix P is used to describe the transitions of the six FBS states (R, F, Be, Bs, S, and D). Equation 5.1 shows the transition matrix of this software design session.

There are only three occurrences of the F event, which result in a round off to zero in many of the rows and columns containing F. The occurrences of R are also low with some round off to zero as well. Any nonzero values of the probability of the transitions from R to

R are therefore less reliable than other transitions. *P* is one characterization of the dynamics of the entire design session. It shows that the most likely transition for this design team is from thinking about the purpose of the design (F) to its expected behavior Be (.67).

$$P = \left(\begin{array}{c|cccccc} & \mathrm{R} & \mathrm{F} & \mathrm{Be} & \mathrm{Bs} & \mathrm{S} & \mathrm{D} \\ \hline \mathrm{R} & .29 & .00 & .12 & .00 & .41 & .18 \\ \mathrm{F} & .00 & .00 & .67 & .00 & .00 & .33 \\ \mathrm{Be} & .00 & .01 & .21 & .23 & .39 & .16 \\ \mathrm{Bs} & .02 & .00 & .47 & .15 & .29 & .07 \\ \mathrm{S} & .03 & .01 & .26 & .20 & .19 & .31 \\ \mathrm{D} & .01 & .00 & .42 & .15 & .43 & .01 \end{array}\right) \tag{5.1}$$

As mentioned earlier, the occurrences of Be and S issues dominated the session, and the probabilities of issues following these two issues characterized this session. The most likely issue that follows a Be issue is S (.39). The transition from Be to S is a synthesis process. Table 5.2 shows three sequences of Be and S issues that are synthesis processes. In the first one from segment 153 to segment 154, Male 1 synthesized the "hierarchy" of rules for the signals from the expected behavior of traffic at an "intersection." In the second sequence, 156–157, two program structure entities were proposed that were synthesized from the expected behavior of "has to synchronize these signals." In the third sequence, 160–161, the expected behavior of "controlling the interactions" translated to "an encapsulated entity."

TABLE 5.2 Examples of S after Be from the Protocol That Are Synthesis Processes

Segment	Name	Transcript	Code
153	Male 1	The intersection needs to say that (gesturing over drawing and words) only one—only safe ones go on at a time so if these are green (gesturing over drawing) these have to be red and ...	Be
154	Male 1	... so there's this kind of this hierarchy where the intersections own (draws ...) the signals and ...	S
155	Male 1	Draws lines and writes S1 S2 S3 that linked to intersections	D
156	Male 1	... so we have this behavior where somebody has to synchronize (drawing circle to enclose S1 S2 and S3) these signals, and so does that occur in the intersection?	Be
157	Male 1	As a containing data entity? Or conceptual entity?	S
158	Male 2	It sounds like more and more like the intersection comes from [inaudible] ...	Be
159	Male 2	... because basically it's going to have given (gesturing over intersections and S1) S1 goes green, it's going to have to delegate the actions of what S2 and S3 are; is it safe from stuff like that	Bs
160	Male 1	Exactly, exactly. Somebody is controlling the interactions	Be
161	Male 1	If you think of this as kind of an encapsulated entity	S
162	Male 1	Draws circle enclose S1	D

The most likely issue that follows an S issue is D (.31), which is a documentation process. An example is the transition from segment 154 to segment 155 in Table 5.2, where the drawn lines represent the hierarchy and the ownership of the "intersections" entity. Another example is from segment 161 to segment 162, where the "encapsulated entity" was drawn as a circle.

The second most expected issue that follows an S issue is Be (.26); segments 157 and 158 in Table 5.2 are an example. We infer from the context that Male 2 was suggesting/expecting the behavior of the intersection based on the proposed entities in segment 157.

Dividing the design session into three equal-sized thirds based on the number of segments allows us to compare the design behaviors during the beginning, middle, and end of the session to determine if there are changes in behavior as the session progresses. P_1, P_2, and P_3 are the probability matrices of the segments from 1 to 214, 215 to 427, and 428 to 640, respectively, for the three thirds of the session.

$$P_1 = \left(\begin{array}{c|cccccc} & R & F & Be & Bs & S & D \\ \hline R & .33 & .00 & .13 & .00 & .33 & .20 \\ F & .00 & .00 & .67 & .00 & .00 & .33 \\ Be & .00 & .02 & .09 & .16 & .42 & .30 \\ Bs & .04 & .00 & .40 & .16 & .28 & .12 \\ S & .07 & .04 & .16 & .14 & .16 & .44 \\ D & .03 & .00 & .40 & .18 & .43 & .00 \end{array} \right) \quad (5.2)$$

$$P_2 = \left(\begin{array}{c|cccccc} & R & F & Be & Bs & S & D \\ \hline R & .00 & .00 & .00 & .00 & 1.00 & .00 \\ F & .00 & .00 & .00 & .00 & .00 & .00 \\ Be & .00 & .00 & .26 & .08 & .49 & .17 \\ Bs & .00 & .00 & .40 & .23 & .34 & .03 \\ S & .01 & .00 & .26 & .24 & .23 & .26 \\ D & .00 & .00 & .37 & .17 & .43 & .03 \end{array} \right) \quad (5.3)$$

$$P_3 = \left(\begin{array}{c|cccccc} & R & F & Be & Bs & S & D \\ \hline R & .00 & .00 & .00 & .00 & 1.00 & .00 \\ F & .00 & .00 & .00 & .00 & .00 & .00 \\ Be & .00 & .00 & .24 & .39 & .31 & .07 \\ Bs & .02 & .00 & .57 & .09 & .25 & .07 \\ S & .00 & .00 & .38 & .21 & .17 & .23 \\ D & .00 & .00 & .53 & .05 & .42 & .00 \end{array} \right) \quad (5.4)$$

These three matrices characterize the transitions in each third of the session and are used to measure differences across the design session. If we examine the transition from F to Be in these parts of the session, we can observe a rapid diminution of this activity: in the first third its probability is .67 and in the second and final thirds it is .00. Similarly, we can trace the behavior of other transitions from the beginning, through the middle, to the end of the design session and use these results to build a quantitative model of designing behavior.

If we assume that each segment is cognitively related to its immediately preceding segment and that the transition probabilities indicate design processes, the first third of the design session had high transition probabilities of formulation (F → Be: .67), documentation (S → D: .44), reformulations I and II (D → S: .43, D → Be: .40), and synthesis (Be → S: .42). The second third of the design session had high transition probabilities of synthesis (Be → S: .49), reformulation I (D → S: .43), and evaluation (Bs → Be: .40). The last third had high transition probabilities of evaluation (Bs → Be: .57) and reformulation II (D → Be: .53).

If we only examine the highest transition probabilities of each third, they are P_1: .67 (F → Be), P_2: .49 (Be → S), and P_3: .57 (Bs → Be). This follows the FBS ontology of designing that starts with formulation (F → Be: .67), then synthesis (Be → S: .49) followed by evaluation (Bs → Be: .57). This also confirms the analysis-synthesis-evaluation model of designing, where the formulation in the FBS model is a more articulated form of "analysis" as described in other models.

If we select the probabilities of those transitions that represent the eight design processes in Figure 5.1, we can plot the design processes of this design session for the beginning, middle, and end of the session (Figure 5.5).

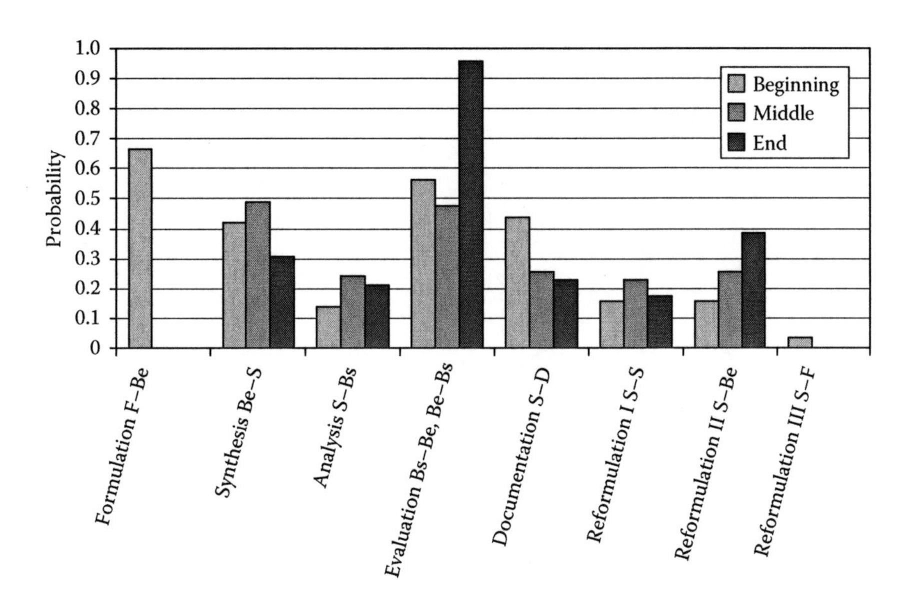

FIGURE 5.5 The transition probability distributions for the eight design processes for the beginning, middle, and end of the session.

From the transition probabilities in Figure 5.5, we can infer the behavior of this design session in terms of design processes. Formulation only occurs at the beginning of the session. Synthesis increases in the middle of the session and then drops off. Analysis is low at the beginning. There is a significant increase in evaluation and reformulation II at the end of the session. The transition probability matrix results are concordant with the qualitative analysis and they provide the basis for further quantitative characterizations of this design session such as the mean first passage time.

The mean first passage time M is the average number of events traversed before reaching a state from other states. The mean passage time can be obtained from the probability matrix. Kemeny and Snell [11] proved that the mean first passage matrix M is given by

$$M = \left(I - Z + EZ_{dg}\right)D \tag{5.5}$$

where

I is an identity matrix

E is a matrix with all entries of 1

D is the diagonal matrix with diagonal elements $d_{ii} = 1/\alpha_i$

Z is the fundamental matrix such that

$$Z = \left(I - (P - A)\right)^{-1} \tag{5.6}$$

Z_{dg} is the diagonal matrix of Z

A is the probability matrix, with each row consisting of the same probability vector $a = (\alpha_1, \alpha_2, \ldots, \alpha_n)$, such that $\alpha P = \alpha$.

Equation 5.7 describes the number of FBS states expected to be passed through from one state before reaching other FBS states.

$$M = \left(\begin{array}{c|cccccc} & \text{R} & \text{F} & \text{Be} & \text{Bs} & \text{S} & \text{D} \\ \hline \text{R} & 49.1 & 168.4 & 3.9 & 6.7 & 2.5 & 5.0 \\ \text{F} & 70.0 & 168.4 & 1.9 & 6.2 & 3.7 & 4.5 \\ \text{Be} & 69.3 & 167.3 & 3.2 & 5.0 & 2.7 & 5.3 \\ \text{Bs} & 68.3 & 168.2 & 2.5 & 5.4 & 2.9 & 5.8 \\ \text{S} & 67.6 & 166.7 & 3.1 & 5.2 & 3.2 & 4.6 \\ \text{D} & 68.6 & 168.1 & 2.7 & 5.4 & 2.6 & 6.0 \end{array}\right) \tag{5.7}$$

Equations 5.8 through 5.10 are the unnormalized mean first passage times matrices of the design session divided into thirds: segments from 1 to 214, 215 to 427, and 428 to 640, respectively. These provide the quantitative bases for an understanding of the character and the extent of the iterative nature of the design session through changes in the mean length of the iteration cycle across the session. For example, the mean length of the passage time from R to R in the first third is 20.1 and it then increases to around 190 in the remainder of the session; the mean length of the passage time from F to F in the first third is 61.3 and it then increases to infinity (meaning there is no return to F) in the remaining two thirds; and the mean length of the iteration cycle of S to Bs starts at 6.9 in the first third, shortens to 5.6 in the middle third, and drops to 4.0 in the final third. These results can then be used to characterize quantitatively this particular design session.

$$M_1 = \left(\begin{array}{c|cccccc} & \text{R} & \text{F} & \text{Be} & \text{Bs} & \text{S} & \text{D} \\ \hline \text{R} & 20.1 & 61.6 & 4.4 & 8.3 & 2.8 & 3.6 \\ \text{F} & 30.4 & 61.3 & 2.1 & 7.7 & 3.6 & 3.1 \\ \text{Be} & 29.6 & 60.0 & 4.1 & 6.7 & 2.6 & 3.2 \\ \text{Bs} & 28.8 & 61.3 & 3.1 & 6.8 & 2.9 & 3.8 \\ \text{S} & 27.9 & 59.6 & 3.8 & 6.9 & 3.3 & 2.8 \\ \text{D} & 29.1 & 61.1 & 3.2 & 6.7 & 2.6 & 4.2 \end{array}\right) \tag{5.8}$$

$$M_2 = \left(\begin{array}{c|cccccc} & \text{R} & \text{F} & \text{Be} & \text{Bs} & \text{S} & \text{D} \\ \hline \text{R} & 192.8 & \infty & 4.3 & 6.6 & 1.0 & 6.4 \\ \text{F} & 194.2 & \infty & 4.2 & 6.9 & 3.4 & 6.8 \\ \text{Be} & 193.9 & \infty & 3.3 & 6.4 & 2.1 & 5.6 \\ \text{Bs} & 194.3 & \infty & 2.9 & 5.6 & 2.5 & 6.6 \\ \text{S} & 191.8 & \infty & 3.3 & 5.6 & 2.8 & 5.4 \\ \text{D} & 194.1 & \infty & 3.0 & 6.0 & 2.3 & 6.5 \end{array}\right) \tag{5.9}$$

$$M_3 = \left(\begin{array}{c|cccccc} & \text{R} & \text{F} & \text{Be} & \text{Bs} & \text{S} & \text{D} \\ \hline \text{R} & 188.8 & \infty & 3.3 & 5.0 & 1.0 & 8.8 \\ \text{F} & 187.7 & \infty & 3.3 & 4.8 & 4.4 & 9.8 \\ \text{Be} & 187.1 & \infty & 2.6 & 3.3 & 3.3 & 9.1 \\ \text{Bs} & 183.8 & \infty & 2.0 & 4.3 & 3.4 & 9.2 \\ \text{S} & 187.8 & \infty & 2.3 & 4.0 & 3.7 & 7.8 \\ \text{D} & 188.2 & \infty & 2.1 & 4.4 & 2.9 & 9.6 \end{array}\right) \tag{5.10}$$

What these mean passage times for this design session show is how formulation decreases over the duration of the session, while at the same time analysis increases. These are the quantitative measures of qualitative design behaviors.

Using different assumptions, we can obtain additional results in other dimensions. If we assume that design processes occur not with consecutive segments but with related segments, we can produce other data for design process behavior. For example, when there is an F segment, it needs a connection to a later Be segment to form a formulation process (F → Be). With the mean first passage time, we can predict when the next Be will appear. Using Equations 5.8 through 5.10, we can measure those average first passage times that form the eight design processes shown in Figure 5.6.

The measured shortest mean passage time: for a formulation process (2.1) it was at the beginning of the session; for a synthesis process (2.1) it was in the middle of the session; for analysis (4.0) and evaluation (3.3) it was at the end of the session; and for a documentation process (3.3) it was at the beginning of the session. These are quantitative measures of qualitative behavior. Considerable detailed information can be derived from the data in Figure 5.6.

Figure 5.6 also shows that there is a change in the behaviors of analysis, reformulation II, and documentation when an S design issue occurs. At the beginning of the session, there was an average of 2.8 steps for a documentation process, which increased to 7.8 steps at the end of the session. Whereas the steps in the analysis process after an S design issue decreased from 5.2 to 4.0; and the steps in reformulation II decreased from 3.8 to 2.3. This matches the qualitatively observed behavior of a large number of structure entities being proposed and documented at the beginning of the session, while toward the end there was more simulation, evaluation, analysis of the proposed structure, and reformulation of the expected behavior.

5.5.3 Comparison with Another Domain of Designing

We compared this software design session with an architectural design session where the task was to generate a conceptual design of a university student union's gallery. The session lasted for 30 minutes. The first 10 minutes was coded as the first third of that design session for comparison with the first third of the software design session. The design issue distributions of the first one third of the two design sessions are shown in Figure 5.7.

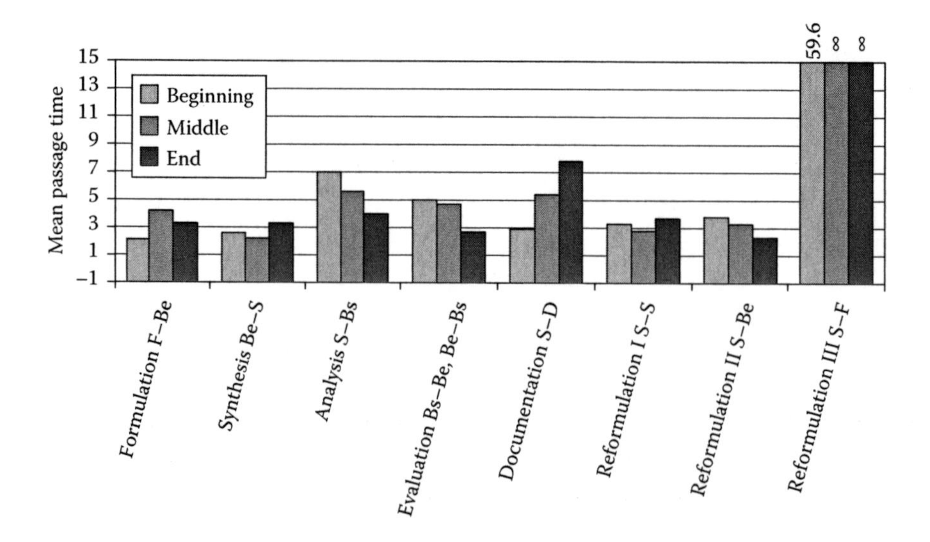

FIGURE 5.6 The distribution of the mean first passage times of the eight FBS design processes for the beginning, middle, and end of the design session.

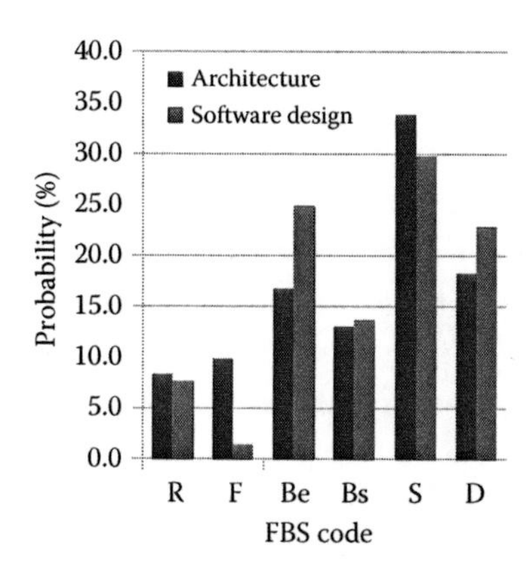

FIGURE 5.7 The design issue distributions of the software design and the architectural design sessions.

Since the same generic coding scheme was used for both domains, the results are commensurable. Figure 5.7 shows that the architectural design session had considerably more F design issues, less Be design issues, and more S design issues than the software design session.

These are interesting observations: they indicate that in the architectural design session, the designers focused more on function, while in the software design session, the designers spent very little time on function-based discussion. A qualitative analysis of the videos of the sessions confirms these quantitative observations qualitatively. It is known that in architectural design, the same structure can lend itself to a variety of function interpretations. Therefore, in architectural design, the relationship between the set of functions being considered (i.e., the main set of purposes for which the structure is being designed) is an open one, as architects continue to create or discover new functions, and continue to reformulate and redefine the already identified ones. In the software design session however, the function space appears much more defined and fully formulated from the start. There appear to be a very small number of main, explicitly defined functions, and the designers' reasoning focuses more on expected behavior, derived behavior, and structure.

We are now in a position to compare quantitative results from a wide range of measurements both within a single domain and between domains.

5.6 CONCLUSION

The results in this study demonstrate that the FBS coding scheme can be used to code and quantify the cognitive behavior of software designers. A wide range of measures of cognitive behavior have been shown to be derivable from the basic design issues that are the consequence of the segmentation/coding. Distributions of design issues and design processes across the design session and any fraction of the session can be measured. When measured across fractions of the session, the resulting differences can be tested for statistical significance. Markov models and mean passage time models can be developed for the session and can be used to develop an understanding of the design cognition of software designers.

In this design session, the designers' cognitive resources were mostly focused on expected behavior and structure issues. These two design issues accounted for 60% of the cognitive effort of this session. Formulation processes only occurred at the beginning of the session. There was an increase in synthesis processes in the middle of the session, which subsequently dropped off. Although the probability of analysis processes was lower at the beginning of the session, this session approximately reflected the analysis-synthesis-evaluation model of designing. There was a significant increase in the probability of evaluation and type II reformulation processes at the end of the session.

Further, since we are using a domain-independent coding scheme, we can develop comparisons with the design cognition of designers in other domains. In the cases presented in this chapter, the software designers were more concerned with behavior compared to the architects who had a greater concern for function and structure (form).

This discussion shows how the FBS ontological framework provides a powerful common ground from which to analyze designing activity in various domains. It aids in developing rich insights into the design cognition of software designers to compare the similarities and the differences in design content, knowledge, and processes in different domains through an analysis of *in vivo* and *in vitro* design sessions.

This ontology-based FBS coding scheme has been applied to statistically significant cohorts of designers and has been shown to be robust [12,13]. This chapter has demonstrated its utility in developing data from design protocols, data that form the basis of an understanding of the design cognition of software designers.

ACKNOWLEDGMENT

This research is supported in part by a grant from the U.S. National Science Foundation grant no. IIS-1002079. Any opinions, findings, and conclusions or recommendations expressed in this chapter are those of the authors and do not necessarily reflect the views of the National Science Foundation. We would like to thank Somwrita Sarkar for assistance with the segmentation/coding.

REFERENCES

1. N. Cross, H. Christiaans, and K. Dorst, Introduction: The Delft Protocols Workshop, in *Analysing Design Activity*, Wiley, Chichester, pp. 1–15, 1996.
2. J. McDonnell and P. Lloyd, Eds, *DTRS7 Design Meeting Protocols: Workshop Proceedings*, Central Saint Martins College of Art and Design, London, 2007.
3. J. McDonnell and P. Lloyd, Eds, *About Designing: Analysing Design Meetings*, Taylor & Francis, London, 2009.
4. T. R. Gruber, A translation approach to portable ontology specifications, *Knowledge Acquisition*, 5(2), 199–220, 1993.
5. J. S. Gero, Design prototypes: A knowledge representation schema for design, *AI Magazine*, 11(4), 26–36, 1990.
6. W. T. Kan, Quantitative methods for studying design protocols, PhD dissertation, The University of Sydney, 2008.
7. J. W. T. Kan and J. S. Gero, A generic tool to study human design activity, in *Human Behavior in Design*, M. Norell Bergendahl, M. Grimheden, L. Leifer, P. Skogstad, and U. Lindemann, Eds, Springer, Berlin, pp. 123–134, 2009.
8. H. Kenner and J. O'Rourke, A travesty generator for micros: Nonsense imitation can be disconcertingly recognizable, *BYTE*, (12), 449–469, 1984.

9. A. N. Langville and C. D. Meyer, Updating Markov chains with an eye on Google's page rank, *SIAM Journal on Matrix Analysis and Applications*, 27(4), 968–987, 2006.
10. M. Farbood and B. Schoner, Analysis and synthesis of Palestrina-style counterpoint using Markov chains, in *Proceedings of the 2001 International Computer Music Conference*, International Computer Music Association, San Francisco, CA, pp. 111–117, 2001.
11. J. G. Kemeny and J. L. Snell, *Finite Markov Chains*, The University Series in Undergraduate Mathematics, D. Van Nostrand, Princeton, NJ, 1960.
12. H.-H. Tang, Y. Y. Lee, and J. S. Gero, Comparing collaborative co-located and distributed design processes in digital and traditional sketching environments: A protocol study using the function–behavior–structure coding scheme, *Design Studies*, 32(1), 1–29, 2011.
13. J. S. Gero and U. Kannengiesser, Commonalities across designing: Evidence from models of designing and experiments, *Design Studies*, 2013 (in press).

CHAPTER 6

Does Professional Work Need to Be Studied in a Natural Setting?

A Secondary Analysis of a Laboratory Study of Software Developers

John Rooksby

University of Glasgow

CONTENTS

6.1 INTRODUCTION

One way to study the work of software developers is to invite experienced developers to participate in a simulation study conducted in a laboratory environment. In such studies, the participants are given a task to work on for a period of time. The task resembles an ordinary activity, and the study participants' workings are recorded. The simulations are contrived, but the participants have relative freedom within these contrivances to work as they see fit. The simulations are not necessarily experimental, although there may be an element of this. The advantages of laboratory simulations are that (a) the task will have a relatively clear start and end point, and it can be limited to a manageable period of time; (b) the researcher has opportunities to make work visible that may otherwise not be (e.g., by asking the participants to talk aloud as they work, or use a whiteboard to write things down); and (c) comparable examples can be generated. However, these same advantages also give rise to criticism, particularly that laboratory simulations (a) are designed according to the researcher's preconceived ideas about what is important in the work; (b) are shaped according to the demands of generating research data; and (c) may lead to acting or unusual behavior among the participants.

Simulation data are sometimes treated with suspicion, with preference given to analyzing examples of work as it is done "in the wild" (e.g., Crabtree et al. 2012; Szymanski & Whalen 2011; Hutchins 1995). For example, in the context of design studies, McDonnell and Lloyd (2009) have recommended that attention be paid to "authentic design activity as it is practiced in the wild." They argue that "credible analysis relies on studying design processes and outcomes which have not been initiated by the researcher and where researcher intervention is minimised." McDonnell and Lloyd do not specifically exclude the possibility of studying design activities in the laboratory, but clearly, if we are to focus on "authentic design activity," then the contrivances inherent in laboratory simulations are problematic.

This chapter is concerned with the issue of whether to study professional work practices, we *need* to see them in a natural setting. To address this issue, the chapter looks directly at what might be said to be unnatural about a particular simulation. This is a secondary analysis, drawing upon conversation analysis (CA) to qualitatively examine the video data from a simulation conducted by other researchers. The chapter confirms that these simulation data are contrived, but it finds no overwhelming cause for concern about these contrivances. The chapter shows that, even if the study is contrived, the experience and the identity of the participants play a key role in their conduct. More broadly, the chapter argues that the naturalness or authenticity of the professional conduct in a laboratory study is not a product of the study design but it is an achievement of the researcher and the study participants in conducting the study.

6.2 BACKGROUND

This chapter presents a secondary analysis of a simulation study of professional software developers. The study, performed by Baker and van de Hoek, asked professional software developers to work on a design problem concerning a traffic signal simulator (see Baker et al. 2012; Petre et al. 2010). The developers were asked to work on this problem for up to 2 hours, to the point where they had a design that they were happy to hand over to programmers. The developers were asked to work in pairs and to use a whiteboard. Baker and van der Hoek's study is certainly contrived; the participants were asked to work in a very particular way (in pairs at the whiteboard), and assumptions were built in about the division of labor in software development (e.g., that there was a clear design phase, with the design being worked up from scratch and then handed over to programmers). These contrivances are useful for the purposes of research, in that they enable the videos to stand for themselves as self-contained episodes of design. The contrivances of the study were not as extreme as they might have been; for example, the participants were studied in meeting rooms within the organizations in which they work, using the whiteboards already located there. But does this level of contrivance mean that the developers were no longer engaging in what McDonnell and Lloyd call an "authentic design activity"?

Baker and van der Hoek made three videos from their study, which are available for other researchers to analyze. To date, more than 10 papers analyzing the videos have been published, and a further 30 working papers were presented at a workshop (see Baker et al. 2012; Petre et al. 2010). These papers generally treat the videos not just as tasks in which professional software developers participated, but as examples of how professional software developers go about designing. One paper is critical of the effectiveness of the approaches that the developers took to completing the task (Jackson 2010), but none have directly addressed the more fundamental issue of whether the tasks and the activities of completing them actually resembled how the developers participating in the study would work in real-world circumstances. This chapter will adopt a skeptical position, exploring how the laboratory constraints affected and shaped the activities of the study participants. Most of the papers do tie their analysis into the broader, empirical and theoretical literature, and many are written by people with experience in design and development, and so major problems with the data were always unlikely (and indeed are not found). However, the ultimate intention of this work is not to attack or undermine the videos, but to address a wider methodological question: in order to understand professional work, do we need to study it "in the wild"?

One paper that draws on Baker and van der Hoek's videos is a qualitative study by McDonnell, examining how a pair of developers manage a disagreement (McDonnell 2012). Elsewhere, McDonnell has pointed to the importance of research "in the wild" (McDonnell & Lloyd 2009). In her paper concerning the simulation data, McDonnell warns that "the 'laboratory' aspects of the setting are a pervasive influence on what we see" (McDonnell 2012, p. 46). However, she proceeds with her analysis, putting the laboratory influences to one side with a comment about the study participants' experience and their institutional identities. This chapter will explore the relationship between the laboratory contrivances and the experience and identity of the participants. Rather than directly addressing how the developers work on the design problem, this chapter addresses how their work is constrained and shaped by the laboratory aspects of the study. To do this, it examines examples of how the participants orient to features of the study, for example, where they talk with the researcher, discuss the camera, and read through the task sheet.

6.3 CONVERSATION ANALYSIS

McDonnell's is one of five papers (McDonnell 2010, 2012; Matthews 2010; Rooksby & Ikeya 2012; Luck & Ikeya 2010) that draw in whole or in part on CA to analyze Baker and van de Hoek's videos. A further five papers (Gross 2010; Détienne 2010; Visser 2010; Nickerson & Yu 2010; Nakakoji et al. 2010) adopt roughly analogous approaches. This chapter also draws upon CA.

CA has its origins in the work of Sacks (1992) and Garfinkel (1967) (see Heritage 2005; ten Have 1999). Garfinkel broke from mainstream sociology with the development of a program of research termed *ethnomethodology*. He argued that social order is not invisibly imposed on people (e.g., through hidden forms of power or underlying cognitive structures), but it is a practical and ongoing achievement necessary in everything we do as members of society. His work focused on the routine methods by which people constitute social order; in the words of Heritage (2005), ethnomethodology rejects "the bucket theory of context" in which the context is anterior to the action; rather "it is fundamentally through interaction that context is built, invoked and managed" (p. 109). CA grew out of ethnomethodology, taking on a more strongly empirical character. While mainstream CA focuses directly on conversation, its methods have found wide applicability in studies of work practice. Matthews (2009) and Housley and Fitzgerald (2000) explain the value and applicability of CA to the study of design. Examples of its application to simulation data include Suchman (1987, 2007), Francis (1989), and Sharrock and Watson (1985).

Following CA convention (as outlined in ten Have 1999), I have collected and analyzed relevant sections of the videos. The convention is to (a) review the available data and build a collection of sequences relevant to a chosen theme, and (b) transcribe and analyze the collection (ten Have 1999). As such, CA involves a basic amount of coding, but it does not involve the development and application of a coding scheme. The aim is essentially to build a very thorough description of some aspect of how people talk and interact. For this study, I began by collecting every sequence in which there was talk between the participants and the researcher. Following this, I went through the videos again to spot the occasions on which the participants said or did something that was overtly task-relevant but where the researcher was not spoken to. In particular, I took an interest in the times at which the participants read or discussed the instruction sheet.

6.4 ANALYSIS OF LABORATORY CONTRIVANCES

I have selected six examples (four transcripts and two figures) for inclusion in this chapter. For the purposes of readability, simplified transcripts are used; they are designed to convey what was said but are less precise about how it was said than were the full transcripts produced for the analysis (e.g., I omit details about laughter, intonations, and overlapping talk, and I do not time the pauses). For simplicity, I also refer to the videos as Studies 1, 2, and 3, and to the participants in each video simply as A_{1-3} and B_{1-3}, and the researcher as R.

6.4.1 Contrivance 1: Requirement to Use the Whiteboard

To begin, I will discuss Figure 6.1 and Transcript 6.1. These examples concern how the researcher's requirement to use the whiteboard is implemented. These examples do not

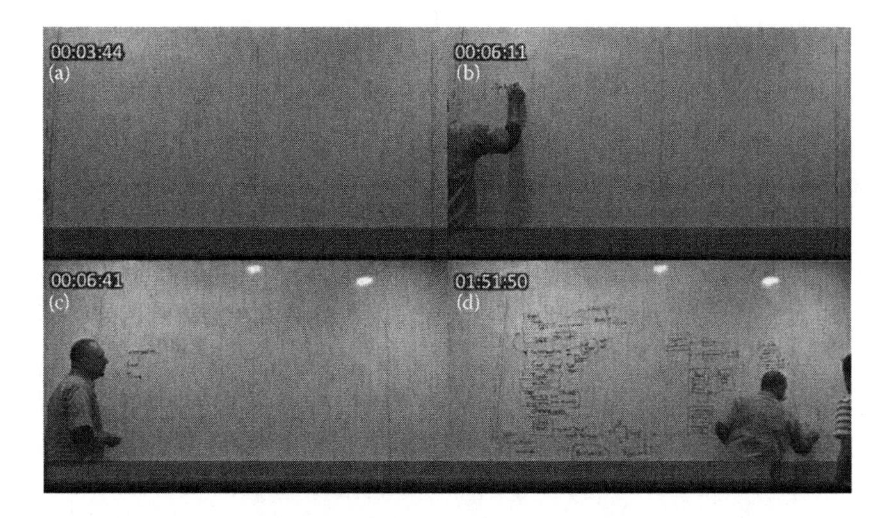

FIGURE 6.1 The designated working area. (a) Lines mark visible area of whiteboard. (b) The participants are not fully on camera. (c) The camera is adjusted mid-session. (d) The participants write outside the area.

show that this requirement is unproblematic or has no effect. Rather, they show that laboratory contrivances are not purely resident within the design of a study, but they are implemented and oriented to in specific and practical ways during each study.

Figure 6.1 gives a sequence of images from Study 2. The meeting room used for Study 2 had a large whiteboard, and so the researcher drew red lines to create a working area that could be captured on camera. Soon into Study 2, the researcher had to adjust the camera because the participants would often stand to the side of these lines. The participants used this working area for most of the study, but at one point when they ran out of room, they began drawing outside this area. This figure shows that the study is contrived, with the participants not just being asked to use the whiteboard, but to use a specific area of it. It also shows that the researcher is sometimes led to make practical adjustments and interventions during the study. So, while the contrivances may be decided in advance, their specific implementation is contingent on the specific setting of the study and the course it takes. The figure also shows that the participants for the most part comply with the contrivances, but they are prepared to break them and make no excuse when they do so.

In Study 1, the requirement to use the whiteboard briefly becomes a joking matter. In all three videos, after the developers participating in the study have spent about 5 minutes reading the instruction sheets, they face the problem of when they will start, who will make the first move, and what this move will be. In the opening turns from Study 1 (reproduced in Transcript 6.1), one of the participants asks the researcher "shall we go?" This and the opening turns from the other studies are points at which the study itself is being discussed. Study 1 stands out because rather than asking her colleague "shall we start?," the participant asks the researcher. This draws the researcher into a revealing discussion (see Transcript 6.1). The researcher seems to be caught off guard when A_1 asks him "shall we go?" The researcher splutters "whenever you're ready" and then repeats this more clearly. Then there is a silence. The researcher, probably worrying that nothing is happening, now begins to direct them to use the whiteboard, "... you should go and use ..."; but he stops abruptly and repairs what he was saying into something less directive "... just try to do it at the whiteboard." The researcher is then met with laughter and is mocked "WE CANNOT WHISPER HERE." The researcher tries to explain but he is laughed down. It is not until A_1 finishes laughing at her own joke that the researcher gets to explain that a previous study failed because a pair wrote on paper. Transcript 6.1 ends where the pair begin discussing the task between themselves.

In this example, the researcher can be seen trying to balance two conflicting priorities: for the participants to work at the whiteboard (visibly, on camera), and for the participants to do things as naturally as they can. It is also interesting that the participants, through their laughter and jokes, show that these concerns are obvious to them. The joke plays on following one obvious thing (use the whiteboard) with another (speak clearly). This is not to say that the participants necessarily know and understand everything the researcher wants (in the example, a pair that tried to work on paper is mentioned, and elsewhere in the video one of the participants appears to be mentioning electronic whiteboards as if he thinks that the researcher is interested in them). The participants also seem awkward about these constraints. This example shows that the developers are not acting blindly or in ignorance of the fact that they are participating in a study; neither the researcher nor the participants are blind to the contrivances of the study.

6.4.2 Contrivance 2: The Task

Transcript 6.2 details a relatively long example in which the developers discuss and disagree about what the task sheet is asking them to do. A_1 reads out a section from the task sheet that says "you may assume you can reuse an existing package to deal with certain mathematical and statistical functionalities necessary for the simulation software." B_1 jokes, "Ok, cos I don't wanna code that," but A_1 is not sure about the scope of what they can assume. A_1 wonders whether they will have to model issues to do with speed and the timing of the signals, but B_1 says, "we can rely on Professor E for kinda the details" and then "we don't have enough information here, but there might be packages out there." I have then omitted several turns from the transcript (in which B_1 changes the subject), and I restart where A_1 steers the subject back to her worry about speed. A_1 is worried if it can be "assumed that all of 'a' have the same speed." By this point, B_1 seems a little irritated and he has started working on something else. He turns to A_1 and replies, "lets assume yes, lets assume its inside a city." Then, to drive home his point that speed is not within the scope, he says, "we'll talk to professor E about it," and gives a pointed grin.

The task sheet points to the scope of the work expected from the participants, but it does not contain a complete description of what must be achieved; the participants must interpret and flesh out what is required (and they might not always agree on the interpretation). Would not the simplest way for the developers to figure out the task, and settle disputes about what is wanted, be to turn and ask the researcher "what do you mean here?" or "what exactly do you want here?." But, the task is treated in terms of what Professor E wants and might say. For the purposes of carrying out the task, it does not matter that Professor E is not really their customer (I am sure the participants know that the study will end after 2 hours and that, even if Professor E exists, there will not be a need or an opportunity to speak to her later). In a sense then, the participants buy into the artificialities of the task, and, arguably, they make things more difficult for themselves by "pretending" that they will speak to Professor E later. But in doing so, they are treating the task according to the logics and rationalities of software design. The participants are not naïvely following instructions or doing what is simplest in that particular situation, but they are approaching the task as designers would. As can be seen in example one, the developers know what is expected of them; but now we can also see that they still draw upon their ordinary methods and logics. This is not just to act a role, but it is also to be involved in things such as settling disputes.

6.4.3 Contrivance 3: 2-Hour Time Limit

The study is run with a 2-hour time constraint. At the point where Transcript 6.3 begins, the participants have had a long discussion about traffic modeling, but without coming to anything they feel necessary to write down. The discussion does not seem to be going anywhere. A question from A_2 is dismissed with the answer "if we even wanted to." A_2 echoes this answer, and then steps backwards shaking his head. A_2 then suggests an idea for what they might work on next, premising it with the words "with half an hour left." A_2 wants them to think about the goal of the task to "communicate this," making specific reference to the fact that "we're playing the role of architect." He picks up the sheet to read through it, and B_2 does likewise. They discuss the stated goals of the task, saying of the first one "which we kind of have been doing here" and the next "which we've done here," essentially checking off what they have done against the study goals. B_2 then states that a goal is to "communicate the interaction UI," and later says, "one way I've found effective in communicating UI interaction is user stories." He suggests, "we could come up with a list of user stories."

For a long period, the developers had "forgotten" the sheet, both in the sense that they did not need it to do what they had been doing, and in the sense that they had forgotten some of the specific instructions. In this example, they find an occasion when it is appropriate to come back to the sheet and read it again. The occasion has two key features: having half an hour left, and requiring a new topic to follow an idea that has fizzled out. The lack of available time is referred to specifically, and this appears to support returning to the goals of the study as a credible next thing to do (as if this would not be preferable were there plenty of time). Returning to the instructions first involves deciding whether the work done so far satisfies the stated goals. In this instance, "kinda" is good enough; they are not concerned whether they have precisely fulfilled what has been asked of them. Returning to the instructions also enables the developers to decide on a relevant next course of action: listing "user stories." This course of action is not stated in the sheet, but it is given by one of the participants as an adequate and reasonable response to a requirement that they "communicate" their design. I also want to note with this transcript that the participants orient to "playing the role of architect." They invoke this role not as something for them to pretend to be as such, but as a referent for what would be the appropriate course of action; *what would an architect do here*?

6.4.4 Contrivance 4: Video Camera

Finally, I will discuss the camera. For the most part in the videos, the participants seem to ignore or even forget that the camera is there. In Figures 6.1 and 6.2, two participants are working together at the whiteboard. The images give a typical scene in that the participants position themselves in such a way that each can see and access the board, and, for the most part, they show little awareness or concern for what is visible to the camera. In the sequence shown in Figure 6.2, one of the participants briefly checks his telephone (which had beeped earlier). The participant times this in such a way that the other participant does not notice. When the first participant is writing something, the second participant quickly removes his telephone from his pocket. As soon as the first participant begins to stand and look back, the telephone is pocketed. We can see that the participants' primary concern is managing face with their partner, not how they appear on video; working together can and does take precedence over any acting. However, as Transcript 6.4 shows, this is not to say that the participants are oblivious to the camera.

In Transcript 6.4, the participants wonder if they can take a shortcut because they know that something they have discussed has already been captured on camera and they think, therefore, that there is no real point in producing a written record. The transcript begins with A_2 stating an idea for a user story, and then, with B_2's encouragement, writing down "I want to create an intersection." They discuss "the way you do that" according to their design. Pointing to the blank area beneath their user story, B_2 says, "do we wanna then annotate" and then "or are we letting the video camera capture our description." Essentially, B_2 is wondering whether it is worth writing down everything in full. The participants laugh about this idea. The researcher coughs to gain their attention and says, "you don't have to write everything down." The participants seem a little surprised at R speaking, and just nod and turn back to the whiteboard.

The participants take a shortcut on the rationale that their verbal description is already on camera. Until now, the camera had never been oriented to or mentioned. It

FIGURE 6.2 The developer on the right briefly checks his phone. (a) Participant B_3 speaks. (b) Participant A_3 moves to board. (c) Participant A_3 glances at phone. (d) Phone to pocket and eye contact is made.

was forgotten for all practical purposes, which is to say, whether or not the participants were aware of the camera, they did not orient to it. But here, the presence of the camera is made explicitly relevant, and it is laughed about as a possible determinant of how they will proceed (specifically, whether they can omit something from their working). As with example one, joking and laughing feature where the participants directly discuss constraints, perhaps to mask their awkwardness or embarrassment. As with example one, the researcher is also dragged into the proceedings. The researcher has to make an on-the-spot decision regarding the study contrivances, choosing this time to intervene. Arguably, the participants themselves minimize the intervention of the researcher, turning to him only briefly. In this example, the participants orient to the contrivances of the study, but in ways that show sensitivity to the researcher's task (as best they understand it), and in ways that appear to trade working through everything in full against doing it in the 2 hours available. Perhaps the participants are being a little lazy, but even if this is so, they are lazy with sensitivity to the study.

6.5 FINDINGS

I have discussed four transcripts and two figures in order to examine how contrivances feature in a laboratory study. These examples confirm that the videos are contrived. I will argue, however, that this is not a reason for dismissing simulation as a method of producing data for studies of professional practice.

6.5.1 Laboratory Simulations Are Contrived

The examples confirm, first, that the simulations are contrived. Simulations are crafted to resemble the design process, but in a tractable way to be captured on camera and to produce examples that can be compared with each other. The examples also show that the researcher and the participants are not always comfortable with the contrivances of a study nor are they confident about how to implement them. The study participants would often laugh when discussing aspects of the laboratory setting (such as the camera, that the researcher wanted them to use the whiteboard, and the time constraints). This laughter, I suggest, does not mean that they find studies funny, but that they are uncomfortable. The researcher also seemed uncomfortable at times, particularly in Transcript 6.1 where he had to make the requirement to use the whiteboard explicit.

To point out that a laboratory simulation is contrived is *not* to point out something that the researcher and the study participants do not know. So, if someone complains that laboratory studies are contrived, he/she is not pointing out something that would be "news" to anyone involved. This shows, or at least implies, that the presence of contrivances is not inherently grounds for dismissal: if researchers and participants know that a study is contrived, why would they proceed with it? I find it more plausible that the researchers *and* the participants recognize contrivances and believe that they are problems that can be overcome, rather than being somehow in denial. I also believe that the uncertainties and awkwardnesses of the researcher and the participants signal a need for research that supports and guides the running of simulation studies.

6.5.2 Contrivances Are Pervasive but Not Controlling

The examples show that the researcher and the participants deal with the contrivances on occasion, but not at all times. The contrivances may be a pervasive influence on the study, but their influence is not constant. The features of the study, such as the task, the camera, and the working arrangement, are always present, but they are not of equal relevance at all times. The study participants predominantly just get on with the job at hand and only on occasion talk or worry specifically about what the sheet says, what the camera means, how much time they have left, and so on. They ignore the camera most of the time, worrying more about how they are seen by their partner, than how they appear on video. When they pick up the task sheet with half an hour left, they go back through the goals and decide to what extent the work they have done satisfies these goals. For the most part, the contrivances are forgotten—both in the sense that the participants really do not remember what is in the sheet or that they are on camera, but also in the sense that the sheet, the camera, and such like, are not, at that moment, relevant to what they are doing. This is not to say that the contrivances do not shape the action, but that there is also room for other factors.

6.5.3 Identity and Experience Also Shape the Simulation

The examples confirm and cast light on McDonnell's (2012) point that the participants call upon their experience and institutional identities. The study participants do not work in ways that are unique to the laboratory context, but draw upon their experiences and ways of designing. The study participants draw from their experiences of what they have found useful (e.g., creating user stories), and they take decisions that are professionally accountable rather than blindly relevant to the situation they are in (e.g., claiming that they will ask Professor E later). The participants are not drawing purely from personal experience, but they can be seen to orient to and talk about institutionally relevant identities; for example, they orient to how a system architect would communicate something (rather than their own specific experiences in systems architecture), what a user would do, and what a customer might want. This reflects Heritage's (2005) remarks about talk at work, that people will select institutionally relevant identities, and that these have procedural consequentiality. This happens in a multifaceted way in the study, with the developers orienting at times to "me as a developer" (e.g., *what I've found useful* or *what I would do in this kind of situation*), at times to what any professional developer would do (e.g., *what architects would do* or *what type of problem this is*), and at times to what characterizes the work of their organization (e.g., *what we do here* or *how we approach things*).

6.5.4 Practices in a Simulation Can Be Both Professional and Contrived

It is not the case that experience and institutional identity feature in ways that are independent of the contrivances. Experience and identity do not, for example, fill the gaps where the contrivances are "forgotten." Rather, the examples show the participants choosing their courses of action with reference to a combination of experience, identity, and contrivance. For example, the task sheet sets a scope for what is required, but the participants figure out for themselves exactly what this scope is and how they will meet it. The task sheet does not give a set of steps that will get the participants through the task; instead, they must select courses of action that will address what is asked of them. These courses of action (e.g., to list user stories, to designate something as a question to be settled by the customer, or to sketch an interface) are chosen with relevance to what is asked in the task sheet, with relevance to the time they have remaining, as well as with relevance to things such as their experience as designers. These courses of action can take the developers away from the task sheet, with them spending time doing the thing they have decided to do. But when they later come back to the sheet, they may talk about the extent to which what they have been doing satisfies what is on the sheet. There are multiple relevant contexts for the participants, including the simulation itself, but also their experience, their own identities, and their wider organizational and professional identities.

Heritage (2005) criticizes what he calls "the bucket theory of context." He argues that context is not something enfolding and determining action, but that it is fundamentally through interaction that context is built, invoked, and managed. In this study, we have seen that neither the professionalism of the developers nor the laboratory constraints are anterior to their practices; both are oriented to, drawn upon, reasoned about, and sometimes thrown out, all in the course of the simulation. From this perspective, laboratory contrivances should not be thought of as things that exist in the design of the study, but as features that are implemented in contingent ways during the course of the study. The professional practices too are implemented in the course of the study, and in ways that are tethered to the contrivances. This means that while the study participants can be seen to draw on professional practices, these practices are drawn upon in ways that are specific to the simulation. Rather than seeing professional practice in a professional setting, we see it in the laboratory. This practice cannot be isolated from the laboratory just by throwing away simulation-specific talk. But this is only a problem insofar as the settings in which developers work may see them drawing upon experience and identity in quite different ways. So, the laboratory simulation can tell us about professional practice, but it does not necessarily tell us about what happens "in the wild," and might, in places, tell us something specific to practice in the laboratory.

6.6 CONCLUSION

Even though the study discussed in this chapter is contrived, it cannot be dismissed as unnatural. The study participants have been shown to act according to their identity and experience. However, they draw upon and implement their experience and identity in ways that are specific to and intractable from the simulation. So, while a laboratory simulation can serve to reveal aspects of professional practice, research must remain sensitive to how such practice is simulation bound. Three general, overlapping recommendations that can be derived from this chapter are (a) the naturalness or authenticity of practices in a simulation is not a direct result of the way a study is designed but it is a practical problem encountered by the participants and the researcher during the conduct of a study, and therefore in order to understand how and why the participants work in particular ways, we need to pay attention to the work itself; (b) simulation-specific talk and action (e.g., researcher–participant interaction) ought not and often cannot be omitted from an analysis of their practices; and (c) methodological support for simulations might better address not just how to design a study, but also how to conduct it.

REFERENCES

Baker, A., van der Hoek, A., Ossher, H., and Petre, M. 2012. Studying professional software design. *IEEE Software*, 29(3): 28–33.

Crabtree, A., Roucefield, M., and Tomie, P. 2012. *Doing Design Ethnography*. London: Springer.

Détienne, F. 2010. Roles in collaborative software design: An analysis of symmetry in interaction. Presented at the NSF Workshop on Studying Professional Software Design, 8–10 February, UC-Irvine, CA.

Francis, D. 1989. Game identities and activities: Some ethnomethodological observations. In Crookall, D. and Saunders, D. (eds), *Communication and Simulation*. Bristol: Multilingual Matters, pp. 53–68.

Garfinkel, H. 1967. *Studies in Ethnomethodology*. Cambridge: Polity.

Gross, T. 2010. Talk and diagrams in software design: A case study. Presented at the NSF Workshop on Studying Professional Software Design, 8–10 February, UC-Irvine, CA.

Heritage, J. 2005. Conversation analysis and institutional talk. In Fitch, K. and Saunders, R. (eds), *Handbook of Language and Social Interaction*. Mahwah, NJ: Lawrence Erlbaum, pp. 103–148.

Housley, W. and Fitzgerald, R. 2000. Conversation analysis, practitioner based research, reflexivity and reflective practice. *Ethnographic Studies*, 5: 27–41.

Hutchins, E. 1995. *Cognition in the Wild*. Cambridge, MA: MIT Press.

Jackson, M. 2010. Representing structure in a software system design. *Design Studies*, 31(6): 545–566.

Luck, R. and Ikeya, N. 2010. "So the big rocks we're putting in the jar": Using situated reasoning to coordinate design activity. Presented at the NSF Workshop on Studying Professional Software Design, 8–10 February, UC-Irvine, CA.

Matthews, B. 2009. Intersections of brainstorming rules and social order. *CoDesign*, 5(1): 65–76.

Matthews, B. 2010. Designing assumptions. Presented at the NSF Workshop on Studying Professional Software Design, 8–10 February, UC-Irvine, CA.

McDonnell, J. 2010. A study of fluid design collaboration. Presented at the NSF Workshop on Studying Professional Software Design, 8–10 February, UC-Irvine, CA.

McDonnell, J. 2012. Accommodating disagreement: A study of effective design collaboration. *Design Studies*, 33(1): 44–63.

McDonnell, J. and Lloyd, P. 2009. Editorial. *CoDesign*, 5(1): 1–4.

Nakakoji, K., Yamamoto, Y., and Matsubara, N. 2010. Understanding the nature of software design processes. Presented at the NSF Workshop on Studying Professional Software Design, 8–10 February, UC-Irvine, CA.

Nickerson, J. and Yu, L. 2010. There's actually a car—Perspective taking and evaluation in software-intensive systems design conversations. Presented at the NSF Workshop on Studying Professional Software Design, 8–10 February, UC-Irvine, CA.

Petre, M., van der Hoek, A., and Baker, A. 2010. Editorial (studying professional software design). *Design Studies*, 31: 533–544.

Rooksby, J. and Ikeya, N. 2012. Face-to-face collaboration in formative design: Working together at a whiteboard. *IEEE Software*, 29(3): 56–60.

Sacks, H. 1992. *Lectures on Conversation*. Oxford: Blackwell.

Sharrock, W. and Watson, D. 1985. Reality constructions in L2 simulations. *System*, 13(3): 195–206.

Suchman, L. 1987. *Plans and Situated Actions: The Problem of Human-Machine Communication*. Cambridge: Cambridge University Press.

Suchman, L. 2007. *Human–Machine Reconfigurations: Plans and Situated Actions* (2nd edn). Cambridge: Cambridge University Press.
Szymanski, M. and Whalen, J. 2011. *Making Work Visible: Ethnographically Grounded Case Studies of Work Practice*. New York: Cambridge University Press.
ten Have, P. 1999. *Doing Conversation Analysis*. London: Sage.
Visser, W. 2010. Contributing to the elaboration of a design artefact according to one's interactional position: Visual and audible evidence. Presented at the NSF Workshop on Studying Professional Software Design, 8–10 February, UC-Irvine, CA.

APPENDIX

These transcripts have been slightly simplified for readability. ((Double parentheses indicate notes)). Underlined text relates to the adjacent screen grab from the video.

TRANSCRIPT 6.1 Study 1 (AmberPoint, 00.05.48–00.06.15)

A_1: Shall we go?
R: Eh? Whenever you're ready. Whenever you're ready
A_1: Yeah
B_1: Okay
((1 second pause))
R: Yeah you should go and use tha-. If you want to write anything, just try to do it at the whiteboard
A_1: h huh huh ((continues laughing))
B_1: huh huh huh ((continues laughing))
R: It helpful for us ((drowned out by laughter))
A_1: WE CANNOT WHISPER HERE huh huh huh.
((brief pause))
R: Some people want to draw on paper and that just makes it difficult for us.
B_1: Yeah
A_1: Uhuh, uhuh
((3.5 second pause))
B_1: Well, I mean, so, so, the end users, seem to be the students and the professor.

TRANSCRIPT 6.2 Study 1 (AmberPoint, 00.22.14–00.25.02)

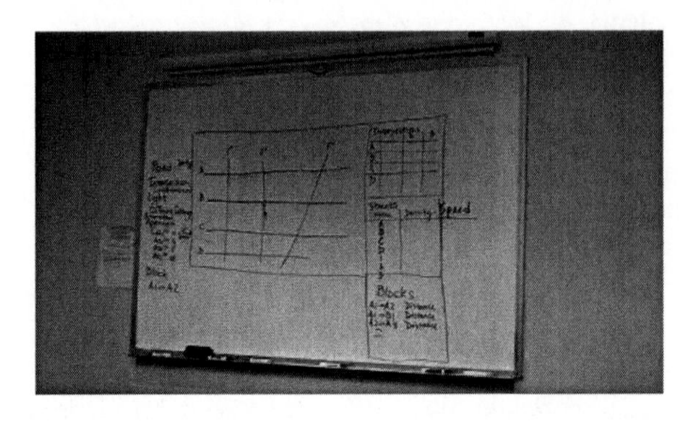

A_1: Yeah ((reading task sheet)) If you wish you may assume that you will be able to reuse an existing software package that provides relevant mathematically functionality such as statistical distributions, random number generators. Yeah but that

B_1: OK cos I don't wanna code that ((laugh))

A_1: No ((laugh)) Yeah but that doesn't really necessarily tell you

B_1: No thats not really a drawing package

A_1: Yeah

B_1: Its more a

A_1: No but it also doesn't say, uhm, you know that if you have two streets that are, you know, a mile long and the average speed is twenty five, the best timing

B_1: Yeah, I think we're gonna have to rely on professor E for kinda the details about, how that, the theory of how that works. I mean, we might have to code whatever her understanding is of, of

A_1: Uhuh

B_1: traffic light theory into it. But we don't have enough information here. But. There might be packages out there, I don't know. But. Does that make sense?

[Omitted 23.09–23.42]

A_1: So the user is dragging, drawing these things, placing intersection, uhm, and now they have, this grid. Uhm. That's the hard part, that I'm not quite

B_1: Which part? Over here? ((points))

A_1: Yeah the part about settings actually. In other words, if you want to simulate it, you have to know exactly the distance between each intersection, you have to know how fast is the maximum speed that the cars possibly, could be going

B_1: Right

A_1: So ha how long does it take to get from one intersection to another, so if you left at a green light, from one point.

B_1: Oh yeah that would be nice so if we ((walks to whiteboard pointing at part of diagram)). That's kind of good down here, if the speed on this street is twenty five miles an hour, how, adding green light, long should it take you to go the distance, this distance?

A_1: So this can get thoroughly complex very fast ((slight laugh)) because you suddenly you know have all these intersections and if the. Hhh. That's the thing, are the speeds to be assumed that all of A has the same speed?

B_1: Which one are you
A_1: You know, and is it the same as B?
B_1: Lets assume yes ((turns to face A_1)), lets assume it's inside a city so ((turns to whiteboard))
A_1: Ok uh
B_1: ((Turns to A_1)) We'll talk to Professor E about it ((grins))

TRANSCRIPT 6.3 Study 2 (Adobe, 01.17.35–01.19.40)

A_2: Yeah right now we've said. Yeah. Right now we've said if you can control ((points at part of their design)) the frequency of the car creation on that road, what about the, how do you, how do you uh

B_2: If we even wanted to

A_2: yeah if we even wanted to that's, that's ((steps back and shakes head to say no))

B_2: ((Waving pen)) I'm wondering with half an hour left if we need to uh, think about how to. Cos I think one of the, ((reaches to pick sheet up, as does A_2)) goal here is to communicate this to, if, if we're playing the roles of architect, how to communicate it to. ((looks at sheet)) so we ((looking through sheet))

((7 second pause, both participants looking through their information sheets))

A_2: So two things, design the interaction, which we've, kind of, have been doing here, design a base, basic structure of the code. Which we've done here ((points)), maybe we should more formally.

B_2: ((reading from sheet)) You should design the basic structure of the code, okay so really the desired outcome right, so we need ta, communicate the interaction UI. But then we need to, design the basic structure of the code that will be used to implement the system.

((3.5 second pause)) So I'm thinking, maybe, uhm, ((3.5 second pause)) so, maybe uh ((1 second pause)) one, one way that I've found effective in communicating, UI interaction is uh, is user stories. So we could come up with a list of user stories

around, uh I as a s:i. Err, uh I as a user of the system, want to uhm, create a, a, a, a road in my simulation, so that I can see traffic flow across it

A_2: Yeah

B_2: And then come up with progressively, start with the simplest user story and then…

TRANSCRIPT 6.4 Study 2 (Adobe, 01.22.18)

A_2: Right, so as a user I want to ((1 second pause, gestures toward whiteboard)) create intersection ((turns to B_2))

B_2: Ok

A_2: ((Writes a user story on whiteboard "I want to create an intersection")) Currently, the way you do that, is, ((pointing at areas of diagram)) you, uhm, click on the edges, to create roads. And then, where the roads meet, you have an intersection.

B_2: Right ((4 second pause)) so. Ok, so, I want to create intersection. So do we wanna, do we wanna then annotate ((points at area beneath where the user story is written)), or are we letting the video camera

A_2: ((laughs))

B_2: capture our description? I think we can assume that we have a video camera rolling ((A laughs)), and we'll go through user stories

R: ((clears throat)) To ((A_2 and B_2 turn to face R)) some extent you should just be able to explain it, you don't have to have everything written down

B_2: Written down ((nods and turns back to whiteboard)) I see. Ok. Right.

II

The Process of Design

CHAPTER 7

Empirical Studies of Software Design Process

Jim Buckley and J.J. Collins

University of Limerick

CONTENTS

7.1 MAKING THE CASE FOR DESIGN PROCESSES

A plethora of software processes can be availed of to support the organization of a software development life cycle and the creation of the resultant software artifacts. Processes such as the rational unified process (Krutchen, 2003) explicitly state how to move from a requirement analysis to design in a structured manner, given a set of architectural decisions. This scoping of the solution space facilitates novices by making software development methodological and repeatable. In addition, high-value, experience-based insights are now leveraged through analysis, architecture, workflow, and design patterns that are subject to periodical renewal. From an architectural perspective, for example, the selection of patterns can be guided by an up-front application of stress analysis techniques from architecture evaluation frameworks such as the architecture trade-off analysis method (Clements et al., 2001). One may therefore be misguided in thinking that the potential for independent decision making is limited.

An implicit assumption in the first paragraph is that validated and verified requirements have been elicited, are correctly scoped, and are of sufficient depth to permit development to proceed in a controlled and orderly fashion. This is increasingly unlikely because of increased ecological complexity (Jarke et al., 2011). Even if requirements are correctly specified, the advances typified by the evolution of the patterns communities illustrate that software developers have to cross a steep learning curve in order to populate the solution space with meaningful alternatives necessary to synthesize a somewhat optimal design. These considerations prompt questions as to the strategies that novices can fall back on in the absence of such wisdom, and the mechanism by which experienced engineers can evaluate their practices. An abstract design process would be an appropriate starting point: one that provides guidance on the types of decision-making strategies and techniques that might be adopted, and their distribution over the development life cycle.

Paradoxically, the opportunities for decision making in software design increase with acquired knowledge. For a simulation, should one use a blackboard architecture or an event-driven loop? Should one use concurrency patterns to simulate independent, autonomous, intelligent entities? Should patterns such as the "visitor" (Gamma et al., 1995) and the "interceptor" (Schmidt et al., 2004) be considered as mechanisms for future proofing? What is the interaction between the user and the system? Even when such questions are framed correctly and therefore should lead to the generation of a set of design alternatives, the candidate solutions are not always evaluated because of developer preferences, an example of autonomous reasoning (Evans, 2003).

The variability in practice and the expansion of the solution space that comes with increased experience compel developers to make complex decisions at all stages in the software development process. It begs the question as to the types of decision-making strategies that might be adopted, their distribution in the design process, their impact on the outcomes, and the implications for the infrastructure that supports collaboration and modeling, to name but a few.

7.2 OVERALL GOALS FOR THIS FORM OF ANALYSIS

The ultimate goal of studying software design processes is to improve the efficiency and effectiveness of the process, toward an improved product quality. It aims to advance knowledge on the fundamental questions of design, such as determining the strategies that can be applied when the designer is selecting from a wide range of design alternatives and is faced with a seemingly combinatorial collection of unranked constraints and requirements. Knowledge such as this enables the evolution of the design craft into a discipline, where designers can rely on clearly defined protocols in times of difficulty and where we know what to teach novices.

The research into software design processes can be deductive or inductive. As Sharif et al. point out, most of the research work in this area has been deductive, where researchers suggest how design should be done. However, it is equally important to determine how experienced practitioners actually do software design, and whether or not a gap exists between theory and practice. This provides a stronger basis for the subsequent evaluation and refinement of software design processes, toward notations, tools, and artifacts that genuinely support designers in their work contexts. Yet, this is not an easy task: it is very difficult to obtain ecologically valid data (Perry et al., 1997) of designers working through design processes. Once obtained, a reliable schema for an analysis must be established that produces meaningful insights into the design process, and these insights must be carefully distilled. This set of papers is directed toward that goal.

7.3 PAPER SUMMARIES

All of the studies presented in this section are based on an analysis of video recordings and transcripts provided by the University of California (Irvine) in 2009. These empirical data were generated by three, two-person design teams engaged in designing a traffic simulation system. The participants were professional designers, who were given a whiteboard to create a user interface (UI), a high-level design, and a high-level implementation of the simulation system. A priori coding schemes were employed to analyze the data, although Sharif et al. did augment their initial coding schema with an additional category, as they progressed. These schemes are all described transparently in the papers, allowing researchers the opportunity to replicate, evaluate, and possibly refine the analyses in future studies.

Two of the papers (Christiaans and Almendra, Chapter 8, this volume; Tang et al., Chapter 9, this volume) based their coding scheme on decision making and decision reasoning. In contrast, Sharif et al. (Chapter 10, this volume) focused more on concrete software engineering activities (such as analyzing whiteboards). At a coarser level of granularity, Baker and Van der Hoek (Chapter 11, this volume) analyzed patterns of intertopic discussions during the design sessions, to see, for example, whether the designers revisited topics repeatedly. Finally, in an attempt to identify possible improvements in tool support for designers, Budgen (Chapter 12, this volume) focused on the "viewpoints" (data, functional, behavioral, and constructional) that the designers adopted and transitioned between during their sessions.

7.4 STRUCTURE OF THE DESIGN PROCESS

Sharif et al. state that their observations on activities in design "demonstrate the iterative process of design." Likewise, Baker and van der Hoek show that the structure of the design process is iterative given that a session can be decomposed into cycles, where a cycle is manifested by a period of focus setting. They decompose the 3 design sessions into 49 cycles where ideas are categorized into orthogonal subjects, and they note that subjects are repeatedly discussed throughout the sessions. In addition, they note repetition at the ideas level with 21% repetition of ideas for Adobe, 31% for Intuit, and 39% for AmberPoint. Tang et al. note that Adobe discussed nine different decision topics throughout the session with issues such as intersections, the simulation of traffic flows, and the system structure occurring repeatedly in an evenly distributed pattern. AmberPoint discussed 12 topics with repetition characterized as clustered.

They further observe that AmberPoint devoted less time to a topic throughout the session when compared to Adobe, and it is postulated that Adobe did so because of their higher frequency of context switching between requirements. Tang et al. state that context switching is a major attribute of the conceptual design process for the two teams studied in their paper. Adobe and AmberPoint made 78 and 69 context switches, respectively, which resulted in the probabilities of 24.9% and 13.2% of doing so, based on the classification of statements made as a percentage of total utterances. Tang et al. postulate that design effectiveness is reduced with increased context switching due to the cognitive load involved when recontextualizing.

Baker and van der Hoek note that the design teams tended to consider two subjects at a time. Low-depth cycles, in which four or more subjects were simultaneously discussed, occurred at the beginning of a session and generally represented a breadth-first investigation of the solution (and problem) space. Low-depth cycles were also detected at the end of the session and corresponded to points in the design session where a holistic view of the solution elements was undertaken in order to coalesce them into a unified whole. Single-subject cycles tended to appear early in the design process and corresponded to an in-depth evaluation of "base components that they will be working with." Baker and van der Hoek further observe that the majority of cycles (28) consisted of two-subject cycles that fell into three categories—containment, communication, and control. They postulate that "by considering two subjects at a time, the designers can consider aspects of the problem in depth without becoming overwhelmed by the problem as a whole." The iterative process in conjunction with the dominance of the two-subject cycles indicates that the designers tended to develop ideas in parallel, and that effort was expended in ensuring that all the elements of the design were kept up to date. Baker and Van der Hoek refer to this as "parallel and even development."

Tang et al. note that the first phase of the conceptual design process was planning, using techniques such as contextualization and scoping in a breadth-first reasoning approach at a high level of granularity. Adobe gave 1.5 minutes to planning while AmberPoint devoted 9 minutes. Adobe demonstrated five instances of scoping and nine examples of key design issue identification. The numbers for AmberPoint were 16 and 18, respectively. Tang et al. additionally observe that while Adobe spent 1% of their time on planning, AmberPoint devoted 9%. However, Sharif et al. note that AmberPoint "did much more planning at the very end of the session with justification provided for each decision made."

Problem solution coevolution followed planning for the majority of the design sessions, in which both the problem and the solution spaces were explored together because of their mutual dependency. Tang et al. note that while AmberPoint generally exhibited this model, Adobe took a solution-driven approach with less time on problem specification, and they relied on assumptions as a means of scoping the process. AmberPoint devoted considerable time to problem exploration and then focused on a single design issue and proceeded to address it. Tang et al. observed that Adobe and AmberPoint made 34 and 120 problem statements, which represent 13.4% and 31.9% of all statements made, respectively. Both teams made a similar number of solution statements, and so the ratio of solution statement generation to problem statement is 32% and 96.7%, respectively, for Adobe and AmberPoint. These numbers indicate that Adobe are solution oriented in their approach to design, whereas AmberPoint are more compliant with the concept of coevolution.

Sharif et al. state that "the team that took a structured approach with systematic planning, and with much consultation and agreement, gave the most detailed solution and covered the most requirements in the deliverable." According to Tang et al., Adobe covered 13 requirements and 14 derived, with the numbers being 19 and 14, respectively, for AmberPoint. However, Sharif et al. state that "we can hypothesise that planning, as well as the high degree of agreement between the two designers of the Adobe session, played a significant role in delivering the most detailed design that covered the most requirements."

7.5 PHASES AND ACTIVITIES

Two distinct but overlapping phases of the design process can be observed from the empirical studies. The first could be referred to as "derivation of design decisions" and, as discussed in the previous section, Christiaans and Almendra refer to three design drivers active within this phase: "problem-based" reasoning, "solution-based" reasoning, and "coevolution," which interleaves problem-based and solution-based reasoning.

Tang et al. relate solution-based reasoning to reasoning based on "fast, intuitive thinking" (Kahneman, 2011) in that there was a less-detailed, less-critical analysis of the problem space in advance. They point to findings in other works (Rowe, 1987; Tang et al.), reinforced by their findings here, suggesting that experienced designers may become tied to the macrosolutions they initially adopt as a basis for subsequent decisions. They thus propose that it is more correct to base initial decisions on extensive, rational critique. However, an alternative perspective comes from the design pattern literature (Gamma et al., 1995) where, in the case of experienced software designers, this "fast thinking" may be reflective of pattern-matching expertise and may provide a valuable and reliable shortcut for designers. Christiaans and Almendra hint at this interpretation in their analysis where they refer to "compensatory (and non-compensatory) rule-based" mechanisms as a prevalent decision-making aid.

The second design process phase is "evaluation," where at least one member of the team took an evaluative role, probing the design decisions for their appropriateness. Most of the reporting of this phase is implicit, in that the authors refer to it in their discussion of the participants' roles. For example, Christiaans and Almendra refer to both members of the Adobe team as becoming more problem oriented and concentrating on evaluation toward the end of their session. However, Tang et al. make it more explicit when they talk of the participants "... evaluating the pros and cons of the available options ...," also suggesting that they used scenarios as a mechanism for this evaluation.

While Sharif et al. focus more at a "software engineering" activity level, there are echoes of these findings in their work too. For example, they found that the designers spent a lot of their time analyzing the whiteboard, suggesting an evaluation of the problem space and of the potential solutions. Additionally, they found that the Adobe team started analyzing the whiteboard only later in their session, supporting Tang et al. and Christiaans and Almendra's assertion that the other teams focused on analyzing the problem space (earlier) to a greater degree.

During the sessions, AmberPoint's focus was largely on user interaction. Sharif et al.'s data show that a large proportion of AmberPoint's team time was consumed with drawing and discussing the UI. Likewise, Christiaans and Almendra note the team's user framing of the problem space and that, unlike the other two teams, the team did not generate any specifications with respect to code structure or code logic, concentrating *only* on the UI specification. They suggest that this UI focus might be a result of the team's expertise or orientation, as one of the participants had "undoubtedly expertise in user interaction." But it is equally feasible that this was just their initial design activity and that the issues they identified (Christiaans and Almendra) caused them to defer other design decisions until they could refer back to their client. AmberPoint's constant focus on the UI contrasts strongly with the other two groups, who focused on data dictionary modeling in the early parts of their sessions and class modeling later. Indeed, even though Adobe concentrated on the UI later on in their session, the Intuit team largely ignored UI issues.

7.6 ARTICULATION AND ROLES

In addition to the points made previously on roles, Tang et al. observe that AmberPoint articulated the context for a problem or a solution more frequently when compared to Adobe, and that the explicit rendering of context is an important facet of an effective design structure. Adobe made 47 cases of implicit reasoning and 52 cases of explicit reasoning. AmberPoint made 33 cases of implicit reasoning and 87 cases of explicit reasoning, and postulated that communicating rationale is an important facet of effective design. Sharif et al. note that Adobe had 414 verbal events, with 448 for AmberPoint. Sharif et al. further state that Adobe had the most agreement between the two designers, and that the consulting activity was scattered uniformly throughout the session, and this observation may have contributed to Adobe's outcomes and their perceived ranking. Baker and van der Hoek observe that there was a slight tendency for the designers to raise ideas that they themselves originally proposed during reiterations. The percentages were 49% for Adobe, 56% for AmberPoint, and 60% for Intuit. Baker and van der Hoek comment on the concept of attachment to a "pet" idea in which an issue was proposed and reiterated by one designer. Finally, Budgen observes that the whiteboard annotations tended to be dominated by one team member for all three teams.

7.7 NOTE ON DOCUMENTATION

Sharif et al. observe that, early in their sessions, the designers spent time constructing the data dictionary, use cases, and a class model, and that considerable time was devoted to discussions on code logic even when constructing the class model. Budgen notes that Adobe exhibited a cyclical pattern of data modeling (entity relationship diagrams) interspersed with behavioral (state transition) modeling and some functional (DFD) clarification. There was very little discussion of classes and program modeling. The Adobe team used the functional form for an analysis of the specification, specifically with respect to the UI requirements. In contrast, AmberPoint exhibited a strong focus on UI throughout their session and concluded it with a drawing of an entity–relationship (ER) diagram.

There was a notable trend, across the groups, of using informal representations. All the teams used notes (quite extensively in the Adobe and Intuit teams) and drew pictures that were not representative of standard software engineering design diagrams. These findings are echoed in Budgen's work where he notes the importance of informal lists and informal diagrams. Both AmberPoint and Adobe drew out potential UIs. In contrast, none of the teams drew out use case diagrams or code structures and, while each team drew class diagrams or ER diagrams, no team drew both; this is possibly a commentary on the degree of overlap between the utility of these representations. The informality of the representations employed could be because the teams were working with just a whiteboard, but it could also be a more prevalent tendency in the practitioners' conceptual design process. Regardless, Budgen notes that informal artifacts are not traditionally supported by software engineering design workbenches, that there is frequent switching between viewpoints that also challenges the available design environments, and that these environments capture outcomes and not the underlying rationale.

7.8 ISSUES

While these papers provide a strong foundation for future work in the field, they also suggest several issues that should be addressed going forward:

- Capturing design quality: while the number of requirements covered was a frequently used metric to guide the evaluation, Sharif et al. state that the availability of a design-quality gold standard should provide better support for a comparison within and across studies.
- Shared ontology: a collective analysis needs to be supported by a collective ontology that maps and relates the terms used in each analysis schema. Currently, a comparison across the analyses is hindered by obfuscated semantic relationships between the various analysis categories.
- Ecological validity: a body of work needs to be carried out to characterize system design *in vivo*, as it is unclear how representative these sessions were of the designers' normal conceptual design context. For example, do conceptual designers work alone or in larger teams? Do they use whiteboards or other media and tools? Building on this characterization knowledge, the preliminary findings reported here direct several future studies.

ACKNOWLEDGMENT

This work was supported, in part, by Science Foundation Ireland grant 10/CE/I1855 to Lero—the Irish Software Engineering Research Centre (www.lero.ie).

REFERENCES

Clements, P., Kazman, R., and Klein, M. (2001). *Evaluating Software Architecture*. Boston, MA: Addison-Wesley.

Evans, J. (2003). In two minds: Dual process accounts of reasoning. *Trends in Cognitive Sciences* 7(10): 454–459.

Gamma, E., Helm, R., Johnson, R., and Vlissides J. (1995). *Design Patterns: Elements of Reusable Object-Oriented Design*. Reading, MA: Addison-Wesley.

Jarke, M., Loucopoulos, P., Lyytinen, K., Mylopoulos, J., and Robinson, W. (2011). The brave new world of design requirements. *Information Systems* 36: 992–1008.

Kahneman, D. (2011). *Thinking Fast and Slow*. New York: Farrar, Straus and Giroux.

Krutchen, P. (2003). *The Rational Unified Process: An Introduction*, 3rd edn. Boston, MA: Addison-Wesley.
Perry, D., Porter, A., and Votta, L. (1997). A primer on empirical studies. Tutorial presented at the International Conference on Software Maintenance, Bari, Italy.
Rowe, P. G. (1987). *Design Thinking*. Cambridge, MA: MIT Press.
Schmidt, D., Stal, M., Rohnert, H., and Buschmann, F. (2004). *Pattern-Oriented Software Architecture*, Vol. 2. New York: Wiley.

CHAPTER 8

Accessing Decision Making in Software Design*

Henri Christiaans

Delft University of Technology

Rita Assoreira Almendra

Technical University of Lisbon

CONTENTS

* Reprinted from *Design Studies* 31 (6), Christiaans, H. and Almendra, R.A., Accessing decision-making in software design, 642–662, Copyright 2010, with permission from Elsevier.

8.1 INTRODUCTION

The study described in this chapter is one of the contributions to the 2010 international workshop, Studying Professional Software Design. The goal of this workshop was to collect a foundational set of observations and insights into software design, drawing on theories and methods from a variety of research disciplines. One of the disciplines that already has a long tradition in design research is product design. Because the authors' backgrounds are in research on product design, they used their expertise in this area to reflect on the data.

The videos and transcripts of three pairs of professional software designers working on the conceptual design of a software system were made available. The video material also included a feedback session with each team in which they gave their perception on their own process. Participant researchers were asked to analyze the videos. A fuller description of the design task and the data capture can be found in the introduction to this Special Issue.

Design processes can be studied in different ways. *Analyzing Design Activity*, by Cross et al. (1996), is a good example of how the same protocols of designers have been analyzed from various scientific and practical perspectives. One of the most relevant aspects in modeling the design process is decision making. In 1969 Simon already noted that decision making and design are so intertwined that the entire decision-making process might be viewed as design. In particular, the conceptual design phase, when the solution space is explored and the product and context have to be observed from multifaceted perspectives, is a decision-intensive process (Rehman & Yan, 2007). Also, Longueville et al. (2003) noticed that in recent years a number of proposals have been advanced for the study of decision-making processes in knowledge areas such as management, cognition, engineering design, and artificial intelligence. Thus, decision making is a field of study that is constantly addressed in all domain knowledge areas; the main driver of those studies is the cognitive assessment of how decision making occurs. Therefore, a decision-making framework will be presented that forms the basis for the analysis of the three protocols. The next section will first give some background information regarding the design processes in different domains. Next, the method used in this study will be explained briefly, followed by an analysis of the three protocols.

The analysis will be twofold, based on (a) the decision-making framework used and (b) the summaries (the way that each participant viewed his own performance along the process). Finally, the results and conclusions are presented.

8.2 BACKGROUND

Research into product development processes has the final aim to understand those processes and to build models that can support design practice in efficiency and quality. It is about designing a better design process. For a number of separate phases in the design process, satisfying methods and tools have been developed and implemented, such as those dealing with information flows, computer-aided design (CAD) tools in the detailing process, and methods of production.

Modeling the conceptual phase of the process, however, is still a very complicated issue because this part of the product development process occurs in a state of uncertainty and ambiguity. It involves the generation and evaluation of design alternatives, and the selection of a single or a set of design solutions to fulfill a particular need or function. The importance of conceptual design to the overall success of the product is crucial. Once a final concept is chosen, the majority of the design decisions relating to the product behavior, cost, and quality have been fixed; and the subsequent product life-cycle activities (manufacturing, assembly, use, and recycle/dispose) are implicitly determined by the concept (Rehman & Yan, 2007). At the same time, this is the most complicated phase because of both the ambiguity of the brief and the consequences that the decisions taken at this phase have for the result. Whitney (1990) uses the term *interactions* to express that particular design activities are not that difficult but rather that they affect each other in circular ways. "Furthermore, some variables or decisions in the design sooner or later are found to dominate others in the sense that these decisions ordain others or make them moot. These dominating variables are often called 'design drivers'. Failing to identify them early in design often results in a failed design or the need to start design over" (Whitney, 1990, p. 10). At the least, designers need to be aware of the consequences of their decisions early in the conceptual design stage to undertake effective and informed life-cycle-oriented decision making.

The nature of modeling methodologies requires the use of precise data (Smith & Morrow, 1999), hence the models so far consider only the most structured parts of the engineering design process while the conceptual and creative aspects have been neglected. Several authors complain that the tools and frameworks hitherto developed do not support the decision-making process from the holistic point of view of the product, its user, and the environment (cf. Rehman & Yan, 2007). Particularly for the conceptual phase, there is still a gap between the academic approach to model building and the applicability of the models in practice.

8.2.1 Domain-Specific "Language"

In discussing the conceptual phase in design, the authors are well aware of the fact that they reason from the perspective of product design and architecture. However, the study described here is about software engineering and interaction design, which raises the question of whether the findings can be generalized from product development processes to these areas of design. Both software engineering and interaction design are rather young disciplines with settled practices. At first face, software design does not differ that much: it involves the integration of multiple knowledge domains: knowledge of the application domain, of software system architecture, of computer science, of software design methods, and so on. Regarding the development of large software systems, the following phases are usually described: requirement analysis; system design (high-level and detailed); implementation (coding, debugging, and unit testing); and system testing and maintenance (Guindon, 1990). "During high-level system design, a designer has to transform an incomplete and ambiguous specification of the requirements into a high-level system design, expressed in a formal or semi-formal notation" (p. 279). It includes low-level component and algorithm implementation issues as well as the architectural view. Furthermore, for the design of systems for the intelligent control of urban traffic, artificial neural networks (ANNs) (learning systems) and expert systems (knowledge-based systems) have been extensively explored as approaches for decision making. The traditional design methodologies in software design usually begin, as Pennington et al. (1995) explain, by identifying a top-level view of the main function of the software system, which is then decomposed into more and more detailed subgoals. The focus of these methods is on the structure of the data and the processes to be performed on the data. The question is whether, in the meantime, design processes have adopted other views.

Nowadays, it is common knowledge that software design should be developed not so much from a constructor's eye view, but rather from a designer's eye view, taking the system, the user, and the context all together as a starting point (Winograd, 1996). However, there still are complaints about the fact that for software engineers and even for computer professionals' design education, the primary concern is to implement a specified functionality efficiently while the user interface (UI) is seen as a cosmetic final stage in the overall software development process and/or as a job for an interaction designer.

A similar discussion was once characteristic for the areas of industrial design and product design as well, probably as part of becoming mature and accepted. That brings us to the question of how software design differs from industrial design more than a matter of maturity. One particular study has to be mentioned although it regards only interaction design. Edeholt and Löwgren (2003) analyzed both industrial design and interaction design on the basis of key dimensions: process, material, and deliverable (see Figure 8.1). For most dimensions, each of the disciplines shows a rather clear orientation toward one of the extremes of the scale attributes. The overlap between the disciplines in this analysis as a whole is only 14%. While in our study, aspects of both software engineering and interaction design are under investigation, it will be interesting to see if these differences in "language" also pertain to the software engineering part.

8.2.2 Design and Decision Making

Different approaches regarding the nature of decision-making processes can be found in the literature. According to Sarma (1994), three main approaches can be distinguished: (1) a *descriptive* approach, which makes use of models and theories to describe and explain human decision-making behavior by studying human beliefs and preferences as they are; (2) a *normative* approach, which utilizes axioms to make optimal decisions, studying mainly the logic of decision making and the nature of rationality in an attempt to suggest how good decisions ought to be made; and (3) a *prescriptive* approach, which develops techniques and aids for supporting and improving human decision making. The analysis in this chapter belongs to the first category, the descriptive approach, in the way that Longueville et al. (2003) define it: as an approach aimed at modeling in order to study, understand, represent, and reuse existing decision-making processes. In our opinion, the most relevant contribution lies in the analysis of the relationship between the decision-making process and the quality of the result. The reason is our belief that product development should solve a profit maximization problem.

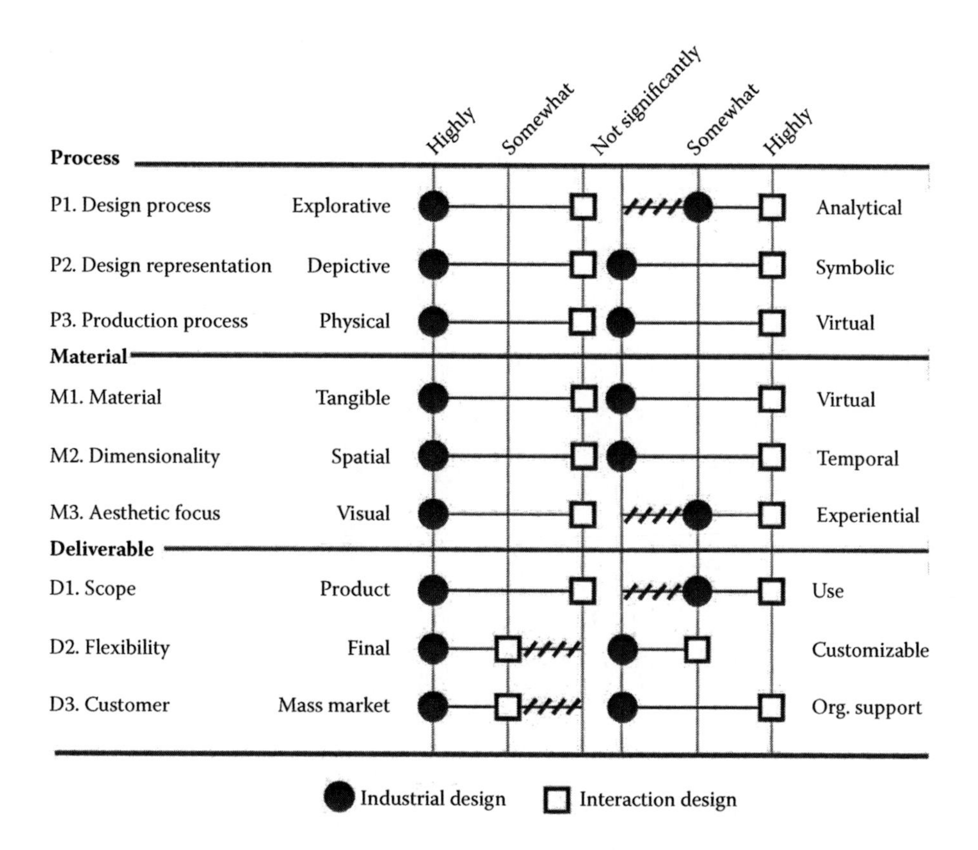

FIGURE 8.1 Comparison of industrial design and interaction design. (From Edeholt, H. and Löwgren, J., *Proceedings of the 5th Conference European Academy of Design*, Barcelona, 2003.)

8.3 DECISION-MAKING FRAMEWORK

Based on a number of empirical studies conducted by the authors, a descriptive framework of decision making was developed and used in the analysis of the three protocols (Almendra & Christiaans, 2009). The framework is presented in Figure 8.2.

The framework consists of two major levels: (A) the mindset and (B) the operationalization of the mindset. In between, the *nature of decisions* forms an intermediate level. Each of these levels can be further detailed:

(A) *Mindset*—macro level. This depends on: (1) the *design strategy*, which is based on Christiaans and Restrepo's (2001) findings where two different orientations were identified in terms of the way that assignments are approached by designers: problem oriented when there are descriptions made in terms of abstract relations and concepts; and solution oriented when from the beginning one or more possible solutions are the drivers of the process. In the framework, these categories correspond, respectively, to *problem driven* (A) and *solution driven* (C), respectively. We added a new category that we called *integration driven* (B), which is defined by the coevolution of the problem and the solution along the design process. (2) *Creative cognitive processes* are identified at two modes: *exploratory*, which has to do with operations

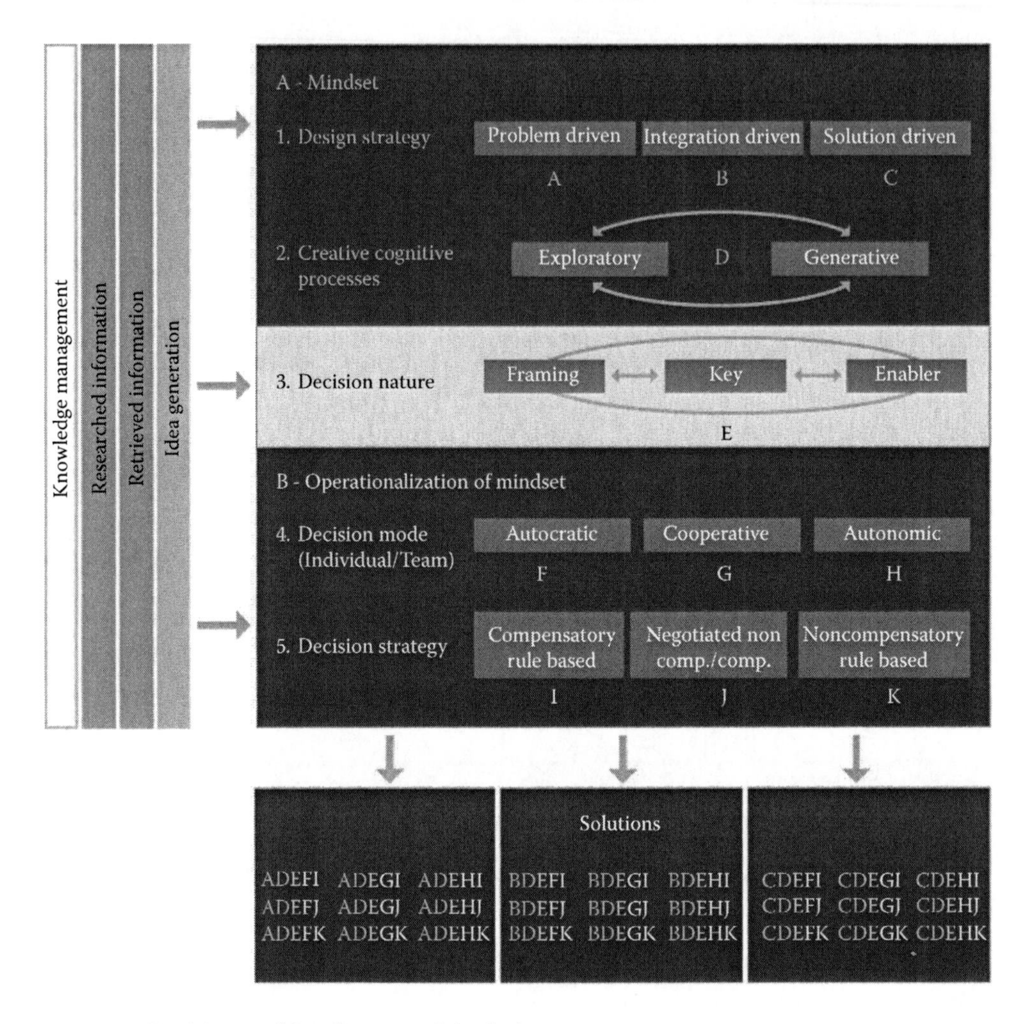

FIGURE 8.2 Decision-making framework in design.

such as contextual shifting, functional inference, and hypothesis testing; and *generative*, which is related with analogical transfer, association, retrieval, and synthesis. (3) *Decision nature*—intermediate level. This is defined by Christiaans and Almendra (2008) as being of three types: *framing decisions*—decisions made during the period when a designer mentally "frames" the object; *key decisions*—those made at moments when the (preparation of the) product creation occurs; and *enabler decisions*—which signify mental object representation instants. Again, since designers always display the three natures of decisions, all three are identified with a single letter, "E."

(B) *Operationalization of mindset*—micro level. Decision-making is defined according to the following descriptors: (4) *Decision mode*, which has to do with group dynamics in relation to decision making, including three modes: *Cooperation* (F): the group leader works together with the team members, seeks integration of people's ideas, and prompts and enables members to decide. *Autocracy* (G): the leader works for people; this mode can be either autocratic or it can have a consultative direction. *Autonomy* (H): the leader delegates to the team members; this delegation can be quite structured, meaning that a procedure or a more broad approach must be followed. (5) *Decision strategy*, encompassing three distinct strategies: A *noncompensatory rule-based strategy* (I): under such a strategy, as defined by Rothrock and Yin (2008), designers generally do not make use of all available information and trade-offs are often ignored. A *compensatory rule-based strategy* (J): information is processed exhaustively and trade-offs need to be made between attribute cues. A *negotiated strategy* (K): designers use both previous strategies, trying to balance their decision constraints between several aspects such as time, expertise, and level of information.

The 27 possible solutions in the framework are no more than distinct descriptions of designers' processes in terms of the parameters listed in the decision-making framework. As an example, we can see that an "ADEGK" designer is A—problem driven; D—engaging in cognitive creative processes (that can be further analyzed to see which are dominant); E—taking the framing, key, and enabler decisions (that can be further explained regarding their content and prevalence); G—displaying a cooperative mode of making decisions; and K—making a prevailing use of the noncompensatory rule-based type of decision strategy.

8.4 METHOD

The study is based on the verbal protocols of the three software design teams. Within limitations, the protocols show the product/system development organization in terms of a sequence of steps that transform customer requirements into a satisfactory design; and in terms of the information flow governed by one decision maker who makes both the design decisions and the development decisions under time and budget constraints. It is a decision production system (Almendra & Christiaans, 2010).

In experiments such as the one described here, the coding of data is particularly relevant. Coding data is a data reduction method that aims to facilitate information management and to keep focus on the relevant issues that should be scrutinized critically. The two authors performed the coding. The analysis of the data in this case involved two types of processes: (a) a process of immersion where both researchers immersed themselves in the collected data by reading or examining some portion of the data in detail, and (b) a process of crystallization wherein the researchers temporarily suspended the process of reading and examining the data in order to reflect on the analyzing experience and to identify and concatenate patterns or themes.

The protocol analysis was done according to four sequential, distinct phases:

1. Independent reading of the three protocols' transcripts
2. Independent reading of the brief (design prompt)
3. Independent reading of the three summaries' transcripts
4. Independent watching of the three protocol videos

In between each phase, the two authors confronted their own analysis and consolidated it.

For the analysis in Phase 1, a supporting spreadsheet was used with the three protocols' names as the columns and the descriptors that integrate the decision-making framework as the rows. The comments made (in the cross cells) were supported by the protocols' speech fragments identified with their time occurrence. The option of accessing the brief after a first overview of all the transcripts was intentional: it gave origin to a second reading and a discussion of the transcripts of the three protocols.

Before addressing the next step in our analysis process, a visual representation synthesis of each solution building was done. The aim of this analysis method was to try to verify if our synthesis was coincidental with the one made by the subjects of the protocols. With the analysis hitherto done on the design processes of the three teams, a next step had to be made by analyzing the feedback that the teams gave in evaluating their own work. This was possible through the assessment of the reflective summaries made by the teams after they had finished the assignment. This analysis of the feedback was compared with the analysis previously made. Finally, the videos were watched as a check on the data gathered.

8.5 ANALYSIS OF THE THREE PROTOCOLS

8.5.1 Analysis of Protocols Using the Decision-Making Framework

8.5.1.1 Protocol "Intuit"

The duration of their design process took 1 hour, half the time that the other two teams took. The team members are both male (M) with different experience: M2 is a recent practitioner; M1 has experience but with completely different applications and systems. Phase 1 of the analysis is synthesized in Table 8.1.

Going into the process of the Intuit protocol by way of the different levels of the decision-making framework shows the following results:

- The *design strategy* never becomes a joint one; the two subjects have different approaches to the brief: M1, who visibly took the leadership of the process, is *problem driven* while M2 is *solution driven*. The process is a continuous "ping pong" between an operational level of analysis and a more broad, abstract one.
- The *nature of decisions* shows all varieties, the *framing* ones being more frequent in the period that precedes the first *key decision*. After this period, the *enabler decisions* occur predominantly and are by and large proposed by M2. It is significant to note that all these decisions mainly concern the subsolution of the structure/coding system. The UI development is almost absent in their decision process.
- The *decision mode* is *cooperative* although it is possible to plainly identify two distinct roles for the two subjects: M1 gives the inputs; M2 develops and enables. M1 raises questions and tests the subsolutions and he encourages M2. Halfway through the process, M2 becomes more inquisitive and tries to take a wider perspective that he adds to the evaluation of M1.

TABLE 8.1 First Phase Analysis of Team "Intuit"

	Levels of Analysis	Intuit: The "Ping-Pong" Case
1. Design Strategy	A. Problem driven	Male 1 (M1): assumes leadership—05:29:07; 06:23:07; 12:55:01; 14:51:2; 17:40:5; 23:54:01; 34:38:07; 43:52:09; 48:31:09.
	B. Integration driven	
	C. Solution driven	Male 2 (M2): 06:47:5; 12:50:1; 35:13.15 (almost all of his are interventions).
2. Creative Cognitive Processes	D. Exploratory (operations such as contextual shifting, functional inference, and hypothesis testing)	The exploratory and generative approaches are always present. In this case, there is no visible dominance of one process over the other; the subjects use both in a continuum along the process; 41:35:03 (example of analogy); 49:33:04 (example of synthesis).
	D. Generative (related with analogical transfer, association, retrieval, and synthesis)	
3. Decision Nature	E. Framing	Examples of *framing decision* moments: 14:51:02; 17:04:07; 19:33:0; 27:19:01; 41:35:3; 43:52:09 (these decisions were taken by reducing and prioritizing the problem variables).
	E. Key	First *key decision*: 29:07:03—after a process of identifying the key variables, a solutions system was identified by the subjects: 38:33:04; 46:24:04; 49:39:09; 56:50:02—moments indicating that finally a model to build the end-user application was found.
	E. Enabler	*Enabler decisions* help the work of reducing the variables, thereby "giving shape" to the possible solution; they occur in almost all of the moves of M2. There was a clear dominance of enabler decisions after *key decision 1.*
4. Decision Mode	F. Autocratic	
	G. Cooperative	From the first moment Male 1 (M1) takes the role of leader. He gives the inputs so that male 2 (M2), trying to answer the question that M1 raises, tests it in terms of subsolutions. M1 incentivizes M2 (49:19:0; 51:52:08) and halfway through the process, M2 becomes more inquisitive and tries to take a wider perspective, seeking confirmation/ agreement with M1 (51:32:03).
	H. Autonomic	
5. Decision Strategy	I. Compensatory rule based	
	J. Negotiated	Decisions are taken based on the subject's own experience (08:17.09) and on the reduction of variables (12.01.1; 21:35:06) and also upon the establishment of a hierarchy of variables (11:01:07; 16:56:04; 23:54:01).
	K. Noncompensatory rule based	

Main Conclusions: In this protocol there is clearly one subject, M1, who is *problem driven* and the second subject, M2, is *solution driven*. This causes a constant "ping pong" between an operational level of analysis and a more broad, abstract one.

In terms of the two desired outcomes of the design brief, these subjects were clearly focused on Outcome 2 (structure of the code); 43:02:03 is one of the rare moments where there is visible concern about Outcome 1 (the user interface); 57:20:05 is another moment where the interface is presented.

- The overall *decision strategy* is considered to be a *negotiated* strategy with both *noncompensatory rule-based* and *compensatory rule-based* characteristics. Some decisions depend upon a rational analysis of the variables in place and a thorough hierarchy of them; others are assumptions based upon experience and/or circumstantial simplifications generated in order to enable the work.

8.5.1.2 Protocol "Adobe"

The process of the Adobe team took 2 hours, the time available for the experiment. Both subjects are male and have experience in this area: M1 has 16 years of experience and M2 has 10 years. The subjects are accustomed to working together although they never worked so directly and on such a type of exercise.

Phase 1 of the analysis is summarized in Table 8.2. The fundamental aspects to retain from the analysis regarding the Adobe protocol are:

- The *design strategy* is clearly *solution driven.* Both team members try to reach a solution from the beginning in a very intensive developing process. The brief is used as a control instrument along the process. The subjects have defined a modus operandi: to start with the simplest structure (ground 0) and progressively add complexity. It is a development that can be seen as a sequence of frames in a time/space perspective.

- Regarding the *nature of decisions*, all three types naturally occur; the *key decision* is traced early and it rules the whole process. The idea is the development of a basic simple unit to which other elements should progressively be added; this should be done first in terms of the structure and later in terms of the UI. Furthermore, being a *solution-driven* process is, by consequence, a basis for very intensive *enabler decision* taking. Both team members contribute in a sequence of reasoning where one completes or corrects the idea of the other or adds more information. Moreover, enabling is mainly done through the use of simulation.

- The *decision mode* is *cooperative* and the roles of both subjects naturally arise at the beginning of the process. The leadership role in the development is taken by M2 (the subject who is designing). M1 acts as a controller of the "line of reasoning," making a synthesis of the ongoing work and raising questions and/or bringing up elements that were either forgotten or not yet thought of. M2 works on the solution from the "inside out" and M1 does it from the "outside in."

- The overall *decision strategy* is, in general, *compensatory rule based.* The team members dissect each subalternative and do rational judgments that are normally supported by a simulation of the alternatives generated. However, it is possible to observe some noncompensatory behavior. This is the case at moment 1:00:43 where a shortcut proposal is made; another example is the one that appears at about 1:48:29 where a simplification has been accepted.

TABLE 8.2 First Phase Analysis of Team Adobe

	Levels of Analysis	Adobe from Unit to System: A Story in Frames
1. Design Strategy	A. Problem driven	
	B. Integration driven	
	C. Solution driven	Male 1 (M) and male 2 (M2) are *solution driven*—M2 is the developer; M1 is the feeder/point of the situation/control element (05:32:9; 49:41.0; 58:15.0). Although the subjects are not *integration driven*, they try to match their structure with a possible UI.
2. Creative Cognitive Processes	D. Exploratory (operations such as contextual shifting, functional inference, and hypothesis testing)	The exploratory and generative approaches are always present. Subjects use both approaches in a very intense way with M2 more in the exploratory field and M1 in the generative one (this has to do with the assumed roles).
	D. Generative (related with analogical transfer, association, retrieval, and synthesis)	
3. Decision Nature	E. Framing	Subjects start framing at 5.51.6—subsequently, they enable (8:51.0, 09:24.7); from enabling they consider again the conceptual framing (13.37.9) and the definition of the structural elements and their hierarchy and role in the system (14:14.1; 21:47.4; 32.18.2; 42:40.4; 48:20.4). Then, the subjects start framing another level of complexity: the framing of the UI starts at 52:44.0.
	E. Key	First *key decision*—05:18.6/05:32.9 (this decision has ruled the process—to develop a basic simple unit and add complexity—this is in terms of the structure); later, the subjects had a similar *key decision* regarding the UI—the interface solution—also starting with a simple base and developing it by adding complexity (1:00:02.5/1:03:28.8); other key decisions: user stories (1:19.09.1); UI process—start-observe-pause (1:25:08.5); structure after the UI (1:33:37.7).
	E. Enabler	Very intense enabler process—decisions are explored by both subjects in a sequence of reasoning—one completes or corrects the idea of the other or adds more information; (06:41:1)—enabling started early in the process; subjects alternate framing (09:37:4) with enabler; enabling is mostly done by using simulation while designing (10:24.0–13.30.1).
4. Decision Mode	F. Autocratic	
	G. Cooperative	Leadership of the development (driver) is assumed by M2 (probably the one who is designing); M1 acts as a controller of the line of reasoning, making a synthesis of the ongoing work and raising questions and/or bringing elements that were either forgotten or not yet thought of (41:20.4; 41:30.7; 59:44.9). M2 works "inside" and M1 works "outside."
	H. Autonomic	

(continued)

TABLE 8.2 (Continued) First Phase Analysis of Team Adobe

	Levels of Analysis	Adobe from Unit to System: A Story in Frames
5. Decision Strategy	I. Compensatory rule based	In general, this is the dominant strategy—subject analyzes each subalternative and does rational judgments—normally supported by the simulation of the alternatives generated.
	J. Negotiated	(1.01:10.6)—an attempt at dismantling the system in order to check its consistency and to simplify
	K. Noncompensatory rule based	(1:00:43)—time for a shortcut proposal; (1:48:29.0)—time for a simplified assumption

Main Conclusions: In this protocol, both males are solution driven but they have different roles in the design process—one is the developer (M2) and the other is the process controller (M1). The brief is used intensely to help that control (45:06.1;57:26.2); the process here does not evolve in a "ping pong" between the big picture and the operationalization of it; the designers have defined a modus operandi—to start with the simplest structure (ground 0) and progressively add complexity (23:38.3; 43:41.8)—assuming stage 2 of the model in terms of complexity; (55:27.1) statement of more sophistication, evaluation (30:55.8; 44:21.8), and correcting/fixing (33:51.0); the designers used this procedure both for the rules system and for the UI; it is a sequence of frames in a time/space perspective.

In terms of the desired outcomes, the structure (Outcome 2) is intensely developed during the first hour and then the designers introduce the user interface (Outcome 1) at 1:03:28.8, trying to translate the subsolution found for the structure into a UI. 25:32.5 was the first moment when the user was broached; 47:07.4; 52:44.0 are moments when the "user" starts to be important in the process; after these moments, the user's viewpoint is confronted with the structure model "put-in-action" and that brings to light some frailties of the rules system. Some moments where the user was considered: 57:26.2; 58:25.7; some examples of the use of the user's perspective to correct the coding: 57:49.9; an example of assessing the dimension of the rule's system: 1.00.36.8; example of an attempt to make a shortcut (1:00.43.0).

From 1:03:28.8 until 1:16:51.6, the subjects intensely developed the interface subsolutions and tried to simulate the visual representation and the use of it; then they changed to the system again with inputs coming from the interaction simulation. Because of the time already spent (1:17:59.5) they felt the need to look at the stage they were at, forcing them to concentrate on reinforcing the structure system. Subjects then iterated between coding and the UI; an example of process concern (1:22.49.9); an example of an attempt to clean the solution (1:39:57.5); an example of the possibility of assuming two alternatives (1:41:22.9). The last 40 minutes were almost dominated by the revision of the developed system, by testing it against user simulation without achieving a clear result.

8.5.1.3 Protocol "AmberPoint"

The process takes the full 2 hours. The team members are one male (M) and one female (F). Of all the teams, they have the longest experience in this area: M has 20 years and F has 26 years. M undoubtedly has expertise in user-interaction issues. The team members were already accustomed to working together and have a friendly and in-control interaction.

Phase 1 of the analysis is synthesized in Table 8.3. Regarding the AmberPoint protocol, the essential aspects that arise from the analysis presented in Table 8.3 are:

- The *design strategy* followed by both team members shows an *integration-driven* approach. This is particularly evident in M. The starting point is the development of the UI in its visual and usability aspects; that gives origin to the description of the structure that again contributes to the detailing of the learning tool, the one asked for in the brief.

TABLE 8.3 First Phase Analysis of Team AmberPoint

	Levels of Analysis	AmberPoint: An Integration Process
1. Design Strategy	A. Problem driven	
	B. Integration driven	Both subjects, male (M) and female (F) display an integrative-driven strategy with this characteristic more visible in M; it is also M who has the designing/sketch skills that allow him to represent the ideas being created. The starting point was the development of the UI in its visual and usability aspects; that gave origin to the description of the structure that again contributed to the detailing of the UI as a learning tool.
	C. Solution driven	
2. Creative Cognitive Processes	D. Exploratory (operations such as contextual shifting, functional inference, and hypothesis testing)	This is the only protocol where we can talk about explicit creative cognitive processes since there is a strong and dominating focus on the visual representation of the interface; drawing as a tool to generate ideas.
	D. Generative (related with analogical transfer, association, retrieval, and synthesis)	
3. Decision Nature	E. Framing	Framing occurs a lot along the process starting with the framing of the UI system (6:27.4; 8:03.1; 08:53.7; 9:31.1; 15:05.3; 17:04.8) followed by the coding system (22:16.7; 29:12.3; 32:52.1; 42:04.0; 44:47.0–45:18.6) and again both are integrated (52:38.054:15.5; 1:04:33.6).
	E. Key	Key decision—22:46.7 is the moment when the subjects decide that the system to be developed depends on new information from Professor E. Some other key decisions: regarding a subsolution (35:24.9); a subsolution related with real-time simulations (1:30:00); again the solution depends upon the goal of this learning tool (56:33.4); needed information (statistics about traffic) to feed the solution (1:23:54.0); subsolution in terms of structure (1:27:55.3); subsolution of the final design (1:36:43:05); subsolution to the structure regarding car ruling and its link to intersections (1:44:05.5); 1:47:53.2 until 1:48:23.3—summary of things needed to ask Professor E to implement the solution.
	E. Enabler	Enabling is very intense since this is a very interactive process with a high density of decisions to enable the partial or the whole solution that is being developed—13:35.5; 20:29.4; 24:22.9; 24:50.2; 26:43.5; 29:28.2; 29:59.9; 34:30.6; 43:55.2; 46:26.8; 53:49.8; 58:13.9; 1:03:15.9; 1:09:28.0; 1:13.09.0; 1:25:56.0; 1:20.06.1; 1:28:21.4; 1:34:33.8; 1:35:09.5 are some examples of enabler decisions undertaken.
4. Decision Mode	F. Autocratic	
	G. Cooperative	The dominant mode is the cooperative one; however, the protocol has two stages of interaction modes: during the first one, F is rather challenging and it looks like she is playing the "devil's advocate"—she has many doubts about the ongoing work, she displays a very pushing dialog with her partner (21:09.3; 24:31.2; 24:57.8; 31:00.6; 31:58.0; 35:24.9; 36:55.0; 41:07:0; 41:49.4; 44:17.6); after 45 minutes, F engages in truly cooperative and more constructive behavior and her reasoning also becomes more fluid and efficacious (45:18.6; 49:38.7; 54:00.2; 58:13.9; 1:01:33.9; 1:04:33.6; 1:13:56.3; 1:16:01.3; 1:19:23.1; 1:29:03.8; 1:33:49.3; 1:49:01.7).

(*continued*)

TABLE 8.3 (Continued) First Phase Analysis of Team AmberPoint

	Levels of Analysis	AmberPoint: An Integration Process
5. Decision Strategy	H. Autonomic	
	I. Compensatory rule based	Some examples of compensatory decisions: 20:50.0; 1:13:25.8; 1:18:13.6; 1:23:54.0; 1:44.05.5.
	J. Negotiated	In general, this is the dominating strategy; there is a balance between compensatory and noncompensatory strategies: F is the one who displays more noncompensatory decisions (examples: 35:51.1–36:35.2).
	K. Noncompensatory rule based	Some examples of noncompensatory decisions: 09:31.1; 19:13.2; 20:29.4; 24:31.2; 25:09.4; 29:12.3; 29:28.2; 30:42.8; 44:19.7.

Main Conclusions: In this protocol, the subjects started with the user (06:08.9); they devised the map of the "drawing tool" by making use of framing decisions (08:03.1); simulation was the method used to evolve the proposed solution and that depended heavily on the drawing skills of M (1:47:33). It is the only protocol where the software engineers questioned the ill-structure of the problem, stating that "they will present a solution that can be worked out by others while they have to get back to the professor E who needs the device to get more precise information (22:46:7; 24:56.5;55:16.6;1:23.54;1:47:53—summary of the doubts to pose to Professor E)."

It is also the only protocol where the question is posed about the usefulness of the tool (52:38.0) and where the tool's goals and methods are questioned: it could be a learning tool where Professor E would like students to learn through trial and error (56:33.4). Correlation seems to be the pedagogical goal supporting the use of the tools (1:33.17.1) The way of reporting the results after using the tool is decisive for this team. Again, the importance of the utility of the proposed solution is stressed (1:35:22.4; 1:35:56.5; 1:48:23.3; 1:49:38.6–1:50:53.6).

The focus on the user's perspective is highly relevant—the whole protocol is the clear expression of a true concern with usability, the ease of use of the tool (18:10.8; 47:32.8; 1:16:01.3), the friendliness of use (18:46.1; 19:13.2; 20:03.6; 40:35.5; 1:07:24.8; 1:09:04.8; 1:11:28.1; 1:11:20.4), the redundancy in the system to make it clear to the user (41:07.0–41:13.3; 1:14:03.9), the customizing aspects of the tool (43:08.7), and the enabling of the visual comprehension of the tool (1:29:10.3; 1:31:05.9; 1:33:17.1; 1:40:33.9).

- Regarding the *creative cognitive process*, this is the only protocol where creativity is explicitly expressed in the strong and dominating focus on the visual representation of the interface. This visual way of displaying and designing the simulations, using drawing as a structuring tool, might in itself call for creativity. That is not to say that creativity was not present in the other protocols; but in the other teams, the explicitly expressed conversation was focused on analyzing, structuring, and coding.
- Regarding the *nature of decisions*, all different types naturally appear; *framing decisions* arise very often along the process, starting with framing the UI system; it is pursued by decisions concerning the structure system and again by decisions where both systems are *framed* in an integrated mode. The *key decision* happens at a moment where they decide that whatever system is developed, it should be based on new information from Professor E (the client). After this assumption, other *key decisions* are taken concerning subsolutions; but all are supported in the conviction that it depends on the goal of this learning tool, the methods used in teaching—trial and error or other. The design solution appears at 1:19:23.1; subsequently, several subsolutions are decided upon with respect to structure. Another *key decision* occurs when a summary of "things to ask to Professor E" is created to implement the solution. Finally, *enabling decisions* are very frequent since this is a very interactive process with a high density of decisions to enable part or whole of the solution being developed.

- The dominant decision mode is *cooperative*. However, the protocol has two stages of interaction modes: in the first one, F is rather challenging and it looks like she is playing the "devil's advocate"—she has many doubts about the ongoing work, and she displays a very pushing dialog with her partner. After 45 minutes, she engages in a truly cooperative and more constructive behavior and her reasoning also gets more fluid and efficacious.
- The overall decision strategy is, in general, a *negotiated* one between *compensatory* and *noncompensatory rule-based decisions*. There is a balance between the two strategies. F is the one who displays more noncompensatory decisions.

8.5.2 Analysis of the Summaries

The summary videos presented the perception that the teams had about their own process and its outcome. On face validity, such an analysis was only possible for the Adobe and AmberPoint protocols because the Intuit protocol was biased by the interference of an external person who asked different questions and made comments. Nevertheless, considering only the team members' appreciation before the intervention of this external person, it is possible to draw the following conclusions:

- *Intuit*—In the words of one of the subjects: "... we took the approach of modelling the whole interaction between everything. That's a weird annotation here but it works." This comment clarifies that the two team members considered their overall process and outcome as very positive and successful. However, their evaluation is not so rigorous if we consider that the design brief asked for two outcomes while they only give one. They were mainly focused on the structure and the coding (Figure 8.3).
- *Adobe*—The following comments of the team members give a concise impression of their evaluation of the overall performance: "We started with a picture, and drew a picture of a very simple case, the simplest intersection and then thought about how to (...). (...) I think probably both of us were looking at that picture and trying to take the simple case but come up with a scheme that we intuitively thought would grow well as we added more complexity to the system. (...) at the end we kind of we had these user stories and those are, I think, quite helpful in walking through a particular path through the design or through a functioning system and seeing, okay if you had tried to do such and such you have (...) is that accommodated with the objects in the processes that we've got." In

FIGURE 8.3 Intuit session.

fact, while analyzing the long and detailed description about the process given by the subjects, it becomes evident that they have a strong and very clear definition of the coding structure, which they defended in an efficient way. It is obvious that this summary underpins the analysis done by the authors (see Figure 8.4).

- *AmberPoint*—In this case it is important to underline the comment that M made regarding their approach: "I'm kind of trying to do is go and immediately figure out what's the end user's goal in doing this, and so the end user and the customer are two different people here, which quite often happens in software. Marketing will sometimes think that the customer, the one who's going to make the purchase is the one who's going to use it day to day, and its not. And actually it's pretty clear in this description that it's not. So we want to make the professor happy, but the students also have to be able to use this as a learning experience and so trying to figure out what are the most important steps that the professor wants the students to go through in order to learn about the traffic civil engineering problem this problem led itself to a very visual solution, you know, its clearly we needed a map. We knew that right away, and so that was kind of the starting point with this. I don't know that I normally would start with the kind of visual solution, but because this one called out for one, we were much more likely to start from there, but we also did kind of come up with like what's the whole list of objects that we need to deal with—both for storage, and that the user and the professors are going to have to talk about together." The description of the process done by these subjects was very detailed and clear, combining explanations on both the UI and the entity relationship (ER). The analysis of their own performance is in line with the one made by the authors (before seeing the summary) (see Figure 8.5).

FIGURE 8.4 Adobe session.

FIGURE 8.5 AmberPoint session.

8.6 CONCLUSIONS

8.6.1 Regarding Design Strategies

A usability approach is absent in the Intuit and Adobe protocols. In these two situations, the user interaction is not an initial variable framing possible subsolutions and whole solutions. The driver of all the decisions in these two protocols is the coding structure, which is intensely explored. Both teams use a strategy of stepwise refinement, in which a complex design problem is decomposed top-down into subproblems. Subproblems are then decomposed until the final layer of subproblems is simple enough to be solved directly (Pennington et al., 1995). The strategy of the Adobe team can be characterized as holistic or integrated. From the beginning, they take both the UI and the ER into account; and at a certain point, the team attempts to make the structure coding fit with the user necessities. This is done by testing it against user needs through a basic simulation of use. The AmberPoint protocol starts with a UI, and having it as a departure point the team could solve the structure in a natural "dialoguing" way. The ability of designing scenarios and the interface, displaying the different elements—maps, tables, menus, and palettes—contributes to the design of the software system.

8.6.2 On Creative Cognitive Processes

In Section 8.1, a comparison was made between product/industrial design and software design. At first, it was suggested that in their processes the two disciplines do not differ that much. The designers have to transform an incomplete and ambiguous specification of the requirements into a system design. However, analyzing the processes of the three software design teams gives rise to a more "nuanced" view. The development of functional solutions, including structure and coding, asks for deep analytic thinking and reasoning, resulting in the implementation of algorithms. But this is only a constructor's eye view, as was mentioned earlier, and quite characteristic for two of the three teams in this study. However, if one really takes a designer's point of view, that is, taking a holistic approach with the user, context, and functionality as integrated aspects, the processes of both disciplines will be more similar. Such an approach shows richer mental processing and gives more chance for creative actions and solutions.

8.6.3 On the Subject of Decision Nature

In general, enabling decisions are intensely taken while designing. It is interesting to observe the role of both partners in this process that enables key decisions. The team members who do the actual design work by writing and drawing are the ones who speak more. On the other hand, the "outsiders" within the team who are not doing the actual drawing/writing tasks, normally take the role of "controller" of the process by stressing the context of the design problem, the variables to be involved, the requirements from the brief that have to be taken into account, and so on. There is no case where only one key decision frames the whole solution. In fact, the final solution is the result of a chain of key decisions that address either the subsolutions regarding the UI or the structure/ruling or tackle the subsolutions for the integrated UI/structure product. The solution-driven protocols displayed a larger number of enabling decisions. This has also been observed in previous studies and it has to do with the fact that solution-driven subjects focus on the operational, technical aspects of the solution, "trying" to make the solution feasible.

8.6.4 Concerning Decision Mode

In terms of group dynamics, the *cooperative mode* was dominant among all three teams. But within this mode, one can observe very different team behavior as a consequence of the different and consistent roles that were taken by the team members. Both in the Intuit and the Adobe cases there was an unambiguous leader. In the Intuit protocol, the leader is the more experienced subject who takes control of the designing and writing while challenging his colleague who takes the role of enabler, operationalizing the framed decisions/reflections. In the Adobe case, the leader is the one who is in charge of drawing and writing the solution on the whiteboard while his colleague is the "controller" of the line of reasoning and the information input. The AmberPoint protocol displays two different moments of cooperation: (a) one where the woman takes the role of the skeptical, challenging partner, while the man is the developer and solver (he is also the one in control of the whiteboard); and (b) one where both subjects engage in a collaboration process where the woman empowers the man and engages in a constructive, supportive reasoning mode.

8.6.5 About the Experiment

The way that the design brief was structured in the experiment is similar to product design briefs in that it presented an ill-defined problem. The time available for the design process as well as the available resources in terms of equipment clearly influenced the outcomes. In fact, being an exercise that calls for interaction design implies the control of numerous variables in a time/space evolving context. The restriction of just using a whiteboard and colored pens introduced difficulties to the subjects. It might be interesting to evaluate whether the presence or absence of drawing software tools would interfere in a dominating way with the results.

The debrief moment (summaries) is essential to access the perception that the subjects had of their own performance. We assumed it was the final phase of the experiment. This is because the brief stated the need to communicate the solution to be developed to a group of engineers' developers. This being so, the way that the subjects communicated their processes and solutions was central to the project. Its analysis made it possible to understand how important the communication of the generated ideas and solutions is. This is consistent with the view of Curtis et al. (1988, p. 1282) when they state that "… developing large software systems must be treated, at least in part, as a learning, communication, and negotiation process." It was also visible that the way the subjects viewed their own performance, especially in what concerned the clarity of the reasoning, had no correspondence with reality.

8.6.6 In Relation to the Outcomes

The data show a relationship between the outcome of the project and the approach in the foregoing process. The decision that has the most profound impact on the (quality of the) outcome is the design strategy chosen at the beginning: (1) A strategy of stepwise refinement with the decomposition into subproblems—starting with coding (structure) and eventually followed by a UI focus. (2) An "integrated" strategy, trying to engage in a coevolutionary process where both desired outcomes (UI and structure) are explored in an integrative mode.

The integration-driven strategy was the one that resulted in the most balanced solution and the only one to fully address the desired outcomes.

8.6.7 Synthesis of this Study

As a synthesis of our work, it seems useful to visually present our analysis (see the diagrams for the three teams in Figure 8.6). This becomes even more significant since it is a visual representation that summarizes both the outcomes and the processes underlying those outcomes. In addition, in this experiment visual representation played a decisive role. As Curtis et al. (1988, p. 1283) noticed "... three capabilities that we believe must be supported in a software development environment are knowledge sharing and integration, change facilitation, and broad communication and coordination." The AmberPoint process better matches these three conditions.

8.6.8 Concerning the Differences between Software Design and Product Design

The level of complexity of the product design process and the software design process is different. Software designing involves more complex processes. Because software problems embrace both structuring and coding and a UI, the holistic view necessary for solving design problems in general, is, in this case, particularly critical.

The software design environment is more complex because it has a systemic nature. The decisions are chained and call for a more in-depth level of decision.

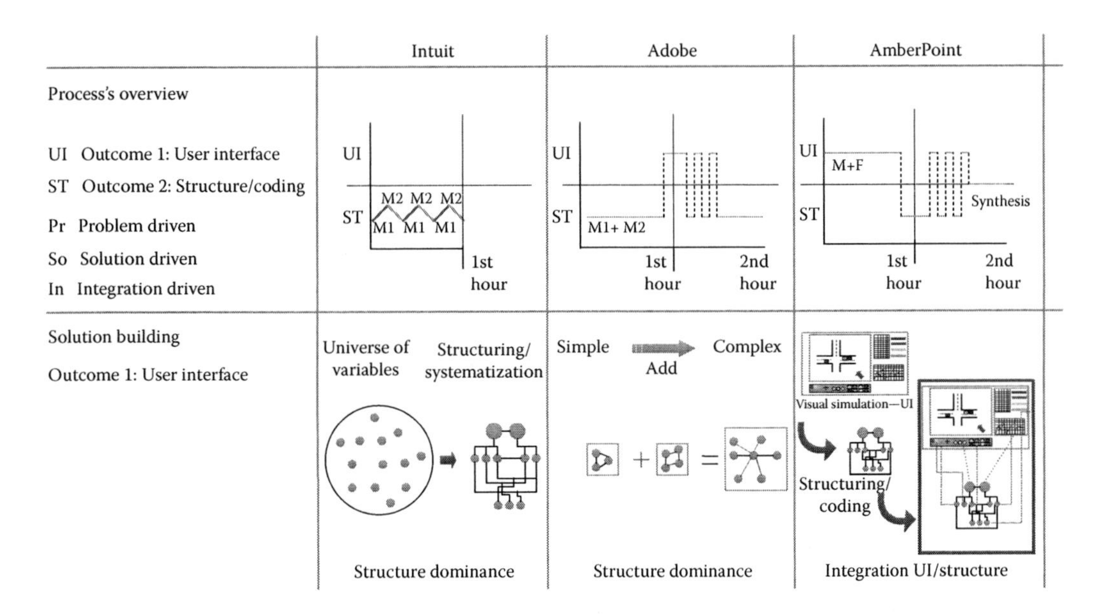

FIGURE 8.6 Synthesis after Phases 2 and 3 of the protocols study.

To put together a logical, rational ruling system and an interface that should be used by users in an intuitive way is very hard since the structuring calls for deductive and inference reasoning while user interaction is ruled by abductive reasoning.

Regarding creative cognitive processes, product designers in general alternate between exploratory and generative tasks, while software designers apparently use these two modes in a sequential way, starting with an extensive use of exploratory tasks such as hypothesis testing and functional inference exploration and through these come up with more generative ones, such as associations and analogical transfer.

The structure and user interaction should be connected all the time and developed in a coevolutionary way. This might indicate that the integration-driven strategy is the adequate one in these processes, since a problem-driven or a solution-driven strategy does not guarantee this ongoing coevolution between the problem and the solutions.

The level of knowledge necessary to develop such systems calls for specific domain knowledge in very distinct areas such as computer science, expert systems, interaction design, perception and representation, cognition, usability, etc. That being so, by all means software design has to be developed by teams where all these knowledge domains should be guaranteed by a high level of expertise.

Finally, concerning the decision strategy, both product and software designers display a more intensive use of negotiated compensatory/noncompensatory rule-based decisions over the use of compensatory ruled-based decisions, that is, designers in general tend to privilege their previous knowledge and to neglect an exhaustive treatment of information in order to take decisions along their processes.

ACKNOWLEDGMENTS

The results described in this chapter are based on videos and transcripts initially distributed for the 2010 international workshop, Studying Professional Software Design, as partially supported by NSF grant CCF-0845840. http://www.ics.uci.edu/design-workshop.

REFERENCES

Almendra, R. & Christiaans, H. (2009). Improving design processes through better decision making: An experiment with a decision making support tool in a company case. In: K. Lee, J. Kim, & L.L. Chen (Eds), *Proceedings of IASDR 2009: Design Rigor & Relevance*. Seoul: Korean Society of Design Science (pp. 2315–2325).

Almendra, R. & Christiaans, H. (2010). Decision-making in the conceptual design phases: A comparative study. *Journal of Design Research*, *8*(1), 1–22.

Christiaans, H. & Almendra, R. (2008). Problem structuring and information access in a design problem: A comparative study. In: *P&D Design 2008, 8° Congresso Brasileiro de Pesquisa e Desenvolvimento em Design*. São Paulo: Centro Universitário Senac.

Christiaans, H. & Restrepo, J. (2001). Information processing in design: A theoretical and empirical perspective. In: H. Achten, B. De Vries, & J. Hennessey (Eds), *Research in the Netherlands 2000*. Eindhoven: Eindhoven University of Technology (pp. 63–73).

Cross, N., Christiaans, H., & Dorst, K. (Eds) (1996). *Analysing Design Activity*. London: Wiley.

Curtis, B., Krasner, H., & Iscoe, N. (1988). A field study of the software design process for large systems. *Computing Practices*, *31*(11), 1268–1287.

Edeholt, H. & Löwgren, J. (2003). Industrial design in a post-industrial society: A framework for understanding the relationship between industrial design and interaction design. In: *Proceedings of the 5th Conference European Academy of Design*, Barcelona.
Guindon, R. (1990). Designing the design process: Exploiting opportunistic thoughts. *Human–Computer Interaction*, *5*(2), 305–344.
Longueville, B., Le Cardinal, J., Bocquet, J., & Daneau, P. (2003). Towards a project memory for innovative product design: A decision-making process model. In: *ICED03 International Conference on Engineering Design*, Stockholm.
Pennington, N., Lee, A.Y., & Rehder, B. (1995). Cognitive activities and levels of abstraction in procedural and object-oriented design. *Human–Computer Interaction*, *10*, 171–226.
Rehman, F.U. & Yan, X.T. (2007). Supporting early design decision making using design context knowledge. *Journal of Design Research*, *6*(1–2), 169–189.
Rothrock, L. & Yin, J. (2008). Integrating compensatory and noncompensatory decision-making strategies in dynamic task environments. In: T. Kugler, J. Smith, T. Connolly, & Y.J. Son (Eds), *Decision Modelling and Behavior in Complex and Uncertain Environments*. New York: Springer (pp. 125–141).
Sarma, V. (1994). Decision making in complex systems. *Systems Practice*, *7*(4), 399–407.
Simon, H. (1969). *The Sciences of the Artificial*. Cambridge, MA: MIT Press.
Smith, R.P. & Morrow, J.A. (1999). Product development process modeling. *Design Studies*, *20*, 237–261.
Whitney, D.E. (1990). Designing the design process. *Research in Engineering Design*, *2*, 3–13.
Winograd, T. (Ed.) (1996) *Bringing Design to Software*. New York: ACM Press.

CHAPTER 9

What Makes Software Design Effective?*

Antony Tang and Aldeida Aleti

Swinburne University of Technology

Janet Burge

Miami University

Hans van Vliet

Vrije Universiteit Amsterdam

CONTENTS

* Reprinted from *Design Studies* 31 (6), Tang, A., Aleti, A., Burge, J. and van Vliet, H., What makes software design effective?, 614–640, Copyright 2010, with permission from Elsevier.

9.1 INTRODUCTION

Software design is a highly complex and demanding activity. Software designers often work in a volatile environment in which both the business requirements and the technologies can change. Designers also face new problem domains where the knowledge about a design cannot be readily found. The characteristics and behaviors of the software and the hardware systems are often unknown and the user and quality requirements are highly complex. Under such a complex environment, the ways that design decisions are made can greatly influence the outcome of the design. Despite such challenges, we understand little about how software designers make design decisions, and how and if they reason with their decisions. We also understand little about the decision-making process and how it influences the effectiveness of the design process.

The University of California, Irvine, prepared an experiment in which software designers were given a set of requirements to design a traffic simulator. Their design activities were video recorded and transcribed.* This gave us the opportunity to study and compare design reasoning and decision making by professional designers working on the same problem. The primary goal of our study is to investigate software design from a decision-making perspective. In particular, we want to understand software design decision making in terms of how software designers use reasoning techniques when making decisions. If different reasoning and decision-making approaches are used, would they influence software design effectiveness?

* The results described in this chapter are based on the videos and the transcripts initially distributed for the 2010 international workshop Studying Professional Software Design, which was partially supported by NSF grant CCF-0845840. http://www.ics.uci.edu/design-workshop.

In this study, we analyze the designers' activities from three perspectives:

- *Design planning*: To understand the ways that software designers plan and conduct a design session and how they influence design effectiveness.
- *Design problem–solution coevolution*: To understand the ways that designers arrange their design activities during the exploration of the problem space and the solution space. For instance, we want to understand if designers identify design problems before formulating solutions, and what would be the implications on design effectiveness.
- *Design reasoning*: To understand the ways that designers reason with their design decisions. For instance, would designers be more effective if they created more design options, or if they were explicit about why they make their design choices.

We postulate that these perspectives in combination influence the effectiveness of design activities. Design effectiveness is defined in terms of requirements coverage and design time utilization. That is, given a certain amount of design time, how can one team address more requirements? Does one team solve a design problem more quickly than another team because of their design decision-making activities? In this study, we carry out protocol encodings of the transcripts. We label each part of the designer conversations to identify the decision-making and reasoning activities that went on during the design. The encoded protocols are analyzed to create visual decision maps. We evaluate the results to investigate the factors that influence the effectiveness of software design. We have found that the ways that decisions are made impact on the effective use of design time and, consequently, the derived solutions.

9.2 PRIOR WORKS IN DESIGN DECISION MAKING AND REASONING

9.2.1 Design Planning

Given a set of software requirements, designers must look at how to structure or plan a design approach. Zannier et al. (2007) have found that designers use both rational and naturalistic decision-making tactics. Designers make more use of a rational approach when the design problem is well structured, and, inversely, designers use a naturalistic approach when the problem is not well structured. Goel and Pirolli (1992) reported that, on average, 24% of the statements made by the architects in their study were devoted to problem structuring and this activity occurred mostly at the beginning of the design task. Christiaans reported that "the more time a subject spent in defining and understanding the problem, and consequently using their own frame of reference in forming conceptual structures, the better able he/she was to achieve a creative result" (Dorst and Cross, 2001). These studies show that design structuring or planning is an important aspect of design.

9.2.2 Design Reasoning

Designers are mostly not cognizant of the process of design decision making. Despite lacking the knowledge about how design decisions are made, which is fundamental to explaining the cognitive process of design, software practitioners and software engineering researchers continue to invent development processes, modeling techniques, and first principles to create software products. In real life, it seems that most software designers make design decisions based on personal preferences and habits, applying some form of reasoning and ending up with varying design productivity and quality.

This phenomenon can perhaps be explained by the way humans think. Researchers in psychology suggest that there are two distinct cognitive systems underlying reasoning: System 1 comprises a set of autonomous subsystems that react to situations automatically; they enable us to make quicker decisions with a lesser load on our cognitive reasoning. System 1 thinking can introduce belief-biased decisions based on intuitions from past experiences; these decisions require little reflection. System 2 is a logical system that employs abstract reasoning and hypothetical thinking; such a system requires a longer decision time and it requires searching through memories. System 2 permits hypothetical and deductive thinking (Evans, 1984, 2003). Under this dual-process theory, designers are said to use both systems. It seems, however, that designers rely heavily on prior beliefs and intuition rather than logical reasoning, creating nonrational decisions. Furthermore, the comprehension of an issue also dictates how people make decisions and rational choices (Tversky and Kahneman, 1986). The comprehension of design issues depends on, at least partly, how designers frame or structure the design problems. If the way we think can influence the way we design and eventually the end results of the design, then it is beneficial to investigate how software designers make design decisions.

Rittel and Webber (1973) view design as a process of negotiation and deliberation. They suggest that design is a "wicked problem" for which it does not have a well-defined set of potential solutions. Even though the act of design is a logical process, it is subject to how a designer handles this wicked problem. Goldschmidt and Weil (1998) describe design reasoning as the relationship between contents and structure. They suggest that the process of reasoning as represented by design moves is "double speared," where one direction is to move forward with a new design, and another direction is to look backwards for consistency.

As software design experience can be application domain dependent, it is different from other engineering design disciplines where the context of the domain is relatively constant. For instance, the issues faced by software designers working in the scientific domain are quite different from those working in the transactional financial system domain, so software designers may use different decision-making processes if they are faced with unfamiliar domains and technologies. Cross (2004) suggests that expert designers appear not to generate a wide range of alternatives. We found in an earlier study that experienced software designers intuitively rule out unviable design alternatives, whereas inexperienced software designers can benefit from explicitly considering design options (Tang et al., 2008).

9.2.3 Design Problem–Solution Coevolution

Although many have researched software engineering design, no studies that we are aware of have addressed the cognitive aspects of software designers. The studies that are closest to this subject address the use of design rationale in software engineering, mostly from the

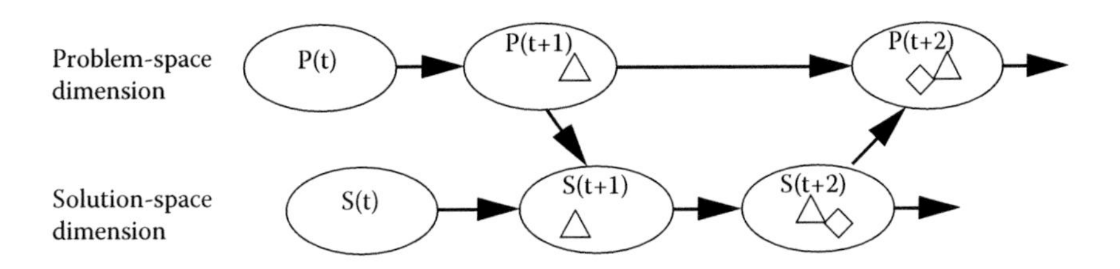

FIGURE 9.1 Problem–solution coevolution.

perspective of documenting design decisions instead of a systematic approach to applying software design reasoning (Tang et al., 2006; Tyree and Akerman, 2005; Dutoit et al., 2006; Curtis et al., 1988). From the problem-solving perspective, Hall et al. (2002) suggest that problem frames provide a means of analyzing and decomposing problems, enabling the designer to design by iteratively moving between the problem structure and the solution structure. These software design methodologies employ a black box approach that emphasizes the development process and its resulting artifact with little exploration of the cognitive and psychological aspects of the designers who use them. Such methods have a resounding similarity with the Mahler model of coevolution between the problem and the solution spaces (Maher et al., 1996), which was also described by Dorst and Cross in 2001. They suggest that creative design is developing and refining together both the formulation of a problem space and the ideas in the solution space (Figure 9.1).

Simon and Newell (1972) suggest six sources of information that can be used to construct a problem space, providing a context or a task environment for the problem space. Defining the context is thus an important aspect of formulating and identifying a problem. Similarly, we suggest that decisions are driven by decision inputs, or the context of a design problem, and the results of a decision are some decision outcomes. The AREL design rationale model (Tang et al., 2009) represents the causal relationships between design concerns, design decisions, and design outcomes. The concerns and the requirements are the context of a design problem. A design problem needs to be analyzed and formulated. When a decision is made to select a solution, a certain evaluation of the pros and cons of each alternative takes place. A design outcome represents the chosen design solution. Often, one design decision leads to another design decision, and the outcome of a decision can become the design input of the next decision. In this way, a design is a series of decision problems and solutions (Tang et al., 2009; Hofmeister et al., 2007).

9.3 STUDYING SOFTWARE DESIGN REASONING AND DECISION MAKING

9.3.1 Design Assignment

The design task set out in the experiment for the software designers was to build a traffic simulation program. The system is intended to be used by students to help them understand the relationships between traffic lights, traffic conditions, and traffic flows. The context of the subject was traffic lights and intersections, which is common knowledge that the designers should have been familiar with. This general knowledge allowed the designers to begin their work. However, building the simulation software required specialized

TABLE 9.1 Number of Requirements in the System

Requirements	Network of Roads	Intersection/ Light	Simulate Traffic Flows	Traffic Density	Nonfunctional Requirements
Explicit	4	6	4	2	4
Derived	3	5	7	0	0

and detailed domain knowledge that general software designers are unlikely to possess, so none of the designers had any such advantage over the others that could bias the results. The experiments involved three design teams, each consisting of two designers. Each team was given 2 hours to produce a design. We have selected to study two teams only: Adobe (Adobe in short) and AmberPoint, which we call Team A and Team M, respectively. Both Team A and Team M took the full 2 hours to design the system and a third team (Intuit) took just under 1 hour and did not design the system fully.* We chose to study Teams A and M only because we cannot compare the effectiveness of the Intuit team due to the short time they had spent on their design.

The designers were given a design prompt or a brief specification of the problem containing 20 explicit functional and nonfunctional requirements, and 15 derived requirements (see Table 9.1). The derived requirements were requirements that were not specified in the design prompt but needed to be addressed to complete the design. They were derived from the analysis by the designers plus the authors' interpretation of the requirements.† For instance, connecting intersections in the simulation program was a derived requirement.

9.3.2 Analysis Method

Software design has certain characteristics that are different from other engineering design disciplines. First, designers often have to explore new application and technology domains that they do not have previous experience with. Therefore, the quality of their design outcomes may not be consistent even for a designer who has practised for years. Secondly, a design is an abstract model and, often, whether it would actually work or not cannot be easily judged, objectively, until it is implemented. Therefore, it is difficult to measure the quality of a design, especially to compare if one design is superior to another design. For these reasons, we do not measure design quality in this study. We note that the draft designs created by the designers make sense and the designers are experienced, so we judge that their designs would have the basic qualities. Our goal is to understand the relationship between software design thinking and its influence on design effectiveness.

Design planning, problem–solution coevolution, and reasoning techniques are factors that influence design decision making. We postulate that software designers use them implicitly and in different ways during a design session. Furthermore, we suggest that designers use reasoning techniques to support design planning and decision making

* A more detailed description of the design task and the data captures is given in the introduction to this special issue.

† This requirements coverage is from the analysis made by the authors, which can be different from the other analyses. Details can be found in a supporting document: http://www.users.muohio.edu/burgeje/SPSD/SPSD_Requirements_Coverage.pdf.

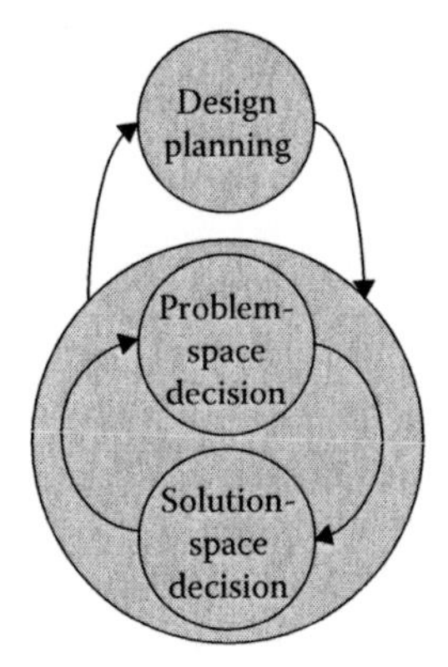

FIGURE 9.2 Iterative design decision-making model.

(see Section 9.4.3 for details). Figure 9.2 describes an iterative design decision-making process model that comprises two stages. Stage 1 is when the designers plan a design session to scope and prioritize the problems. Designers typically commence their design by summarizing the requirements, contextualizing and scoping the problems, and prioritizing the issues. This kind of problem planning is breadth-first reasoning at a high-level of abstraction (Schön, 1988).

Stage 2 is when designers resolve design problems in a coevolving manner. This is typically an iterative process of identifying and solving design problems, that is, problem–solution coevolution. During problem–solution coevolution, designers may carry out further design replanning (see Figure 9.2). Table 9.2 lists the activities of design planning and problem–solution coevolution. All of these activities involve decision making. Design planning involves identifying the scope of the requirements, identifying the key design issues and their relative priorities. The problem-space decision making involves problem analysis and formulation. The solution-space decision making is about generating solutions and other potential options, evaluating between different options, and selecting the most appropriate solution.

TABLE 9.2 General Design Activities and Reasoning Tactics

Design Planning	Reasoning Techniques
1. Scoping	• Conte(X)tualize situation
2. Identify key design	• (S)implify situation
3. Prioritize design problems	• Sc(E)nario generation
	• (C)onstraints consideration
Problem-Space Decisions	• (T)rade-off analysis
1. Analyze problem	• (O)ption generation
2. Formulate problem	• (I)nduction
	• (D)eduction
Solution-Space Decisions	
1. Generate solution	
2. Evaluate solution	
3. Select solution	

Table 9.2 also lists the general reasoning techniques that a designer uses to support these decision-making activities (Tang and Lago, 2010). A reasoning technique can be used by multiple design decision steps. For instance, a designer may identify the contexts (i.e., contextualizing) for design scoping as well as for problem analysis. In making a decision, one or more reasoning techniques can be used in conjunction, that is, they are orthogonal. A designer may use a scenario to analyze and illustrate how a solution may work, and, at the same time, use inductive reasoning to generate a new design idea. Also, a designer may simultaneously consider the context of a situation and the constraints that exist in order to formulate a design problem. Many examples of applying multiple reasoning techniques together are evident in this study.

In this study, we encode the protocol using the activities list and the reasoning techniques shown in Table 9.2. By analyzing the protocols, we aim to understand different design thinking and how it influences design effectiveness.

9.3.2.1 Protocol Analysis and Coding Schemas

We use two protocol coding schemes to analyze the designers' dialog, using a process similar to that of Gero and McNeill (1998) and Yin (2003). This coding scheme allows us to analyze the reasoning and decision-making processes of the designers. Table 9.3 is an example of the first reasoning protocol encoding scheme. We codify the transcripts into problem space (PS) and solution space (SS) (see column 3 in Table 9.3). This enables us to study the coevolution of problem and solution. Columns 4 through 6 show the decision activities that the designers were taking. They correspond to the design planning and design problem–solution coevolution process activities described in Figure 9.2. The encoding classifies each statement made by a designer to indicate the type of process activity that took place. Column 7 shows the reasoning techniques that were used. Column 8 shows whether the designers implicitly or explicitly performed reasoning. Explicit reasoning took place when the designers explicitly explained their decisions.

In parallel, we also created a second protocol to study how each design team made decisions with respect to different decision topics. This protocol represents a set of design rationale for each design using the rationale-import mechanism in the SEURAT (Software Engineering Using RATionale) system (Burge and Kiper, 2008). The SEURAT analysis

TABLE 9.3 Example of the Reasoning Coding Scheme

Dialog Time	Conversation	Design Space	Planning Decisions	Problem Decision	Solution Decision	Reasoning Technique	Explicit/ Implicit Reasoning
[0:09:24.7]	We have some constraints on the flow of the queues because we can't have lights coming in a collision course	PS		Analyze problem		Conte(X)tualize; (C)onstraint; (D)educe	(E)xplicit

TABLE 9.4 Example of the Coding Scheme

Decision ID	Time	Level	Description
1	5	1	| | (Dec) Category: Car
1,1	5	2	| | | (Dec) IMP: See car status
1.1. (Alt)	5	3	| | | | (Alt) And it might be that we want to select a car and watch that particular car snake through the simulation (Adopted)

classifies the decisions and the design alternatives into a hierarchical structure based on the decision topic. A sample of this hierarchy is shown in Table 9.4. All decision topics are time stamped for analysis. This coding scheme enables us to analyze decision making based on decision topics and time; it also allows us to analyze if decisions have been reached to address the requirements.

9.3.2.2 Decision Map

After encoding with the first protocol, we built a decision map to represent the design activities. The decision map is based on the Architecture Rationale Element Linkage (AREL) model, where design decision making is represented by three types of nodes, that is, *design concern*, *design decision*, and *design outcome*, and the causal relationships between them. A design concern node represents information that is an input to a design problem, such as requirements, contexts, or design outcomes. A design decision node represents the design problem. A design outcome node represents the solution of a decision. A decision map shows the progression of a design by connecting the problem space (decision node in rectangular boxes without double vertical lines) to the solution space (design outcome in ovals). The contexts of the design are represented by rectangular boxes with double vertical lines. The nodes in the decision map are linked together by (a) the sequence of discussions, that is, time; and (b) the subject of the discussion. Figure 9.3 shows Team M's design planning in the first 5 minutes of their discussions. The decision map depicts the following: (a) structuring of a design approach; (b) sequencing of design reasoning; and (c) design problem and solution.

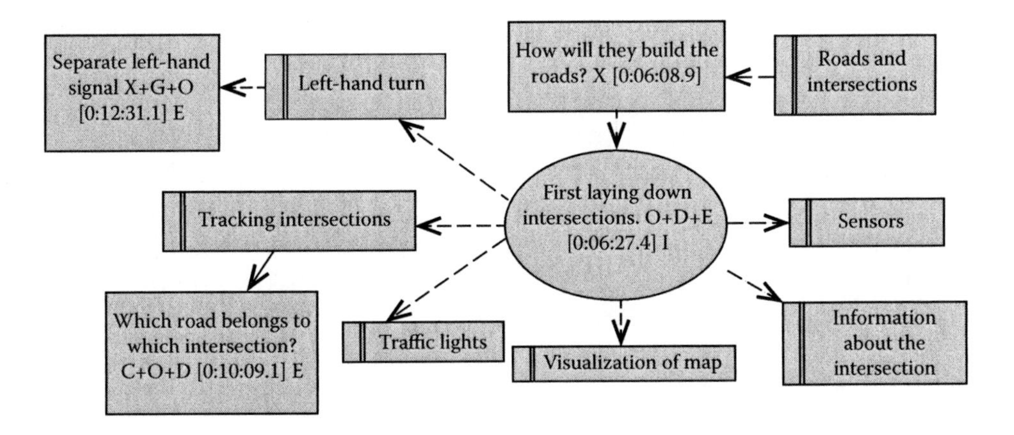

FIGURE 9.3 Extract of a decision map.

9.4 ANALYZING THE DESIGN ACTIVITIES

Using the two coded protocols and the decision maps described in the previous section, we analyze how Teams A and M make decisions and how they reason with their designs. The analysis is based on the three perspectives described earlier: (a) design planning—the question is how the designers plan their design sessions and its impact on design effectiveness; (b) design problem–solution coevolution—the question is how the designers create their design solutions and if they follow a pattern of problem–solution coevolution, and how that impacts on design effectiveness; and (c) the application of design reasoning techniques—the question is what reasoning techniques are used in (a) and (b) and how they impact the decision making and its effectiveness.

9.4.1 Design Planning

We suggest that the decisions on how to plan a design session have bearings on the results and its effectiveness. The considerations are the scope of a design and the priorities that should be given to each design problem. The decisions on these matters dictate the design content and therefore influence the effectiveness of the design.

9.4.1.1 Initial Design Planning

Both teams started off by planning their design approach, which involved scoping the requirements, identifying and structuring high-level design problems, and prioritizing the problems. Team A spent about 1.5 minutes ([0:05:11.0] until [0:06:41.1]) on planning and they identified the representation and modeling of traffic flows, signals, and intersections as their key design issues. Team M spent about 9 minutes ([0:06:08.9] until [0:15:11.7]) on planning and they identified 13 design issues. The two teams spent 1% and 9% of their total time, respectively, on problem structuring. Furthermore, there were 16 instances of scoping and 18 instances of identifying key design issues during Team M's planning, as compared to 5 and 9 instances, respectively, in Team A's planning.

The difference in the way they approached the design planning had a bearing on how each team structured the rest of their design. The up-front identification of specific areas where design was needed dictated the structure of the rest of the design discussions, even though neither team explicitly said that they were planning. For instance, Team A started off by asserting that the solution was a model-view-controller (MVC) pattern. Their focal points were data structure, modeling, and representation. They spent the next 40 minutes ([0:08:28.4] until [0:48:02.1]) on queue management before moving on to the other discussion topics. Team A carried out the design replanning during the design. At [1:17:59.15], Team A assessed their design status and then realigned their focus on the design outcomes, and the rest of the design discussions followed this direction. Team M spent considerably more time exploring their plan. As a result, they first designed the layout of the road [0:16:09.7] and then they considered the other design issues.

9.4.1.2 Structuring Design Activities

Using the discussion grouping by design topics, we observe that the way each team explored the design space followed different reasoning structures. Team A used a *semistructured approach*. They started with a design discussion that focused on the premise of a queue and the push/pop operations, and they explored this solution concept extensively to design the traffic simulation system. Further, they continued the discussions about roads and cars ([0:42:40.4] until [0:52:15.7]); however, without reaching either a solution or a clear identification of the next design problem, they moved to the topic of sensors ([0:54:33.6]). Such inconclusive design discussions continued until [1:17:59.15] when Team A stopped and replanned their design approach. At that point, the designers reassessed their design problems, which guided them till the end of the design session. From that point, three topics, namely, road modeling, simulation display, and code structure, were discussed.

Team M used a *major-minor structuring approach*. Team M tackled the major design problems first and when they concluded the design of those major problems, they addressed the minor associated issues afterward. The major and minor issues are concepts in terms of the difficulties of resolving them, not in terms of their importance to the system or the end users. For instance, Team M started their design discussions by addressing the issue of traffic density [0:13:35.5]. At the end of tackling the high-level design, they investigated the relevant but simpler issue regarding the simulation interface [0:26:43.5]. Then they moved to another high-level critical problem on traffic lights, and at the end of the discussions, they again considered the simpler interface issue [0:41:13.3]. In doing so, the designers started with a major design problem, solved it and then tackled other minor but related design problems. In summary, the initial structuring and planning provided guidance to how the design proceeded in the remaining time.

9.4.1.3 Design Decision Topic Context Switching

The designers in Teams A and M had to deal with many design decision topics and issues that are interrelated. The focusing or switching of the designers' discussions on these topics influences the way that they explore the problem and the solution spaces. We call this *context switching*. Context switching takes place under two circumstances. First, the designers think of related issues during a discussion and switch topic to pursue a new issue instead of the first one. Secondly, the designers may have solved an issue and then switch to a new topic.

Context switching is most noticeable in the semistructured approach employed by Team A. For instance, during the queue management discussion, the designers focused and defocused on traffic lights several times at [0:20:36.2], [0:39:17.8], and [0:49:41.0]. Then this topic was raised again at [0:55:28.8] and [1:49:44.0], but Team A did not have a clear and complete design for all of the issues related to traffic lights. It appears that the context switching affected the resolution of the design issues. The more context switching that takes place, the less the designers follow a single line of thought to explore a design issue thoroughly because the related context of the previous discussions may be lost. Therefore, even when the same issue was discussed many times, a clear solution was not devised. In contrast, Team M addressed traffic lights [0:28:16.4] in a single discussion. There is limited context switching during the discussions of these closely related design issues, for example, sensors [0:28:54.1]. The designers acknowledged the other design issues but quickly returned to the main topic.

We observe context switching in decision making by using the decision topic protocol generated from SEURAT. In this protocol encoding, all statements made by the designers are grouped by the decision topics, for instance, cars and road networks. Each topic has subtopics organized hierarchically in multiple levels, for instance, the subtopics of *car* are *car status*, *wait time per car*, and so forth. Context switching happens when the discussion changes from one decision topic to another decision topic at the top level. We plot the first two levels for Team A and Team M, each symbol in Figures 9.4a and 9.4b representing some discussions of a topic or a subtopic. A subtopic appears in a slightly elevated position than its corresponding topic. The *x* axis is the time line when the decision topics were discussed.

We have analyzed the average number of context switches. Team A and Team M made 312 and 518 total statements, respectively. Out of the statements that they made, we found that Team A and Team M had switched discussion topics 78 and 69 times, respectively. Represented as the percentages of switching topics using the total design statements as a basis, Team A has a 24.9% chance of switching to a new topic and Team M has a 13.2% chance of switching to a new topic. Therefore, Team M is more focused on its discussion topics and performs less context switching compared to Team A.

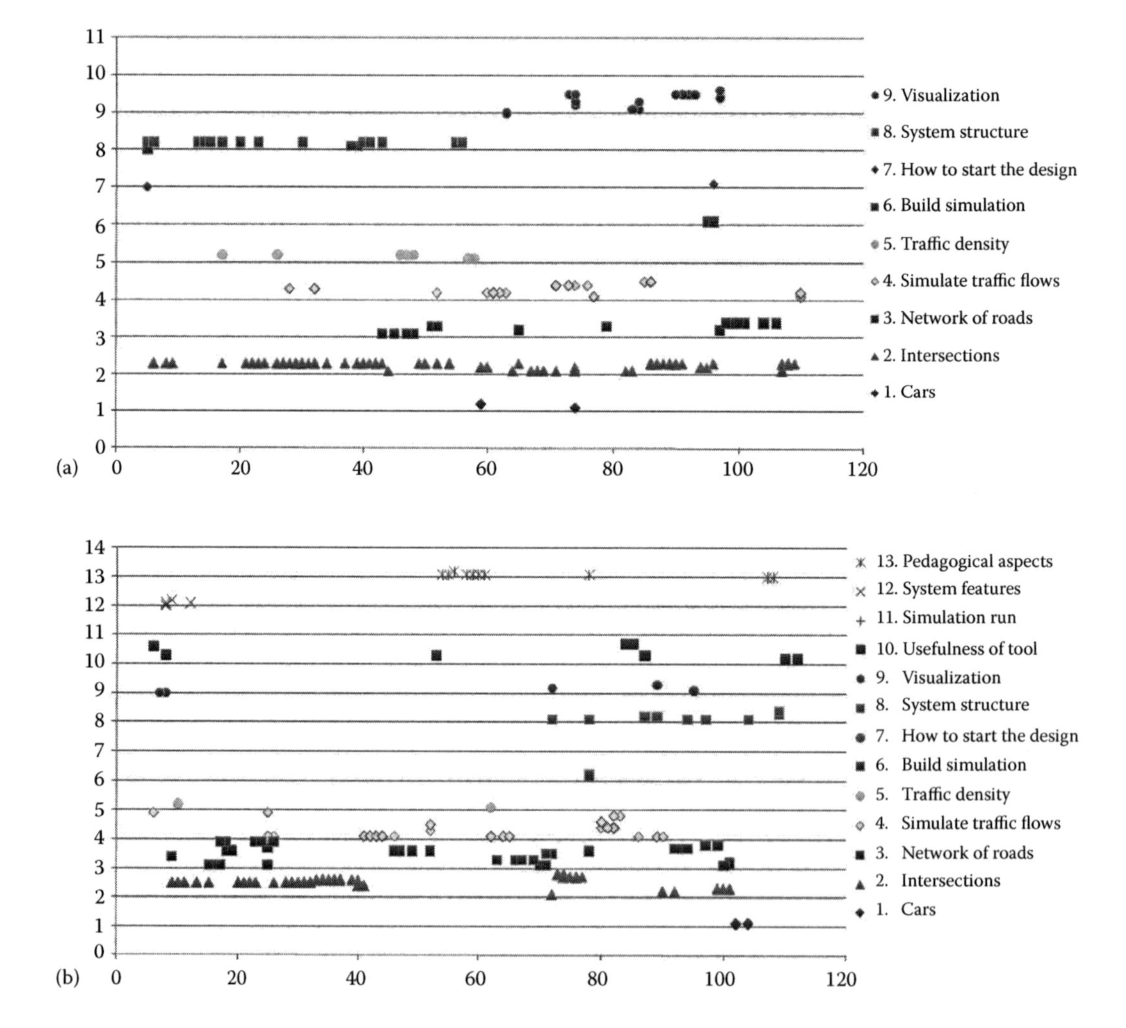

FIGURE 9.4 Decision topics time distribution: (a) Team A and (b) Team M.

9.4.1.4 Time Distribution of Design Discussions

Figure 9.4a shows that Team A discussed nine different decision topics and some of the discussions spanned across the entire design sessions, especially topics 2, 4, and 8. Team M discussed 12 topics. The time distribution of the discussions on topics 2–4 are quite different between Team A and Team M. Team M's discussions on each topic are clustered together in a time line (Figure 9.4b). To enable us to quantify the distribution of time on the discussions of each decision topic, we have computed the standard time deviation (in minutes) for each decision topic (see Table 9.5). As shown in Figure 9.4b, discussion topic 2 has three clear clusters. This means that the designers focused on this topic in three different periods. In this case, the standard deviation is computed for each cluster to reflect the concentration of the discussion. If there is no obvious clustering, then all the data points are used in computing the standard deviation.

The blank cells in Table 9.5 indicate either the team did not discuss this topic or we dismissed the data because there were too few data points to compute the distribution meaningfully. For instance, topic 7 was discussed by Team A only twice, and the two data points give a standard deviation of 64 minutes. The results show that Team A spent considerably more elapsed time discussing topics 2, 4, 8, and 9. When there are many topics that have a high time distribution, it indicates that the topic discussions overlap each other in the same time line. The overlapping of the discussions increases the switching between topics.

One may consider that a decision topic is particularly difficult and therefore requires a longer time to solve. However, if we compare the standard distribution of the discussion time between the two teams on all topics, Team M consistently has smaller time distributions than Team A, indicating that Team M finished dealing with the decision topics more quickly. If we examine the requirements coverage (see Section 9.5.2) together with the time distribution, we see that Team M has addressed more requirements than Team A in total and each topic has a shorter time span. These data suggest that the frequent switching of discussion topics by Team A has made the discussion less effective. Team A spent more time on a topic because the discussion was switched back and forth before the issues were fully resolved. When this happens, it is possible that the context and the issues of a previous discussion have to be reestablished. An example is when a designer in Team A said, "I'm sort of reminded of the fact that we're supposed to allow lights to have sets of rules ..." at [0:49:41.0]. This was the third time that they discussed the same design topic.

9.4.2 Problem–Solution Coevolution

Dorst and Cross (2001) have observed the coevolution of design between problem and solution spaces. The way that the designers in the two teams conducted their designs was largely aligned with the coevolution model. For instance, Team M demonstrated a coevolution process where a problem was identified and solved, leading to the next problem–solution cycle (Figure 9.5). However, this model presumes a logical design reasoning behavior in

TABLE 9.5 Standard Deviation in Time by Decision Topics

	1	2	3	4	5	6	7	8	9	10	12	13
Team A		13.9	2.8	21.9	5.6	0.5		12.5	7.9			
Team M	1.0	5.2	3.3	1.9				10.6	2.5	1.2	2	1.4

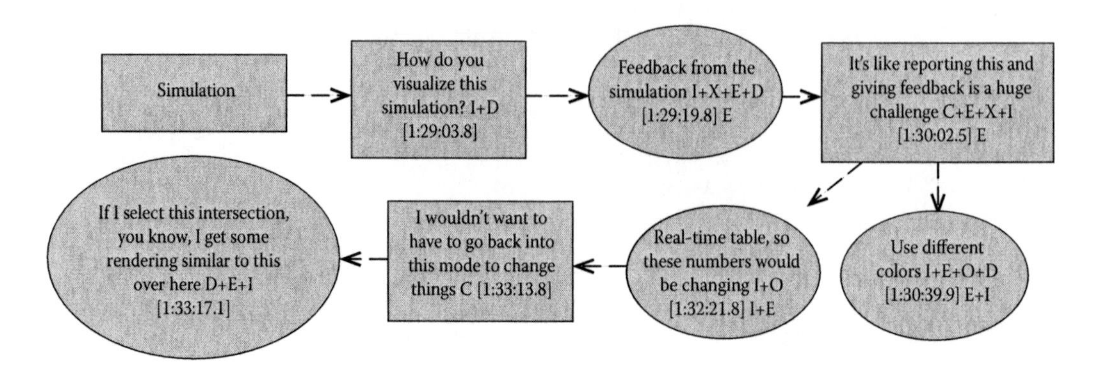

FIGURE 9.5 Example of a problem–solution coevolution in Team M design session.

which a solution is derived from well-defined design problems. We have observed design behaviors in this study that differ from this model.

At a very early stage of design, Team A identified traffic flow and signals as design issues [0:06:10.6] and they decided that a queue should be used in the solution [0:08:28.4]. From this point onward, Team A continued to make a series of decisions surrounding the operations of the queue. Some examples are [0:08:51.0], [0:38:05.6], and [0:39:17.8]. The idea of the queue drove the discussions from [0:08:28.4] to [0:50:43.1]. The design plan and all subsequent reasoning are predominantly based on this solution concept that was decided up-front, without questioning whether this is the right approach or if there are alternative approaches. It is also evident that Team A did not spend much time exploring the problem space and they may have made implicit assumptions about what the solution would provide. This case demonstrates that when designers take a solution-driven approach, the preconceived solution becomes the dominant driver of the design process, and the rest of the solutions evolve from the preconceived solution. Such design behavior hints at the dominance of cognitive System 1, where designers autonomously selected the MVC and queue as the solution, over System 2, the reasoning system.

There are possible reasons for this situation. First, it has been observed that when people make estimates to find an answer from some initial values, they tend not to adjust to another starting point but instead are biased toward the initial values. This is called the anchoring and adjustment heuristic (Tversky and Kahneman, 1974; Epley and Gilovich, 2006). Similarly, Rowe (1987) observed that a dominant influence was exerted by the initial design even when severe problems were encountered. Designers are reluctant to change their minds. Tang et al. (2008) also found that for some software designers, the first solution that came to mind dominated their entire design thinking, especially when the software designers did not explicitly reason with their design decisions. Secondly, a designer may be forced to make premature commitments to guess-ahead a design, but the design commitment remains (Green and Petre, 1996).

9.4.2.1 Problem-Space Decisions

Deciding on what design problems need solving can be done implicitly or explicitly. Team A generally took a solution-driven approach, defining their problem space implicitly.

Team M, on the other hand, explicitly stated their problem statements; they typically analyzed the relevant information to the problem, such as the goals, resources, and current situations of a system. Designers who work explicitly in the problem space typically determine what information is known and what is unknown, and based on that they decide what to design for (Goel and Pirolli, 1992). We encode explicit problem statements, distinguishing between problem analysis and problem formulation. Problem analysis is a step to identify the information relevant to a problem. It provides the context of a problem. Problem formulation is a step to articulate what problems need to be solved or how to proceed to solve them.

Team A and Team M made 49 and 46 analysis statements, respectively. However, Team A only formulated 34 problem statements whereas Team M formulated 120 problem statements, which represent 13.4% and 31.9% of all the statements made by the two teams, respectively. In a study of general problem solving, Goel and Pirolli (1992) found that problem structuring accounts for between 18% and 30% of all design statements made in a problem-solving exercise. Furthermore, Teams A and M made 153 and 173 solution statements, respectively. We plot the problem formulation and solution generation across time in a graph for both design sessions (Figure 9.6). The "problem formulation" columns indicate the frequency of problem formulation statements in 1 minute; the "solution generation" columns indicate the frequency of solution generation statements in 1 minute. The concentration of problem analysis and formulation by Team M is high (Figure 9.6b), especially in certain periods, for example,

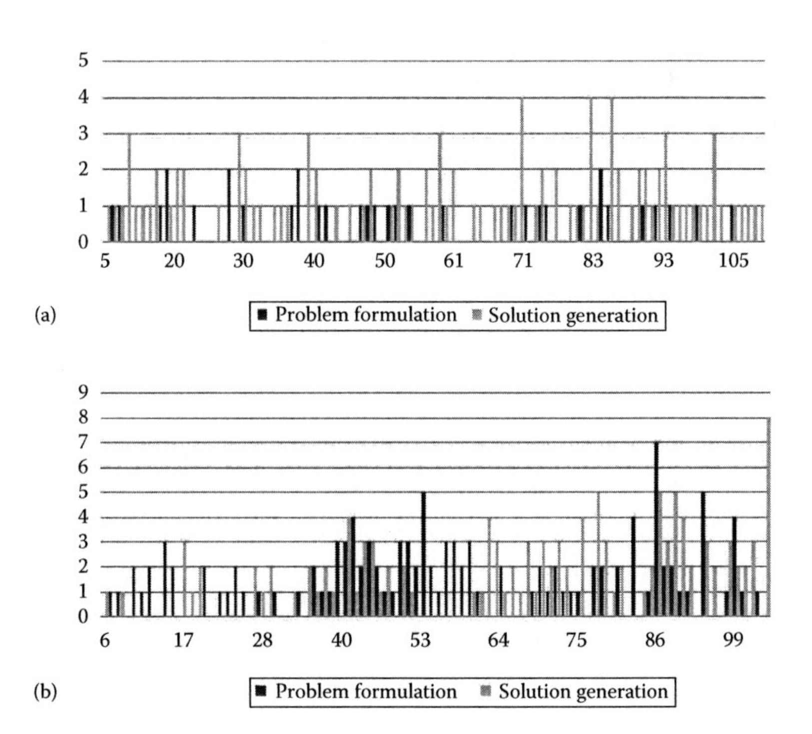

FIGURE 9.6 Problem formulation and solution generation in a design session: (a) Adobe and (b) AmberPoint.

between [0:12] to [0:16] and [0:30] to [1:02]. On the contrary, Team A has fewer problem statements (Figure 9.6a).

9.4.2.2 Solution-Space Decisions

To analyze decision making in the solution space, we divide the decision process into three general steps: generating a solution option, evaluating the pros and cons of the available options when there is more than one solution option, and selecting a solution from the available options. In encoding the protocols, Teams A and M made 106 and 124 solution generation statements, 17 and 16 evaluation statements, and 30 and 32 selection statements, respectively. The numbers of solution statements made by the two teams are very similar. However, the processes that the two teams used are quite different. Problem and solution statements are encoded in the protocol analysis (see Table 9.2).

An analysis of the problem and solution statements that the two teams made shows that the ratios of all the solution statements to all the statements made by Teams A and M are 60.2% and 45.7%, respectively. It means that for Team A, 6 out of 10 statements are solution statements. In total, Team A made 153 solution statements out of a total of 254 statements, whereas Team M made 172 solution statements out of a total of 376 statements. Given the 2 hours of design time, Team A made a similar number of solution statements to Team M, but their solution statements represent a higher proportion of their design dialog, implying that they are spending more time on devising the solutions.

Additionally, the ratios of solution statement generation to problem statement formulation made by Teams A and M are 32% and 96.7%, respectively. Team A has a lower ratio of design problem formulation to solution generation, indicating that they generate solutions with less explicitly identified problems. Although there is no benchmark on the sufficiency of problem-space formulation, it is suggested that a clearly defined problem space is important (Maher et al., 1996). In the case of Team A, they have a lower ratio of formulated problem and a higher ratio of generated solution. We typify this kind of design strategy as a solution-driven strategy. On the other hand, Team M has a ratio that is close to 1, indicating that, on average, for every problem they posed, they also generated a solution.

9.4.2.3 Interleaving Problem–Solution Decisions

The Maher et al. (1996) and Dorst and Cross (2001) coevolution models suggest that design evolution takes place along two parallel lines: the problem space and the solution space. Tang et al. (2009) suggest that software design evolves linearly between the problem and the solution in a causal relationship. These models follow a fundamental premise that solutions are generated from problems. However, the evidence from the Team A design session has suggested that some software designers employ a solution-driven decision-making approach in which problem formulation plays a lesser role (see Figure 9.6a). Designers can devise solutions even though the problems are not well articulated. This may be due to the nature of the software design in which the goals and subgoals of a design can be abstract. Therefore, designers can arbitrarily decide how much of the problem space to explore before delving into the solution space.

The coevolution models also suggest that the development of the problem space and the solution space progresses in parallel. However, the behavior of the designers depicted in Figure 9.6b suggests that other activities are also involved. First, there are extended periods when Team M formulated a lot of questions, at [0:12] and [0:53]. This indicates that the designers explored the problem space intensively for some time without producing any solutions. The designers analyzed and formulated interrelated problems to gain a comprehensive perspective of the domain. The decision topics protocol analysis (Figure 9.4b) shows that Team M was focusing on intersection arrangement from [0:12] and traffic flow simulation from [0:53]. Secondly, after intense explorations of the problem space, the designers followed the general coevolution model in which the problem decisions and solution decisions statements are interleaved, indicating that the designers had decided on what problems to solve and then proceeded to solving them.

The activities carried out by Team M demonstrate that problem–solution coevolution would happen at different levels. When the designers are at the exploratory stage, all related problems are explored to gain a better understanding of the problem space. At this stage, few solutions are provided. When all the design issues are well understood, the designers would then hone in on a single design issue and proceed to creating a design with problems and solutions coevolution. This approach seems to serve Team M quite well in terms of exploring the problem space effectively.

9.4.3 Design Reasoning Techniques

Design is a *rational problem-solving* activity—it is a process in which designers explore the design issues and search for rational solutions. It is well established by many that design rationale is one of the most important aspects in software design (Shum and Hammond, 1994; Tang et al., 2006). However, the basic techniques that lead to rational design decisions are not well understood. Therefore, it is necessary to explore the different design reasoning techniques that are used by the design teams. The initial list of reasoning techniques (Table 9.2) is derived from the design rationale literature (Tyree and Akerman, 2005; Tang et al., 2006; Tang and Lago, 2010) and various literature that describe how software design is to be conducted. First of all, there is a general creative process and some reasoning process that goes on. We encode the protocol with inductive and deductive reasoning. Inductive reasoning, especially analogical reasoning, is a means of applying the knowledge acquired in one context to new situations (Csapó, 1997). It promotes a creative process in which issues from different pieces of information can be combined to form a new artifact (Simina and Kolodner, 1997). There are different techniques of inductive reasoning (Klauer, 1996) that software designers can use.

For instance, one technique is a scenario-based analysis (Kazman et al., 1996). It describes a process of using scenarios to evaluate system architecture. Another technique is contextualization, where identifying the context at the right level of abstraction helps to frame the design issues (Simon and Newell, 1972). We conjecture that reasoning techniques are fundamental to the design decision-making process, so they can be used in design planning, and problem and solution decision making. In this section, we discuss the reasoning techniques and how they are used by the two teams.

9.4.3.1 Contextualizing Problems and Solutions

Contextualization is the process of attaching a meaning and interpreting a phenomenon to a design problem. Well-reasoned arguments state the context of the problem, that is, what and how the context influences the design. All teams contextualize their problems in different ways; we counted the number of statements where they discuss the context of the problem and state the details that can influence the potential solution. We identified 62 such statements, problems, and solutions, by Team A and 97 statements by Team M. This means that Team M would identify the information that surrounds a problem or a solution as part of their analysis before they proceed to decide on what to do with the information. Team M is more thorough in articulating the context. An example is when Team M examines the traffic lights [0:28:16.4]. The designers spell out three relevant contexts of the problem, that is, the timing of the cycle, the impact of the sensors, and the left-turn arrow. Their discussion of the design problem using those contexts happened in the same time frame and, as such, it provides a focal point for investigating a design problem comprehensively.

On the other hand, Team A contextualizes their design problems in an uncoordinated way. For example, when Team A discussed traffic signals, they did not consider all the relevant contexts at the same time. The designers discussed signal rules at [0:12:31.1] and they revisited the subject of traffic signals with respect to lanes at [0:25:37.4]. Later, they decided that it was the intersection that was relevant to traffic signals [0:30:41.9]. At [0:56:06.7], they discussed how signals could be specified by the students. Twenty-one minutes later [1:17:58.4], their discussion went to the subject of sensors and left signals. Contextualization influences how a design problem can be framed; the difference between how the two teams use contextualization affects what information is used for reasoning with a design problem. The lack of thorough contextualization in the problem space may be one reason why Team A's context switches more frequently. If the subject matter is not investigated thoroughly and the design gaps are identified later, the designers would revisit the same design topic.

9.4.3.2 Scenario Construction

The use of scenarios enhances the comprehension of a problem and facilitates communication between designers. Constructing a scenario is a way to instantiate an abstract design problem and evaluate if the abstract design is fit for the real-world problem (Kazman et al., 1996). Carroll suggests that scenarios help designers to understand needs and concerns, and then make generalizations of that knowledge into a solution (Carroll, 1997). For example, Team A built a scenario of how the simulation could work, talking about the possibility of having a slider to control the timing of lights and car speed [0:48:20.4]. Using this scenario, the designers highlighted a number of related design problems, for example, the average wait time of a car traveling through the city streets, car speeds, traffic light control, etc. Team A and Team M made 93 and 80 scenario statements, respectively. This indicates that the scenario is an important reasoning technique that both teams use extensively. In this study, we do not find any significant relationship between how the designers use scenarios as a reasoning technique and the effectiveness of design.

9.4.3.3 Creating Design Options

Creating design solution options is an important reasoning technique used by expert designers (Cross, 2004). It has been found that by simply prompting designers to consider other design alternatives, designers with less experience create better designs (Tang et al., 2008). Team A stated 37 options for 8 problems, a 1:4.5 problem to option ratio. Team M stated 58 options for 20 problems, a 1:3 problem to option ratio. In total, Team M generated 36% more options than Team A, probably because of the higher number of design problems identified.

We have observed certain characteristics about generating design options: (a) Problem refinement—designers contemplated the problem and the solution together to gain a better understanding of the design problem. Designers would use phrases such as "But you would have to deal with this issue and you may also do this." Designers would generate solution options to validate the design problem. (b) Constraint specification—when a constraint(s) is considered, a new solution option is sometimes generated to overcome the constraint. Designers would use phrases such as "but it depends on this" or "it cannot do this." (c) Scenario driven—a new design option can be generated when designers consider different scenarios. Designers would use a phrase such as "Maybe we have ... a scenario ... you would want to do this." Although these phrases can be observed during design, they were used subconsciously by designers. Designers did not explicitly ask "What other design options may we have?" Such an observation suggests that the designers did not consciously consider option generation as a helpful technique in finding an appropriate design solution.

9.4.3.4 Inductive and Deductive Reasoning

Inductive reasoning is a type of reasoning that generalizes specific facts or observations to create a theory to be applied to another situation, whereas deductive reasoning uses commonly known situations and facts to derive a logical conclusion. While inductive reasoning is exploratory and helps develop new theories, deductive reasoning is used to find a conclusion from the known facts.

Let us consider first how both teams have used inductive reasoning. We have counted 71 and 112 induction statements from Team A and Team M, respectively. Team M has used 36% as many inductive reasoning statements as Team A. This suggests that Team M employs a more investigative strategy, inspecting the relevant facts to explore new design problems and solutions. As an example, Team M discussed the relationships between the settings of traffic lights to prevent collisions [0:34:30.6]. From the observation that cars can run a yellow light and in order to prevent collisions, they induce that there should be a slight overlap when all traffic lights facing different directions should be red simultaneously. Then, they reason inductively about the manual settings for the timing of the lights and whether the users should be allowed to set all four lights manually. Since the users should not be allowed to set the traffic lights to cause collisions, the designers reason that there should be a rule in which the setting of a set of lights would imply the behavior of the other traffic lights. Team M, in this case, found new design problems from some initial observations.

On the other hand, deductive reasoning is used more times by Team A than Team M, with 56 and 51 statements, respectively. For instance, Team A started off by predicating on the basic components used in the solution, that is, queue and cop, etc., and they deduced the solution from that. They used deductive reasoning at [0:30:19.3]. They said that a lane is a queue, so multiple lanes would be represented by multiple queues. Since they also predicated on the use of a traffic cop, so they deduced that the cop would look after the popping and pushing of multiple queues. Team A was fixed in the basic design of the system from the beginning without exploring other design options.

9.4.3.5 Explicit Reasoning

Explicit reasoning is the use of explicated arguments to support or reject a design problem and a solution choice. Implicit design reasons are decisions made without communicating an explicit reason. When designers use explicit reasoning, they often use the words *because* or *the reason*. We suggest that explicit reasoning helps designers to better communicate their ideas. When explicit reasons are absent, people other than the designers themselves make assumptions about why the decision takes place. In this study, explicit reasons can be identified with the word "because" in the protocol and the arguments to support the explanations. Team M made 33 cases of implicit reasoning and 87 cases of explicit reasoning. Team A made 47 cases of implicit reasoning and 52 cases of explicit reasoning.

9.5 DESIGN DECISION MAKING AND DESIGN EFFECTIVENESS

Software design experience is very important but it is also domain specific. In this experiment, the designers are highly experienced in software design and they possess general knowledge about the problem domain, but they do not appear to have detailed design knowledge in this area. So, the designers have relied on their general design knowledge and reasoning to create design solutions. All the designers were given 2 hours to design, which was insufficient time to exhaustively explore all the possible solution branches to provide a complete solution. Therefore, their design is measured by how effective the designers have been in analyzing the problems and formulating the solutions. We do not verify or validate each team's design solution or evaluate their quality because: (a) the criteria we use in design quality reviews may bias the research findings; (b) the solutions are incomplete; and (c) the designers in some cases have not sufficiently explained a design decision. In order to understand how design reasoning and decision making influence their design outcomes, we investigate time utilization and requirements coverage.

9.5.1 Design Time Effectiveness

The ways that Team A and Team M spent time on organizing their designs are different. This can be summarized in two ways. First, Team M spent more time on planning than Team A did. This seems to have influenced how Team M's design discussions were conducted, which have more structure than Team A's design discussions. Secondly, Team M had fewer context switches than Team A. Team M's discussion topics were clustered together, focusing on the same line of thought and solving most problems in the same design topic before moving onto another topic. In computer processing, context switching between tasks involves computation overheads for preparing the next task. We suggest that cognitive context switching in design also involves certain overheads such as realigning the context of the new discussion topic. The efficiency and the focus of design discussions are lower if extensive context switching is involved.

9.5.2 Requirements Coverage

Requirements coverage is defined as the number of definitive solutions that are provided to address the requirements, derived requirements, and design outcomes. Tables 9.6 and 9.7 show the requirements and the derived requirements coverage. Requirements coverage is the results of the analysis of the two protocols. We classified the design decisions by their status: mentioned (M), discussed (D), proposed (P), implied resolution (I), and resolved (R). I and R indicate that a solution has been reached. The other statuses indicate that the topics have been visited but no decisions have been made (see Table 9.6).

The number of requirements and derived requirements are shown in a format of [*x*,*y*], where *x* indicates the number of requirements coverage and *y* indicates the number of derived requirements coverage. Table 9.6 shows the number of requirements and derived requirements coverage that each of the two teams has achieved. For instance, Team M has resolved three network of roads requirements and two of its derived requirements (see Table 9.6, row 1 right-hand cell). After adding all the requirements and the derived requirements that each team has accomplished (i.e., implied and resolved), we have found that Team A has achieved 13 requirements and 14 derived requirements, whereas Team M has achieved 19 requirements and 14 derived requirements (total column in Table 9.7). This indicates that, given the same time, Team M managed to cover more requirements and derived requirements than Team A, and they are more effective in their design approach.

9.5.3 Differentiating the Design Approaches

Based on the previous analysis, we differentiate how the two teams approach their designs. We compare and contrast them in Table 9.8 to characterize the different

TABLE 9.6 Requirements Coverage Based on Decision Status

	Team A					**Team M**				
Requirement Category	**M**	**D**	**P**	**I**	**R**	**M**	**D**	**P**	**I**	**R**
Network of roads		1,0		1,0	2,2				1,1	3,2
Intersections/lights			1,0		5,5		0,1	1,0		5,4
Simulate traffic flows	1,0				2,7					4,7
Density					2					2
NFR					1					4

TABLE 9.7 Requirements Coverage of Team A and Team M

	Network of Roads [4,3]	**Intersection/ Light [6,5]**	**Simulate Traffic Flows [4,7]**	**Traffic Density [2,0]**	**NFR [4,0]**	**Total [20,15]**
Team A	3,2	5,5	2,7	2,0	1,0	13,14
Team M	4,3	5,4	4,7	2,0	4,0	19,14

TABLE 9.8 Decision Making and Reasoning Approaches of Team A and Team M

	Team A	Team M
Design Planning		
Design planning at the start of the design session	Brief	Extensive
Structure of the design discussions	Semistructured: the discussions did not follow any plan.	Major-minor structure: the discussions followed the initial scope and addressed the difficult issues followed by the easier issues.
Decision topics context switching	More frequent: moving from one discussion topic to the next before a decision topic was fully explored.	Less frequent: exploration of a topic was thorough.
Decision topics time distribution	Widespread over time: some decision topics were revisited many times and distributed throughout the design session.	Limited time distribution: decision topics discussions were focused and were not spread out.
Problem–Solution Coevolution		
Problem statement formulation	Low: the design problems were often not stated. The problems statements were sporadic.	High: problem statements were stated, and they interleaved with solution statements. There were two periods when problem statements were predominantly made to explore the situations.
Problem–solution approach	Solution driven: the concept of using MVC solution and the queue dominated the design thinking.	Problem driven: explored the design questions before providing a solution.
Reasoning Techniques		
Contextualize the problems and the solutions	Medium: less exploration of what factors might have influenced a design decision.	High: explored the contexts of the problems and the solutions more thoroughly.
Use of scenarios	High: used scenarios as an example to understand the needs and concerns.	High
Creating the design options	Less frequent in discussing design options. Higher problem to option ratio.	More frequent in discussing design options. Lower problem to option ratio.
Inductive reasoning	Medium: the level of generalization as evident in the requirements coverage.	High: generalized specific facts or options from a set of observations. The designers generated new ideas using various reasoning techniques.
Explicit reasoning	Low: often the designers did not explain the reasons why certain solutions were chosen.	High: the reasons of a choice were explicitly stated, and this mutual understanding between the designers created opportunities to explore the problem space and the solution space further.

TABLE 9.9 Design Time Effectiveness and Requirements Coverage

	Team A	Team M
Time effectiveness	More context switching; less design problem statements were made; revisited same design topics many times.	Less context switching; focused discussions and more problem statements were made; systematic exploration of problem and solution spaces.
Requirements coverage	Less requirements are covered.	More requirements are covered.

approaches. We suggest that the characteristics listed in Table 9.8 collectively contribute to design effectiveness. However, we cannot delineate the relative importance of individual factors. All these factors contribute to software design effectiveness in one way or another. Effective time utilization depends on good design planning, minimal context switching, and well-organized problem exploration. Additionally, it seems that a problem-driven design approach supported by the effective use of reasoning techniques is also essential in software design. In evaluating time effectiveness and requirements coverage, we note that both designs are potentially viable solutions. They both make sense from a software design point of view. On this basis, we study the effectiveness and the reasoning of the designers.

Spending design time effectively can help designers focus on design issues and analysis. Additionally, good decision-making approaches and design reasoning techniques can also contribute to software design effectiveness. A summary comparison is shown in Table 9.9.

9.6 RESEARCH VALIDITY

One of the major threats to research validity is the potential subjectivity of the researchers in interpreting the protocols. In this work, we have used two separate protocol encodings to analyze the design activities. Although the two protocols have different perspectives, the data obtained from them are consistent, providing triangulation for our interpretation. The first protocol coding was refined over a period of 6 months, in no less than ten iterations, by two researchers. This investigation is different from previous works and the protocol encoding is new; the interpretation of the protocols, especially on concepts such as inductive and deductive reasoning, required deliberations by the researchers. Therefore, we adjusted our interpretations of the reasoning techniques according to our improved understanding in each encoding iteration. The second protocol encoding was performed over a period of 2 months, in three iterations, by a single researcher. This encoding was not new and the researcher had experience using it on prior projects in multiple domains.

The findings from analyzing the two protocols and the decision maps have highlighted different aspects of design reasoning. Our findings in areas such as problem structuring and problem–solution coevolution are consistent with other design theories, which provide external validation to this work.

It emerged in discussions with the designers at the workshop that the design sessions represented a very different situation from those that the designers normally encounter in their work, notably in being supported by far less specialized knowledge of the problem setting and context. This lack of domain knowledge may have influenced the way that the designers acted during the design sessions and thus the resulting design.

Without assessing the quality of a design, we only note that both of the designs are potentially viable solutions. The threat to this research is that we may judge a team effective when they produce a solution that does not fit the requirements. The need for making such an assumption is because, given the short design session and the draft design, it is impossible to evaluate design quality fairly. Nevertheless, this assumption allows us to have a baseline for evaluating design effectiveness.

9.7 CONCLUSION

Design decision making and reasoning play a major role in software design, but there is limited understanding on how they work, whether one way of doing things can help designers to design more effectively than another way. Two teams of designers were given nontrivial software design problems. We studied the ways that they make design decisions and reasoning from the perspectives of planning, time utilization, and reasoning. We have found that the proper planning of design discussions improves the design issue identification and facilitates the searches for solutions. The approach to spending time on decision topics also impacts on time utilization. Excessive context switching of decision topics would defocus the design discussion, and thus designers require more time to address all the relevant design issues. An example is that the repeated discussions on the same topic made by Team A are less effective compared to the single and focused design discussion made by Team M on the subject of traffic lights.

We have found that software designers do not always use the problem–solution coevolution approach. Some designers autonomously assume that the first solution they come across is the right solution and they anchor on that solution all the way without justifying their initial decision. Some designers follow a problem–solution coevolution and a rational reasoning approach. It appears that in dealing with an unfamiliar domain, a reasoning approach (System 2) is better than an autonomous approach (System 1) because of better problem-space exploration and a thorough consideration of the solution alternatives.

We have also found that reasoning techniques, such as the appropriate contextualization of design problems, the explicit communication of design reasoning, explicit design reasoning, and the use of inductive reasoning contribute to the effectiveness of software design. We have found examples to demonstrate that the exploration of design problems may be related to the level of inductive reasoning, suggesting that inductive reasoning can play a role in defining the design problem space. For instance, Team A generated fewer inductive reasoning statements (Section 9.4.3.4), they also explored relatively fewer problems (Section 9.4.2), and they employed a solution-driven approach (Section 9.4.2.3). If inductive and deductive reasoning is complemental, and if problem and solution formulation is complemental, the use of a predominantly deductive and solution-driven approach may have a negative influence on how much of the problem space a designer explores. Although this is an interesting observation, the evidence that we have here to support it is insufficient and will require further research.

From this study, we have preliminary evidence to show that design decision making based on good design planning, minimal context switching, well-organized problem exploration, and good reasoning can help effective software design. As this is an exploratory study, further in-depth empirical and experimental studies of these aspects can improve the way we design software. This is fundamental to the software development process.

ACKNOWLEDGMENTS

The design sessions and the corresponding workshop were funded by the National Science Foundation (award CCF-0845840). We would like to thank the designers who participated in the sessions that provided the input data to this project and the workshop organizers André van der Hoek, Marian Petre, and Alex Baker for granting access to the transcripts and video data. This work is supported by NSF CAREER Award CCF-0844638 (Burge). Any opinions, findings, and conclusions or recommendations expressed in this material are those of the author(s) and do not necessarily reflect the views of the National Science Foundation (NSF). We are grateful to the anonymous reviewers for their valuable suggestions.

REFERENCES

Burge, J. & Kiper, J. (2008) Capturing decisions and rationale from collaborative design. In Gero, J. S. & Goel, A. S. (Eds) *Proceedings of Design, Computing, and Cognition*, Atlanta, GA, July 2008, pp. 221–239. Eindhoven: Springer.

Carroll, J. (1997) Scenario-based design. In Helander, M., Landauer, T. K., & Praphu, P. (Eds) *Handbook of Human–Computer Interaction*, pp. 383–406. Amsterdam: Elsevier.

Cross, N. (2004) Expertise in design: An overview. *Design Studies*, 25, 427–441.

Csapó, B. (1997) The development of inductive reasoning: Cross-sectional assessments in an educational context. *International Journal of Behavioral Development*, 20, 609–626.

Curtis, B., Krasner, H., & Iscoe, N. (1988) A field study of the software design process for large systems. *Communications of the ACM*, 31, 1268–1287.

Dorst, K. & Cross, N. (2001) Creativity in the design space: Co-evolution of problem-solution. *Design Studies*, 22, 425–437.

Dutoit, A., McCall, R., Mistrik, I., & Paech, B. (Eds) (2006) *Rationale Management in Software Engineering*. Heidelberg: Springer.

Epley, N. & Gilovich, T. (2006) The anchoring-and-adjustment heuristic: Why adjustments are insufficient. *Psychological Science*, 17, 311–318.

Evans, J. (1984) Heuristic and analytic processes in reasoning. *British Journal of Psychology*, 75, 451–468.

Evans, J. S. (2003) In two minds: Dual-process accounts of reasoning. *Trends in Cognitive Sciences*, 7, 454–459.

Gero, J. & McNeill, T. (1998) An approach to the analysis of design protocols. *Design Studies*, 19, 21–61.

Goel, V. & Pirolli, P. (1992) The structure of design problem spaces. *Cognitive Science*, 16, 395–429.

Goldschmidt, G. & Weil, M. (1998) Contents and structure in design reasoning. *Design Issues*, 14, 85–100.

Green, T. R. G. & Petre, M. (1996) Usability analysis of visual programming environments: A "cognitive dimensions" framework. *Journal of Visual Languages and Computing*, 7, 131–174.

Hall, J. G., Jackson, M., Laney, R. C., Nuseibeh, B., & Rapanotti, L. (2002) Relating software requirements and architectures using problem frames. In *IEEE Joint International Conference on Requirements Engineering*, pp. 137–144. Washington, DC: IEEE Computer Society.

Hofmeister, C., Kruchten, P., Nord, R. L., Obbink, J. H., Ran, A., & America, P. (2007) A general model of software architecture design derived from five industrial approaches. *Journal of Systems and Software*, 80, 106–126.

Kazman, R., Abowd, G., Bass, L., & Clements, P. (1996) Scenario-based analysis of software architecture. *IEEE Software*, 13, 47–55.

Klauer, K. J. (1996) Teaching inductive thinking to highly able children. In Cropley, A. J. & Dehn, D. (Eds) *Fostering the Growth of High Ability: European Perspectives*, pp. 175–191. Westport, CT: Greenwood.

Maher, M. L., Poon, J., & Boulanger, S. (1996) Formalising design exploration as co-evolution: A combined gene approach. In Gero, J. S. (Ed) *Advances in Formal Design Methods for CAD*, pp. 3–30. London: Chapman & Hall.

Rittel, H. W. J. & Webber, M. M. (1973) Dilemmas in a general theory of planning. *Policy Sciences*, 4, 155–169.

Rowe, P. G. (1987) *Design Thinking*. Cambridge, MA: MIT Press.

Schön, D. A. (1988) Designing: Rules, types and worlds. *Design Studies*, 9, 181–190.

Shum, B. S. & Hammond, N. (1994) Argumentation-based design rationale: What use at what cost? *International Journal of Human–Computer Studies*, 40, 603–652.

Simina, M. & Kolodner, J. (1997) Creative design: Reasoning and understanding. In Leake, D. B. & Plaza, E. (Eds) *Case-Based Reasoning Research and Development*, pp. 587–597. Berlin/Heidelberg: Springer.

Simon, H. & Newell, A. (1972) Human problem solving: The state of the theory in 1970. Carnegie-Mellon University.

Tang, A. & Lago, P. (2010) Notes on design reasoning techniques. Swinburne FICT Technical Reports. Swinburne University of Technology.

Tang, A., Han, J., & Vasa, R. (2009) Software architecture design reasoning: A case for improved methodology support. *IEEE Software*, March/April, 43–49.

Tang, A., Barbar, M. A., Gorton, I., & Han, J. (2006) A survey of architecture design rationale. *Journal of Systems and Software*, 79, 1792–1804.

Tang, A., Tran, M. H., Han, J., & Van Vliet, H. (2008) Design reasoning improves software design quality. In Becker, S., Plasil, F., & Reussner, R. (Eds) *Quality of Software Architectures (QoSA 2008)*, pp. 28–42. Berlin/Heidelberg: Springer.

Tversky, A. & Kahneman, D. (1974) Judgment under uncertainty: Heuristics and biases. *Science*, 185, 1124–1131.

Tversky, A. & Kahneman, D. (1986) Rational choice and the framing of decisions. *The Journal of Business*, 59, S251–S278.

Tyree, J. & Akerman, A. (2005) Architecture decisions: Demystifying architecture. *IEEE Software*, 22, 19–27.

Yin, R. (2003) *Case Study Research: Design and Methods*. London: Sage.

Zannier, C., Chiasson, M., & Maurer, F. (2007) A model of design decision making based on empirical results of interviews with software designers. *Information and Software Technology*, 49, 637–653.

CHAPTER 10

Identifying and Analyzing Software Design Activities

Bonita Sharif

Youngstown State University

Natalia Dragan

Cleveland State University

Andrew Sutton

Texas A&M University

Michael L. Collard

The University of Akron

Jonathan I. Maletic

Kent State University

CONTENTS

10.1 INTRODUCTION

How do we design software? This very important question has implications for virtually all aspects of the software development life cycle, the participating software developers, and even the project managers. Good software design should promote programmer productivity, lessen the burden of maintenance, and reduce the learning curve for new developers. Despite the seeming simplicity of the question, the answers tend to be anything but simple. An answer may touch on aspects of the problem domain, the expertise of the designers, the quality and completeness of the requirements, and the process models and development methods that are followed. Much work has been done on how software *should be* designed, but there are comparatively few studies of how software *is actually* designed (Guindon et al. 1987; Curtis et al. 1988; Maia et al. 1995; Robillard et al. 1998; Detienne and Bott 2001; Ko et al. 2006; Friess 2007).

To address the aforementioned question on how software is designed, we introduce specific research questions based on the types of activities that designers engage in during the early stages of design. These questions were chosen to cover the activities observed in the context of the requirements met by the design. We are also interested in designer interaction and how similar/different the team interactions were. The research questions we attempt to address are:

RQ1: What specific types of activities do the designers engage in?

RQ2: What design strategies (activity sets) are used together?

RQ3: Are all the requirements met or discussed in the design?

RQ4: What is the level of interaction between the designers?

RQ5: What are the similarities and differences between the different design sessions?

Since this is an *observational* study, we do not generate hypotheses for the research questions. Instead, we try to qualitatively assess the design sessions in a structured manner that is reproducible by others. In order to answer the earlier questions, two issues have to be addressed. First, we need data on the design practice of professional designers; such data are extraordinarily difficult to obtain. Individuals and groups inside companies do not always perform all design activities using a formal tool, but they often rely on paper or whiteboards, especially in the early interactive stages. This, coupled with the collaborative interaction between designers, is very challenging to capture for later study. Second, once the data have been collected, the additional questions of how the data should be analyzed and/or processed arise.

The opportunity to conduct an in-depth study of how professional designers engage in design presented itself in September 2009. A group at the University of California, Irvine (2010) made available video recordings and transcripts of three two-person teams engaged in design for the same set of software requirements. Given these data and our research questions, we decided to use an approach rooted in discourse analysis (Gee 2005): the analysis of language with the goal of identifying problem-solving and decision-making strategies.

In order to do this, we first need to generate, refine, and analyze the data from the design sessions. This process involves generating an XML-structured version of the transcripts, the manual annotation of the XML-formatted transcripts, and its visualization on a time line. For each activity undertaken by the designers, we identified the requirement being addressed. We annotated each design session data set with metadata corresponding to the activities (such as drawing and agreement/disagreement between designers) taking place along with the requirement(s) being addressed. We also analyzed the annotated data for trends and patterns in the design process across the teams.

The results indicate that the types of design activities map nicely to the main phases of software development, such as requirements, analysis, design, and implementation. However, the sequence and iteration of these steps were different for each team. The decisions about the logic of the code, discussions of use cases, and the interaction between the designers played major roles for all three teams, and all teams spent the most time on the same three requirements. The team that took a structured approach with systematic planning, and with much consultation and agreement, gave the most detailed solution and covered the most requirements in the deliverables. A software designer engaged in these types of activities, whether an expert or a novice, may benefit from observing what does and does not work. The future goal is to develop theories (Bryant and Charmaz 2007) on how software is designed based on our observations and results.

10.2 METHODOLOGY

In order to address the research questions of the study, we developed a process and a suite of tools to support the extraction and analysis of the data from the source videos and their transcripts. We now give a brief overview of the steps involved in the process.

1. *Preprocessing*: In order to conduct a structured analysis on the data, we first transformed the transcripts from a simple text-based listing into an XML format for easy processing. This allowed us to clean up the transcripts, annotate them with additional data, and perform queries.

2. *Comprehension and Cleanup*: We reviewed the videos to generate a set of codes related to the activities we observed. This step also attempted to fix any ill-transcribed or inaudible statements within the XML transcripts. Almost all cases were fixed by listening to the audio multiple times or at a higher volume level. Very few inaudible sections remained.

3. *Annotation*: Once the data were cleaned up and in an initial annotated format, we reviewed the videos again, this time annotating the XML transcripts with codes indicating the activities being performed and the requirements being addressed at particular points in time.

4. *Visualization*: A time line visualization of the annotated XML transcripts was generated that allowed us to better see how the designers engaged the problem over the course of the sessions. These time line visualizations were manually studied to answer the research questions we posed.

10.2.1 Problem Statement and Source Material

The three teams were charged with designing a traffic signal simulation. They were given a two-page problem statement on paper and access to a whiteboard where they could design a solution. They were asked to produce the user interface (UI), a high-level design, and a high-level implementation. The design sessions ran from 1 to 2 hours. The audience for the design was listed in the problem statement as a team of software developers whose competency level could be compared to a basic computer science or software engineering undergraduate degree holder. Each team was debriefed at the end of their design session.

Our analysis was conducted on the videos of the three teams, each of which depicts a pair of professional software designers designing the traffic signal simulation. Summary information about the videos is presented in Table 10.1. The transcripts of the designers' discussion were also analyzed. Our approach is rooted in discourse analysis where we examine the videos, code the transcriptions, and analyze the coding. In prior related work, Robillard et al. (1998) used protocol analysis to build a cognitive model, devise a hierarchical coding scheme, and look at the influence of individual participant roles (d'Astous and Robillard 2001). We concentrate on finding the design activities and strategies that the designers used to address the requirements.

Typos, inconsistencies, or omissions between the videos and the transcripts were fixed prior to analyzing the videos. For example, if the transcript stated "[inaudible]" but we were able to recognize the spoken word from the video, the transcript was updated with the corrected text. There were also instances where the transcript changed the meaning of the spoken sentence completely. For example, on page 5 of the Adobe transcript (at 0:14:14.1), Male 2 uses the word "interface" which is transcribed as "beginner phase." Such incorrect transcriptions were also fixed. For those cases where we were not able to hear the speech, we left the text as inaudible. The counts of inaudible text are also shown in Table 10.1. Furthermore, additional time stamps for periods of silence greater than 20 seconds were introduced into the transcripts.

10.2.2 Activity and Requirement Coding

After reviewing all three video sessions, two of the authors jointly developed two sets of codes that can be attributed to the speech and actions in the design sessions. The first set of codes identifies the requirements from the problem domain that the designers addressed. The second set of codes corresponds to various kinds of software engineering activities that the designers engaged in during the sessions. These codes are used during the transcript annotation process.

TABLE 10.1 Video Information

Video	Number of Words	Total Time	Total Number of Inaudible
Adobe	10,043	1 hour 52 seconds	36
AmberPoint	12,328	1 hour 53 seconds	25
Intuit	5,917	~1 hour	19

As the designers conversed, the main topic of conversation was the problem given. Their discourse can be categorized into different parts of the software development process, including requirements, analysis, design, and implementation. For each of these parts, specific artifacts and parts of artifacts were discussed, including use cases, actors, objects, entity–relationship (ER) or class diagrams, control and data flow, data structures, patterns, and architecture. The different teams of designers used different aspects of the design process with varying degrees of detail. From the problem statement, we first identified a set of basic requirements and assigned them a code. The fine-grained requirements identified are shown in Table 10.2.

Additionally, an examination of the videos and transcripts was performed to define a set of activities in which the designers engaged. We differentiated between two categories of activities: verbal and nonverbal. All speech belongs to the verbal category, while the nonverbal activities include drawing, reading, writing, silence, and analyzing the whiteboard. We also identified and annotated the activities that were irrelevant to the design process. For example, in the Adobe video there was a 2-minute break with no activity due to one of the designers leaving the room.

The verbal category consists of activities related to the design process and miscellaneous interpersonal communication, shown in Tables 10.3 and 10.4, respectively. The verbal activities related to the design process in Table 10.3 map to the engineering activities of requirements, analysis, design, and implementation. The activities presented in Table 10.4

TABLE 10.2 Requirement Codes Identified from the Problem Statement

Code	Description
map_roads_intersect	Map should allow for different arrangement of intersections
road_len	Map should allow for roads of varying length
intersect	Map should accommodate at least six intersections if not more
light_seq	Describe and interact with different light sequences at each intersection
set_light_timing	Students must be able to describe and interact with different traffic light timing schemes at each intersection
set_lights	The current state of the lights should be updated when they change
left-hand_turn	The system should be able to accommodate protected left-hand turns
no_crash	No crashes are allowed
sensor_option	Sensors detect car presence in a given lane
sensors_light	A light's behavior should change based on input from sensors
simulate	Simulate traffic flows on the map
vis_traffic	Traffic levels conveyed visually to the user in real time
vis_lights	The state of traffic lights conveyed visually to the user
spec_density	Change traffic density on any given road
alter_update	Alter and update simulation parameters
observe	Observe any problems with their map's timing scheme and see the result of their changes on traffic patterns
wait_time	Waiting is minimized by light settings for an intersection
t_time	Travel time
t_speed	Travel speed for cars

TABLE 10.3 Verbal Activity Codes Related to the Design Process

Verbal Code	Description
v_asu	Making assumptions about the requirements
v_req_new	Identifying other new requirements not specified in the design prompt
v_UC	Define/refine use cases
v_DD	Define/modify the data dictionary/domain objects
v_DD_assoc	Add associations to domain objects
v_DD_attrib	Add attributes to domain objects
v_class	Adding/modifying a class
v_class_attrib	Adding/modifying a class attribute
v_class_ops	Adding/modifying a class operation
v_class_rel	Adding/modifying a class relationship
v_ER	Work on the ER diagram
v_arch	Identify architectural style. Identify design patterns
v_UI	Defining/modifying the UI
v_DS	Identifying data structures
v_code_logic	Specific implementation logic. Includes specific algorithms used, data flow, and control flow
v_code_structure	Designing code structure

TABLE 10.4 Verbal Activity Codes Related to Interpersonal Communication

Miscellaneous Code	Description
v_consult	Consulting with other designers, asking questions, answering questions. No distinction is made between who asked or who answered the question.
v_agree	Agreement.
v_planning	Planning. This activity involves one or both designers deciding on the next step of the design process.
v_justify	Justification for a certain design decision.

include general conversational aspects such as consulting, asking and answering questions, and agreeing or disagreeing with the other participant.

The nonverbal activities are mainly related to reading, drawing or writing on the whiteboard, and analyzing the information on the whiteboard. These activities are shown in Table 10.5 and include reading the requirements; drawing pictures, UIs, and diagrams; and writing the code structure. While viewing the videos, we decided to include a separate code for analyzing the whiteboard as a nonverbal activity because this activity occurred with sufficient frequency. The following sections describe how the XML transcripts were created and how the annotations were performed.

10.2.3 Transcript Annotation

The transcripts were annotated with the requirement and activity codes identified in the previous section. First, the time-stamped events from the simple text-based transcript were converted into an XML format. The events were then annotated with the codes for verbal and nonverbal activities by reading the transcript and reviewing the videos, respectively. The result is an annotated XML transcript with activity and requirement codes as

TABLE 10.5 Nonverbal Activity Codes

Nonverbal Code	Description
nv_read	Initial reading of problem statement.
nv_revisitreq	Revisit requirements by rereading the design prompt.
nv_analyze_board	Both designers analyze the whiteboard. This is accompanied by nv_silent or other activity codes where appropriate.
nv_drawpic	Draw a picture.
nv_UC_draw	Draw/modify use case.
nv_CD_draw	Draw/modify class diagram.
nv_ER_draw	Draw/modify an ER diagram.
nv_UI_draw	Draw/modify a sketch of the UI.
nv_notes	Writing notes or rules.
nv_code	Writing code structure.
nv_silent	Silent. Usually combined with one or more of the above codes.

metadata. An example of an event element is given in Figure 10.1 in nonbolded font. Each such event element corresponds to one time-stamped event. The XML element records the time down to a tenth of a second, a label for which participant was speaking, the number of inaudible tags in the text, and the transcription of the designer's speech.

Annotations were then added to the generated transcript based on a manual analysis of the text and video. As can be imagined, just entering the annotations is very labor intensive and prone to errors. To help alleviate this, we developed a custom annotation tool, *TransAnn*, to simplify the process of entering the annotations. A screenshot of TransAnn is shown in Figure 10.2.

```
<timestamp hr="0" min="06" sec="27" sec_tenth="4" person="Female" num_inaudible="2">
<sentence req="map, arr_intersect"
        image_name="vlcsnap-00003.png">
  <activity>
    <phrase>So it's like a drawing tool that can allow them to lay down these
intersections.</phrase>
    <code type="verbal">v_UC </code>
    <code type="verbal">v_UI</code>
  </activity>
  <activity>
    <phrase>And also some mechanism by specifying these [inaudible] distances between
intersections.</phrase>
    <code type="verbal">v_UC</code>
  </activity>
  <activity>
    <phrase>I suppose there's some traffic speeds at which... [inaudible]</phrase>
    <code type="verbal">v_asu</code>
  </activity>
</sentence>
</timestamp>
```

FIGURE 10.1 XML format for an event in the video. The nonbolded text was generated from the original text-based transcript that associates each sentence with the time spoken, the speaker, and any inaudible words or phrases. The bolded text shows the added annotations to the transcript that decomposes sentences into phrases and includes requirement codes, activity codes, and linked still frames.

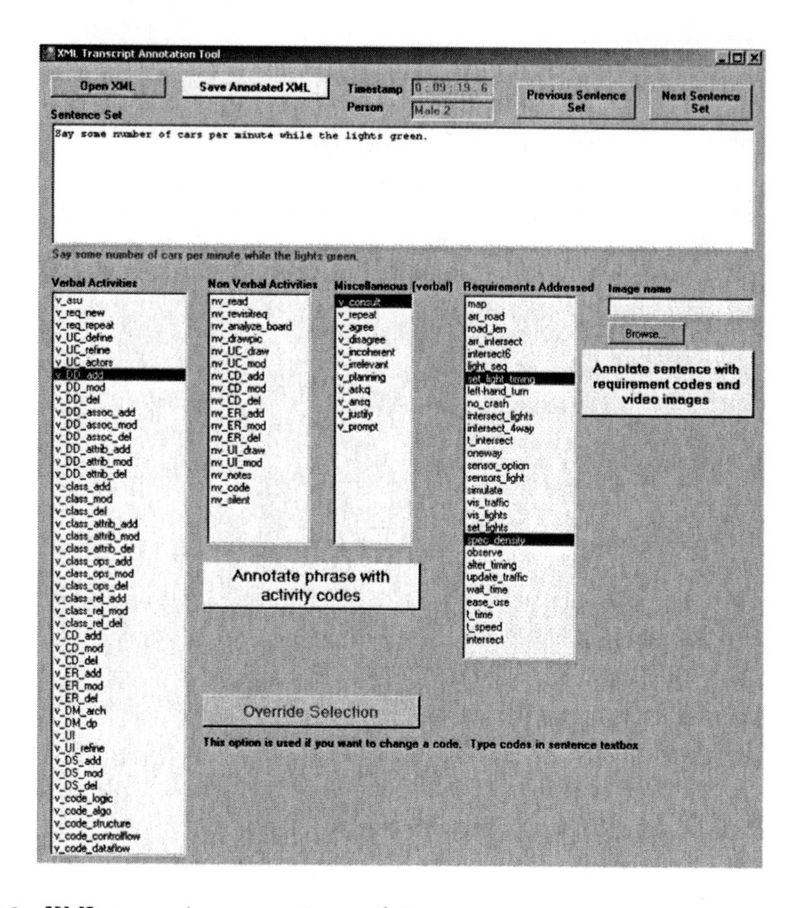

FIGURE 10.2 XML transcript annotation tool *TransAnn* was created to enable the tagging of events with activity and requirements codes.

The process of annotating was performed in the following manner (see Figure 10.3 for the structure of the XML annotations). First, each event (time stamp) was associated with a *sentence*. Each sentence consisted of one or more *phrases*. Each phrase was coded with one or more activity codes. Some phrases did not fit into any activity and were not coded. The resulting set of codes for the sentence is the union of the codes of all phrases in that sentence. Each sentence was additionally annotated with requirement codes, and a relevant

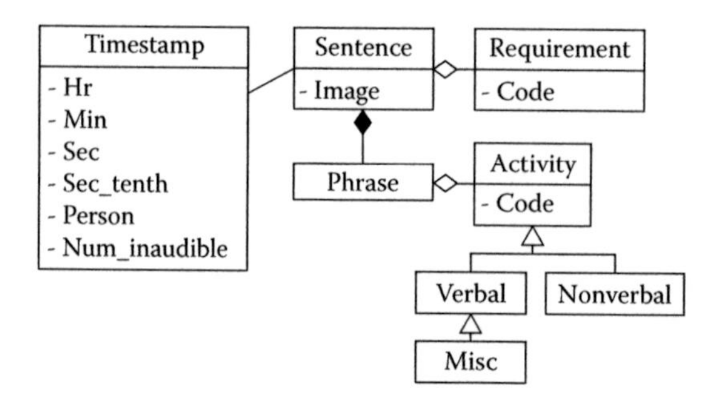

FIGURE 10.3 Class diagram for the annotated XML structure.

image such as a class diagram or a still frame from the video (if available). An activity code based on key words used in the phrase was also assigned. These key words (in the problem and the solution domains) were identified during the initial review of the videos. The annotation decision process was done manually for each time stamp using the annotation tool. An example of the annotated XML is shown in Figure 10.1.

The key words we identified can also be used to query the XML transcripts and this is left as a future exercise. For example, one query could be, "When did the first instance of the word *queue* occur?" The key word *queue* is part of the solution domain. The key word set also includes similar-word mappings. For example, the words "signal" and "light" mean the same thing in this domain context. We do not list the key words here however; they are implicit in the phrases that are annotated. Additionally, since we do not lose any information in the XML annotation process, we can easily query the XML document for these key words.

Some events and activity codes were inserted manually into the XML transcript, specifically events having to do with silence and reading activities, the nonverbal activity codes *nv_silent* and *nv_read*, respectively. Events for both of these were added before the first utterance, that is, before the first entry in the transcript since the designers started the session by reading the problem statement. In the Intuit team video, time stamps were also added for long periods of silence because this was unique to their interaction.

Even with the annotation tool, the analysis of the videos and transcripts and their annotation were quite time consuming as it took 30, 32, and 15 hours to code and annotate the Adobe, AmberPoint, and Intuit sessions, respectively. One of the authors annotated the Adobe video session and another author annotated the AmberPoint and Intuit video sessions. Initially, both annotators worked together to agree on a common coding rules convention. Other challenges included such things as when one designer was drawing and the other was talking, which occurred quite frequently. These instances were put into the same event. Since each video was coded by only one person, we did not calculate an intercoder reliability measure (Artstein and Poesio 2008). However, the two coders consulted each other on coding that seemed unclear and they reached a consensus, which reduced the threat to the validity of the coding scheme. The fully annotated XML transcripts were used for subsequent analyses, queries, and visualizations.

10.2.4 Visualization

In order to help study and further analyze the data, a time line presentation for the annotated XML transcripts is used. Discussion and activities are visualized as a *time line matrix* in which actions and speech are mapped to the event codes described earlier. Examples of engineering activities are shown in Figures 10.5 through 10.7. The duration of the session is shown on the *x* axis divided into 5-minute chunks. The activities and speech are colored blocks indicating the duration of participation. The color of the block indicates the speaker or actor engaging in the activity or requirement.

To generate the visualization, we processed the XML transcripts via a suite of Python scripts, extracting each event and associating it with both a requirement code (from Table 10.2) and an activity code (from Tables 10.3 through 10.5). Recall that events are frequently annotated with multiple activity and requirement codes. Events associated with each code are ordered sequentially by start time. Note that the transcripts are structured so that there are no overlapping time segments for a single event. In the event that an overlap does occur, the most recent activity simply "interrupts" the previous one.

Each of these mappings (event-to-activity and event-to-requirement) is stored in a *time line* file that describes the sequence of events associated with each code. A second Python script renders the time line file into the scalable vector graphics (SVG) format. These images can then be rendered in a browser or converted to bitmap graphics for manual inspection.

To further improve readability, we manually ordered the *y* axis of the time lines in the source data sets. This ordering attempts to group logically related activities. We also inserted missing activity or requirements codes with an empty time line. These modifications are intended to facilitate visual comparisons between design sessions. For example, one design team might not have touched on a specific requirement (thus not appearing on the time line), while another team did. A second set of time lines was generated in which the activity and requirement codes were sorted such that the most engaged activity or the most addressed requirement would appear at the top of the matrix. We refer to these as the *sorted time line matrices.*

10.3 RESULTS AND ANALYSES

In this section, we present selected time line matrices generated from the data analysis. The fully annotated XML transcripts and the time line matrices (including sorted matrices) are published online.* Figure 10.4 depicts the time line matrix for requirements addressed in the Adobe session. From this matrix, we can see that the Adobe team was most actively engaged in addressing problems related to the intersections (*intersect* and *map_roads_intersect*) and the altering or updating of simulation parameters (*spec_density*, *alter_update*, and *t_speed*).

Figure 10.5 depicts the time line matrix for engineering activities in the Adobe session. It is evident from this diagram that, early on, these designers spent time constructing the data dictionary (*v_DD_**), use cases (*v_UC* and *nv_UC_draw*), and class model for the problem (*v_class_**). They spent a great deal of time discussing code logic (*v_code_logic*) even while they were building the class model. They were also consistent in analyzing the whiteboard (*nv_analyze_board*), after designing the class structure and the UI (*nv_UI_draw* and *v_UI*).

The next sections attempt to address the research questions we posed in Section 10.1. From the annotated transcripts and their time line matrices, we can make a number of observations and comparisons about the processes and strategies employed by each design team.

10.3.1 Activity Analysis

Our first research question (RQ1) seeks to determine the types of activities that the designers engage in while solving a problem. The mapping and visualization of phrases to activity codes directly supports answering this question. The types of activities that the designers engage in map clearly to the main phases of the software development, such as requirements, analysis, design, and implementation (see Table 10.6). This mapping was not

* Annotated transcripts and time line visualizations are published online at http://www.sdml.info/design-workshop.

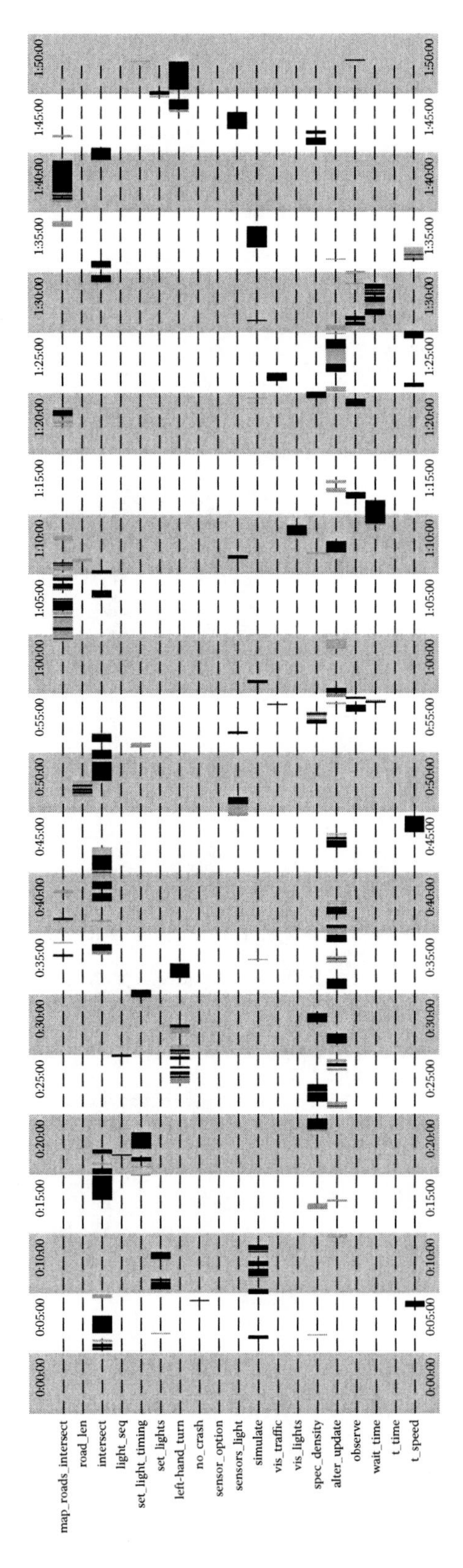

FIGURE 10.4 Requirements addressed during the Adobe session. Events are colored blocks indicating the duration of the participation, with the color of the block indicating the speaker.

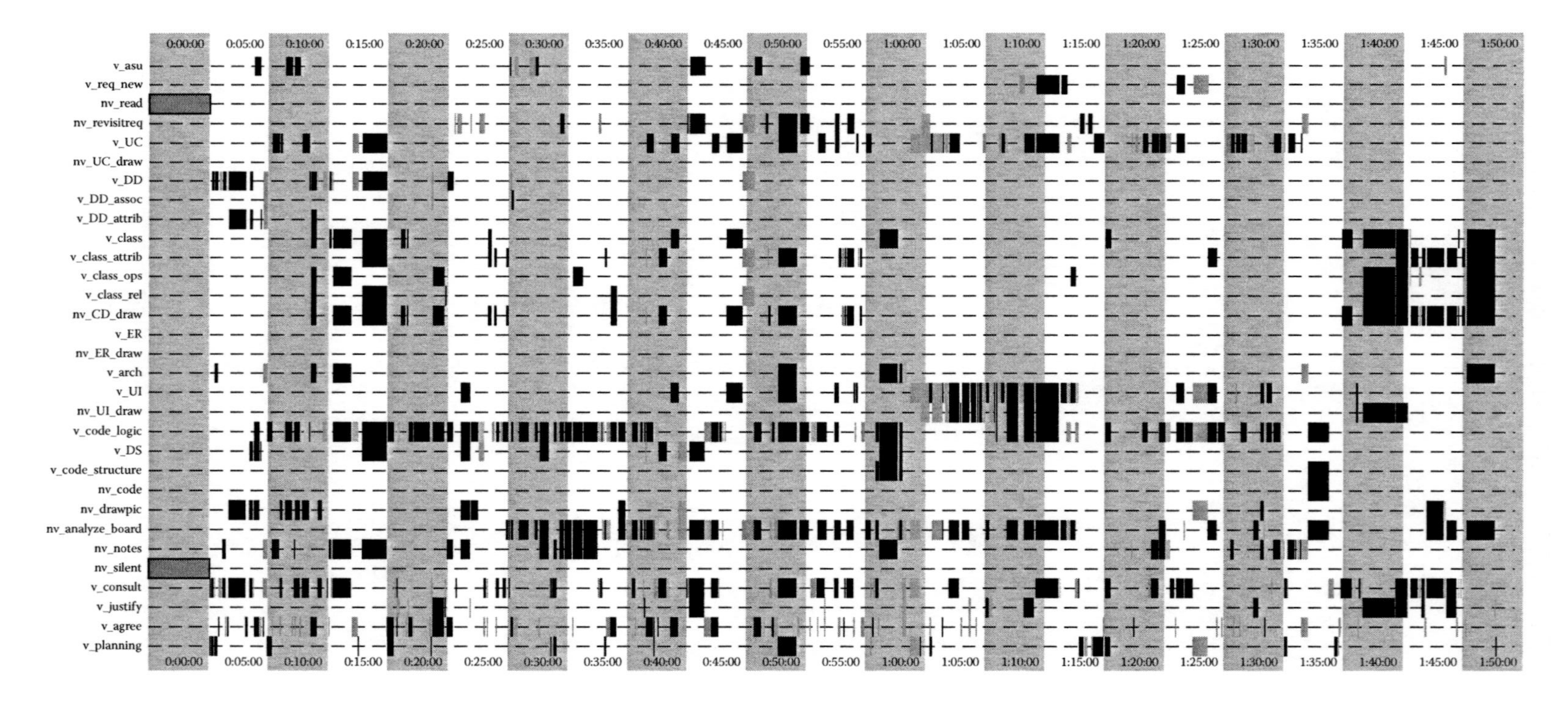

FIGURE 10.5 Engineering activities engaged in during the Adobe session. Events are colored blocks indicating the duration of the participation, with the color of the block indicating the speaker.

TABLE 10.6 Activity Codes Mapped to Phases of a Traditional Software Engineering Process

Phase	Activities
Requirements	v_asu, v_req_new, nv_read, nv_revisitreq, v_UC, nv_UC_draw
Analysis	v_DD, v_DD_assoc, v_DD_attrib
Design	v_class, v_class_attrib, v_class_ops, v_class_rel, nv_CD_draw, v_ER, nv_ER_draw, v_arch, v_UI, nv_UI_draw
Implementation	v_code_logic, v_DS, v_code_structure, nv_code, nv_drawpic, nv_analyze_board, nv_notes, nv_silent

determined a priori, rather it was observed during the analyses of the design sessions. The main phases are shown circled in the time line matrices of Figures 10.6 (Intuit activities) and 10.7 (AmberPoint activities).

We observed that the sequence and iteration of these steps were different for the different teams. The AmberPoint team spent a lot of time designing the UI, whereas the Adobe and Intuit teams spent more time designing the object model. The AmberPoint team engaged in more verbal activity (448 events) compared to the Adobe team (414 events). The Intuit team finished in half the time of the other teams and had the least verbal activity. The participants from only one team (AmberPoint) discussed aspects of verification or validation with regard to their solution. Figures 10.6 and Figure 10.7 group together

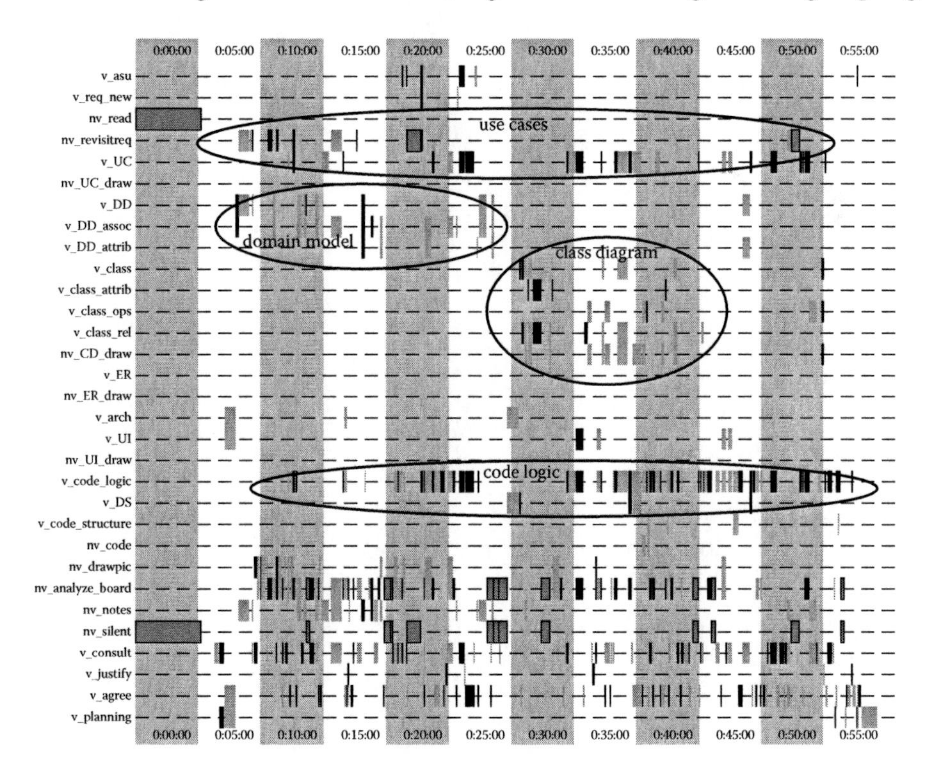

FIGURE 10.6 Time line matrix of activities for the Intuit session shows actions and speech mapped to various event codes. The events are colored blocks indicating the duration of the participation, with the color of the block indicating the speaker. The use cases, domain model, class diagram, and code logic are circled.

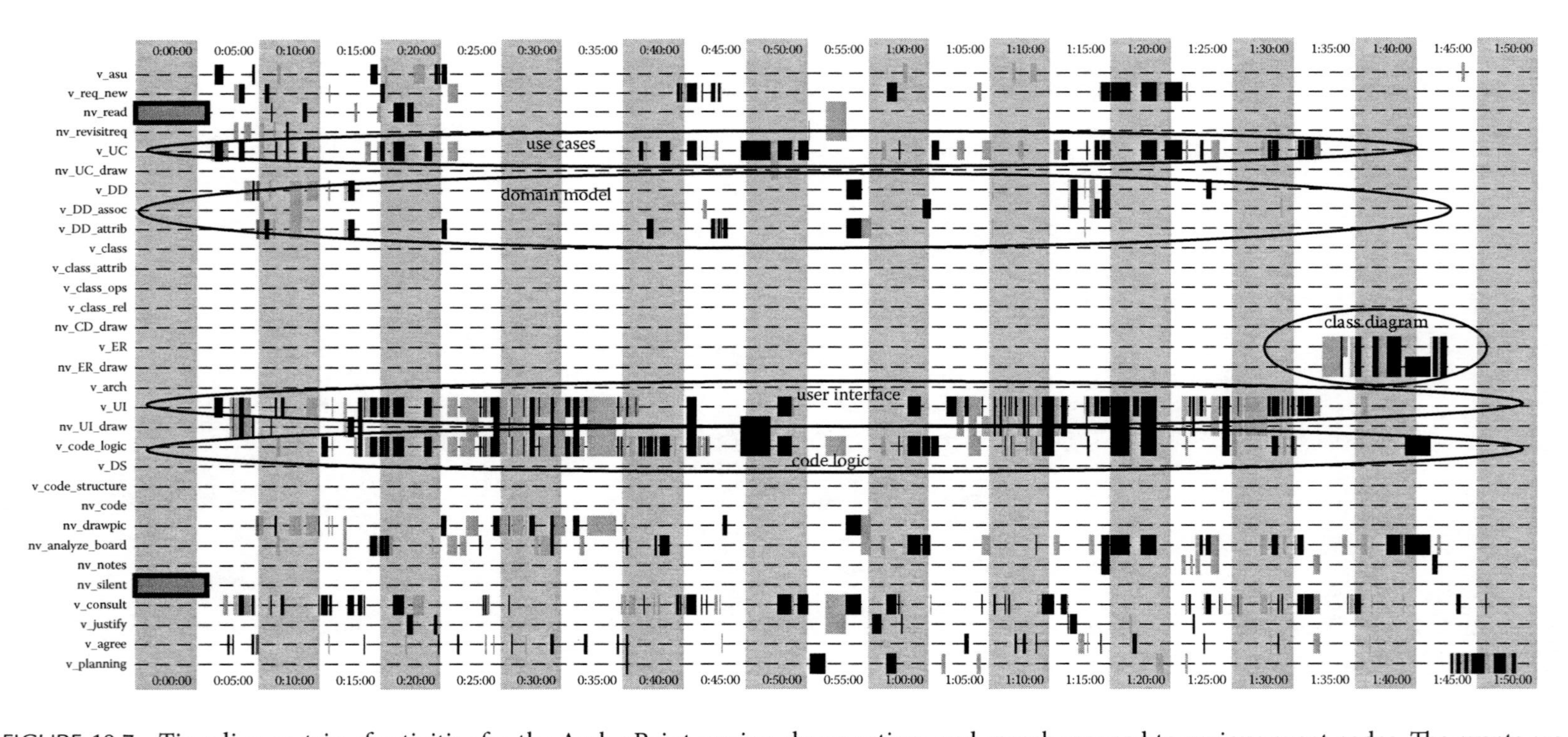

FIGURE 10.7 Time line matrix of activities for the AmberPoint session shows actions and speech mapped to various event codes. The events are colored blocks indicating the duration of the participation, with the color of the block indicating the speaker. The use cases, domain model, class diagram, code logic, and UI are circled.

chunks of activity on the *x* axis, as shown by the circled areas. In Figure 10.6, we observe that the designers discussed code logic (*v_code_logic*) starting at 20 minutes through to the end of the session; however, it is important to see how the logic evolved. At 20 minutes, they were discussing the logic with respect to the data dictionary (as seen by the overlapping domain model oval), whereas at 30 minutes, they had progressed to building the class model. During the entire session, they focused on the requirements, referring back to them at regular intervals.

Our second research question (RQ2) seeks to determine the design strategies or activity sets used by the designers. Examining the distributions of the activities in the time line matrix provides some insights into answering this question. We can determine the design strategies or activity sets by asking the following subquestions. We address each of these subquestions next.

RQ 2.1: Which design activity took the longest?

RQ 2.2: What was the sequence of activities for each session?

RQ 2.3: How many of the activities were interleaved?

RQ 2.4: Did the designers add additional requirements/features and make valid assumptions where appropriate?

The first five activities that took the most and least time are shown in Table 10.7. These are the first five and last five in the sorted time line matrices. Three activities (*v_code_logic*, *v_consult*, and *v_UC*) were common in all three sessions but they did not appear in the same order. In other words, in all three sessions, decisions about the logic of the code, discussions of use cases, and interactions between the designers played major roles. Other top time-consuming activities (RQ 2.1) included analyzing the whiteboard, drawing class diagrams, and talking about and drawing the UI (*nv_analyze_board*, *nv_CD_draw*, *v_UI*, and *nv_UI_draw*, respectively). Silence was much more prevalent in the Intuit session, where they spent the time analyzing the whiteboard.

TABLE 10.7 Top Five and Bottom Five Activities on Which the Most and Least Time Was Spent

	Adobe	AmberPoint	Intuit
Most time spent	*v_code_logic*	v_UI	nv_analyze_board
	nv_analyze_board	*v_code_logic*	*v_code_logic*
	v_consult	nv_UI_draw	*v_consult*
	v_UC	*v_UC*	nv_silent
	nv_CD_draw	*v_consult*	*v_UC*
	:	:	:
Least time spent	nv_code	nv_CD_draw	v_req_new
	v_DD_assoc	v_arch	nv_UC_draw
	nv_UC_draw	v_DS	v_ER
	v_ER	v_code_structure	nv_ER_draw
	nv_ER_draw	nv_code	nv_UI_draw

Note: Activities in italic occurred in all sessions.

TABLE 10.8 Sequence of Main (Not Necessarily Interleaved) Activities across Time Shown in 30-Minute Increments (15-Minute Increments for Intuit)

	Interval	Activities
Adobe	30 minutes	Identifying objects, drawing pictures
	1 hour	Attributes of objects, revisiting requirements, use cases
	90 minutes	UI and use cases
	2 hours	UML class diagram, code structure
AmberPoint	30 minutes	Drawing and talking UI, identifying objects, use cases
	1 hour	Drawing UI, use cases, scenarios
	90 minutes	UI and use cases
	2 hours	UI and ER diagram
Intuit	15 minutes	Identifying objects, drawing pictures
	30 minutes	Identifying objects, drawing pictures, silence
	45 minutes	Drawing diagram attributes and relationships, use cases
	1 hour	Use cases, domain objects, UI

The sequence of activities and the interleaving of those activities for each session varied, which can be attributed to individual differences in approaches to problem solving, expertise, or prior experiences. In order to analyze the sequence of activities (RQ 2.2), we split the time line into quarters (see Table 10.8). For the Adobe and AmberPoint sessions, each quarter was 30 minutes, while in the Intuit session each quarter was 15 minutes since they took only an hour for the complete session. We excluded the top five activities in this analysis in order to focus our attention on more specific tasks; the top five tasks were fairly general activities and were common to all sessions.

For the Adobe session, the first quarter was spent identifying the objects in the system (data dictionary and domain model) and drawing pictures of the road intersection. The next quarter was spent revisiting the requirements (*nv_revisitreq*), use cases (*v_UC*), and specifying the properties of the objects. In the third quarter, they concentrated on the UI and use cases. Finally, they focused on drawing the unified modeling language (UML) class diagram and writing the structure of the code.

For the AmberPoint session, the first quarter was spent talking about the UI, drawing pictures of the UI, and introducing domain objects and use cases. The second quarter focused much more on the UI with some discussion about the use cases and scenarios. The third quarter also focused on the UI and use cases. Finally, in the last quarter, they drew an ER diagram.

For the Intuit session, the first quarter was spent identifying domain objects and drawing pictures. In the next quarter, they continued with long periods of silence. The focus of the next quarter was on drawing a diagram, specifying attributes, relationships, as well as use cases. In the final quarter, use cases were discussed, including some discussion about the UI.

Observing the activity chart for Adobe in Figure 10.5, we get a visual indication of the types of interleaved activities in each quarter (RQ 2.3). The interleaved activities are often related. For example, in the Adobe session, when they talked about the UI, they also discussed user interaction (use cases). The activities of drawing pictures and identifying domain objects were also interleaved in all three sessions. This is evidence that drawing a visual representation actually helps in identifying the data dictionary and conveying the thought process to the designers. In the Adobe session, the logic of the code was discussed almost consistently throughout and hence it interleaved the most with all other activities.

Next, we address RQ 2.4: coming up with new requirements and making valid assumptions. With respect to coming up with new requirements (*v_req_new*), the AmberPoint team did this more than the Adobe team, while the Intuit team addressed the subject very little. In the Adobe session, one of the new requirements was the ability to resize the simulation window. We consider this to be an important feature of the UI and the ability of the team to think ahead in terms of simulation usability.

In all three design sessions, assumptions (*v_asu*) were made about the requirements. For example, in the Adobe session, the assumption was made that roads have two lanes to address the left-hand turn requirement. The Adobe and AmberPoint teams made more assumptions than the Intuit team, some of which were critical to moving the design forward. One prevalent behavior in the Adobe session was to concentrate on the current topic and if something was not clear or if they had trouble with it, they went on to the next point and returned to it at a later stage. This observation demonstrates the iterative process of design.

10.3.2 Requirements Analysis

We now address RQ3: whether the designers took into account all or only some of the requirements. The answer can be determined from the annotations for the requirements in the transcript. For example, the requirements coverage over time for the Adobe session is presented in Figure 10.4. Similar time line matrices showing the requirements coverage for the AmberPoint and Intuit sessions can be found on our companion website. Furthermore, we can determine which requirements were discussed most frequently (in terms of the amount of time spent). There might be a connection between the time taken for a requirement and the inherent complexity involved. However, a requirement that was never discussed is not necessarily less complex or easy. Due to the lack of a gold standard, we do not relate completeness with quality.

The top and bottom five requirements that took the most time and the least time are shown in Table 10.9. Three requirements (*intersect*, *alter_update*, and *light_seq*) were common in all three sessions but did not appear in the same order. The *intersect* requirement dealt with the approach used to describe and represent the intersection of the roads. It also included constraints such as *no one-way roads* or *T-junctions*.

In the Adobe and Intuit sessions, the intersection was the top requirement discussed. Since the AmberPoint session focused more on the UI, their top activity shows up as *alter_update*. This requirement involves interaction by the student using the software (as indicated in the problem statement), and deals with setting simulation parameters and updating the simulation (see Table 10.2). Finally, the *light_seq* requirement was common but all teams spent the least time on it: a scheme for updating the lights.

The Intuit and AmberPoint sessions both discussed the *wait_time* requirement, which states that ideally the wait time needs to be minimized. This was the sixth most talked about activity for the Adobe session. The Intuit team did not pay much attention to this

TABLE 10.9 Top Five and Bottom Five Requirements on Which the Most and Least Time Was Spent

	Adobe	AmberPoint	Intuit
Most time spent	*intersect*	*alter_update*	*intersect*
	alter_update	map_roads_intersect	set_lights
	map_roads_intersect	set_light_timing	*alter_update*
	left-hand_turn	*intersect*	left-hand_turn
	spec_density	wait_time	set_light_timing
	:	:	:
Least time spent	vis_lights	simulate	*light_seq*
	light_seq	*light_seq*	no_crash
	no_crash	set_lights	vis_lights
	sensor_option	road_len	observe
	t_time	t_time	wait_time

Note: Requirements in italics occurred in all sessions.

nonfunctional requirement. Since they finished earlier than the other two teams, they might have glazed over nonfunctional requirements as well as some functional requirements as stated in the following text.

The Adobe session addressed all the requirements of the design problem. The AmberPoint session failed to address the *t_time* requirement; however, they did address the related requirement *t_speed* (seventh highly discussed requirement). The Intuit session did not address the *no_crash*, *vis_lights*, and *observe* requirements. The *vis_lights* requirement states that the state of the lights should be visually conveyed to the user. The observation (*observe*) requirement deals with the students observing problems with traffic patterns and timing schemes. Since the Intuit team did not touch the UI aspect in detail, they missed these requirements. The requirements distribution across time can be compared side by side (or overlaid) with the activity distribution across time, to determine the activities involved to satisfy a given requirement. We discuss some instances of this next.

In the Adobe session, the first 10 minutes of the session was spent on the *intersect* requirement, which involved a lot of consultation between the designers as well as the identification of domain objects. The *intersect* requirement was revisited after a period of time, but then the designers described the objects and their properties in greater detail and even started sketching out a preliminary UML class diagram. We also see that the *left-hand_turn* requirement was interleaved with a discussion of the logic involved in its implementation.

In the AmberPoint session, the *intersect* requirement was talked about intensively in the second part of their discussion for ~20 minutes, which also overlapped with drawing an ER diagram. The first part of the session dealt with traffic light timing schemes (*set_light_timing*), which maps to the nonverbal activities of talking or drawing the UI (*nv_UI_draw* and *v_UI*) and discussing use cases and code logic. During the middle of the session, they discussed the altering of the simulation parameters (*alter_update*) for ~20 minutes, which maps nicely to analyzing the whiteboard, defining use cases, and discussing code logic. It was also interesting to note that they came up with new requirements (building multiple simulations and measuring traffic complexity) during the analysis of the *alter_update* requirement.

In contrast, the Intuit session's talk about the *intersect* requirement spread throughout the session for a total of ~20 minutes. This requirement maps to creating use cases, a domain model, and a class diagram; analyzing and drawing pictures on the whiteboard; and code logic. In the beginning, they discussed updating the current state of the lights (*set_lights*) simultaneously with the *intersect* requirement. These requirements map to creating use cases and the domain model. Finally, they talked about altering the simulation parameters (*alter_update*) for ~10 minutes at the end of the session, which maps to creating a class diagram and discussing code logic. It is important to note that even though each team worked in a different way, there was some commonality in the requirements met and the activities discussed, as presented in the previous sections.

10.3.3 Interpersonal Communication

We also examine the interactions between the designers (RQ4), and their effects on the resulting design. Again, we derive this information from the time line matrices by examining the amount of time spent consulting, planning, questioning, or answering each other. These are shown as the last five rows in the activity time series (Figures 10.5 through 10.7).

The Adobe session (see Figure 10.5) contained the most agreement between the two designers. It also had consulting activity scattered almost uniformly. They engaged in a lot of planning in the last three intervals and also justified their choices. The AmberPoint team also consulted throughout their session with a lesser level of agreement between the designers. They did much more planning almost at the very end of the session with justification provided for each decision made. The Intuit team also consulted throughout their session and they were consistently in agreement. However, there were few justifications made for their decisions and little planning involved. There were no silent episodes in the Adobe or AmberPoint sessions, while the Intuit session had periods of silence spread throughout. To conclude, we can hypothesize that planning, as well as the high degree of agreement between the two designers of the Adobe session, played a significant role in delivering the most detailed design that covered the most requirements.

10.3.4 Similarities and Differences in Design

We will briefly point out the similarities and differences between the approaches used between the design teams (RQ5). The comparison of the sessions is based on the two main deliverables asked for in the design problem: (1) the design interaction/UI and (2) the basic structure of the code.

The Adobe team took a very structured approach to the problem. This can be seen from the density of the activity graphs. They systematically planned what their next moves would be and followed through. They designed both the interaction and the code structures. There was a lot of consulting and agreement between the designers. The AmberPoint team focused on the UI and did produce the deliverable of "an interaction scheme for students." However, they did not spend any time on the structure of the code or even identify associations between the main objects in the system. The Intuit team went through the session very quickly. Their design activity time line matrix is much less dense compared to the other two teams. While they did not design a UI or outline a code structure, they did discuss the code logic (*code_logic*).

The problem that the designers were given was to complete enough of a design such that a student who had recently graduated could implement it. The results of the sessions for teams Intuit and AmberPoint did not achieve this. Conventional wisdom leads us to believe that a software developer would be likely to produce a correct and conforming implementation based on the design provided by the Adobe team because it gives the most detail and covers the most requirements.

10.4 OBSERVATIONS

The three teams took very different approaches to problem solving. Based on the time line matrices, we observed clusters that clearly correspond to the design process stages, that is, requirements, business modeling, design, and implementation; which are shown in Figures 10.6 and 10.7. Some of the observations based on the foregoing analysis are as follows.

- The requirements and use cases were spread throughout the session.
- The data dictionary and the domain model were done at the beginning for the Adobe and Intuit teams and were scattered throughout for the AmberPoint team.
- The class diagram/design model was discussed next by the Adobe and Intuit teams while the AmberPoint team focused on the ER diagram.
- The discussion about the UI was done mostly in the second part of the design sessions for the Adobe and Intuit teams, but it was scattered throughout the AmberPoint session.
- Code logic/implementation was also scattered throughout the session for all teams.

Overall among the three teams, the most time was spent on the requirements: *intersect*, *alter_update*, *map_roads_intersect*, and *left-hand_turn*. An average amount of time was spent on *sensor_option* and *sensor_light*, and the least amount of time was spent on *light_seq*.

Without a gold standard for the design, it is not possible to state which was the overall best or worst design. However, we can make some general points. Two teams (Adobe and Intuit) did a better job at the code logic and system design, while the third team (AmberPoint) did a better job describing and outlining the UI. Curiously, only one team (AmberPoint) discussed the aspects of verification or validation with regard to their solution. We also observed that only two teams (Adobe and Intuit) identified the fact that they had to deal with queues in a traffic simulation problem. We consider the queue as an important part of the design of this system. AmberPoint did not mention this at all, possibly due to a lack of prior experience with simulation systems. In other words, the design teams with different experiences and backgrounds focused on different aspects of the design.

One open question is the relationship between the problem statement and the design process. Did the object-oriented design process help or hinder the final design? Also, did the domain affect the design? Note, the teams were asked to come up with a high-level design, they were not specifically asked to come up with an object-oriented design.

10.5 CONCLUSIONS AND FUTURE WORK

We presented an analysis of three design videos and their associated transcripts. Analyzing video recordings is a time-consuming process and requires an organized approach. As such, we needed to choose a guiding principle for our investigation and develop tools to alleviate some of the manual drudgery. In order to support the analysis, we developed an XML format for embedding dialogs and dialog metadata and we developed an annotation tool to assist in the manual annotation of speech with metadata. We also presented a means to visually compare the data with a time line matrix that was used to analyze and draw inferences about transcribed and videotaped conversations.

We posed a number of research questions on how developers actually design and we addressed them individually. We conducted a fine-grained analysis of the requirements as well as the design activities engaged in by the designers. Another contribution was the generation of various codes for both design and requirements (albeit problem specific). The design codes, however, span across various domains and may be reused.

As future work, we plan to further develop the visualizations, and create an interactive time line exploration matrix that supports drill-down capabilities and overlapping event structures. We also plan to continue developing this method of dialog analysis and the tools used to support it. We believe that a finer-grained analysis of these interactions is both possible and valuable. We are also interested in using these analysis methods to evaluate the effectiveness of certain problem-solving paradigms. This could provide a method of empirically evaluating problem-solving strategies or mental models (Gentner and Stevens 1983) of software engineering and program comprehension.

REFERENCES

Artstein, R. and M. Poesio (2008). Inter-coder agreement for computational linguistics. *Computational Linguistics* **34**(4): 555–596.

Bryant, A. and K. Charmaz, Eds (2007). *Grounded Theory*. Sage, Los Angeles, CA.

Curtis, B., H. Krasner, and N. Iscoe (1988). A field study of the software design process for large systems. *Communications of the ACM* **31**(11): 1268–1287.

d'Astous, P. and P. N. Robillard (2001). Quantitative measurements of the influence of participant roles during peer review meetings. *Empirical Software Engineering* **6**: 143–159.

Detienne, F. and F. Bott (2001). *Software Design: Cognitive Aspects*. Springer, London.

Friess, E. (2007). Decision-making strategies in design meetings. In *CHI '07 Extended Abstracts on Human Factors in Computing Systems*. ACM Press, New York, pp. 1645–1648.

Gee, J. P. (2005). *An Introduction to Discourse Analysis*. Routledge, New York.

Gentner, D. and A. L. Stevens, Eds (1983). *Mental Models*. Erlbaum, Hillsdale, NJ.

Guindon, R., H. Krasner, and B. Curtis (1987). Breakdowns and processes during the early activities of software design by professionals. In G. M. Olson, S. Sheppard, and E. Soloway (Eds), *Empirical Studies of Programmers: Second Workshop*. Ablex, Norwood, NJ, pp. 65–82.

Ko, A. J., B. A. Myers, M. J. Coblenz, and H. H. Aung (2006). An exploratory study of how developers seek, relate, and collect relevant information during software maintenance tasks. *IEEE Transactions on Software Engineering* **32**(12): 971–987.

Maia, A. C. P., C. J. P. Lucena, and A. C. B. Garcia (1995). A method for analyzing team design activity. In *Proceedings of the 1st Conference on Designing Interactive Systems: Processes, Practices, Methods, & Techniques*. ACM Press, Ann Arbor, MI, pp. 149–156.

Robillard, P. N., P. d'Astous, F. Détienne, and W. Visser (1998). Measuring cognitive activities in software engineering. In *Proceedings of the 20th International Conference on Software Engineering*, Kyoto, Japan. IEEE Computer Society, Washington, DC, pp. 292–299.

University of California, Irvine (2010). Studying professional software design. An NSF-Sponsored International Workshop. University of California, Irvine, CA.

CHAPTER 11

Ideas, Subjects, and Cycles as Lenses for Understanding the Software Design Process*

Alex Baker and André van der Hoek

University of California, Irvine

CONTENTS

11.1 INTRODUCTION

As designers work to understand a problem and solve it, they generate many ideas. As the design process progresses, these ideas are weighed and compared to one another, and some notion of good and bad ideas is developed. Eventually some ideas are used to compose a

* Reprinted from *Design Studies* 31 (6), Baker, A. and van der Hoek, A., Ideas, subjects, and cycles as lenses for understanding the software design process, 590–613, Copyright 2010, with permission from Elsevier.

final design (Jones, 1970). We are interested in understanding how this progression takes place in software design. What strategies and patterns govern the kind of ideas that designers generate, how are those ideas evaluated, and what process is used to arrive at the set of accepted ideas?

To build up our understanding of how designers address *software* design problems, we have studied the sessions recorded for the Studying Professional Software Design 2010 Workshop. This workshop was based on video recordings of three pairs of professional software designers, each of which was asked to solve the same provided design problem. The designers were given 1 hour and 50 minutes to design the basic code structure and user interface for a traffic simulation program, and were provided with a whiteboard and pens, but no other tools. A more complete description of the design task being studied can be found in the introduction to this volume.

As a first step, we analyzed the three videos and tracked the individual *ideas* that the designers generated over the course of each session, recording the times when ideas were built upon one another. Then, in order to gain a higher-level understanding of these sessions, we categorized the ideas that the designers generated according to their *subject* and we divided the sessions into clusters of ideas called *cycles.*

With these notions in hand, we observed that the sessions were characterized by the repeated discussion and recontextualization of a set of core ideas and subjects. Specifically, as the designs evolved, previously stated ideas were frequently restated and reconsidered. On a higher level, the designers tended to consider two subjects at a time and over the course of the sessions, they considered the same small number of subjects many times, in different pairings. The result was a process that had characteristics of breadth-first design and incremental design, but that delved into details differently; we found that there was a focus on finding key, organizing ideas and adjusting them to work relative to one another to create a cohesive whole.

This work follows in the tradition of researchers, such as Goldschmidt (1990) and Ennis and Gyeszly (1991), who looked closely at the problem-solving process used by designers in nonsoftware design fields. It also builds on and expands on the comparatively small number of studies of software designers in action, such as Guindon (1990) and Zannier (2007). Its primary contributions are its description of the designers' two-subject discussions and its explanation of the design process as containing frequent repetition and recontextualization of ideas.

11.2 APPROACH

In order to understand the *ideas* that the designers generated, it was necessary to develop two higher-level concepts to use in organizing and categorizing them. Firstly, we observe that each of the six designers structured their thinking about the design problem in terms of a number of distinct *subjects* of discussion. This concept of subjects provides us with a way to categorize the designers' ideas by their content. Secondly, the designers' progress was not steady; they set out to discuss a given aspect of the problem, made progress, and then when that progress dwindled, they picked up a new aspect of the problem to focus on. We describe each such sequence as a design *cycle* and this concept provides us with a way to categorize the ideas according to their temporal location in the design process. In this section, we will describe these notions of ideas, subjects, and design cycles, as well as the techniques we used to define and analyze them in the sessions.

11.2.1 Ideas

The initial step in analyzing the corpus was to move through the sessions and record each occasion where an idea was generated. An idea, in this case, refers to a statement by one of the designers that (a) characterized the provided problem or (b) provided some suggestion about how to solve part of the problem. For example, a statement about what real-world phenomena might be modeled describes the design problem, while a suggestion about a feature to be added provides a suggestion about how to solve the problem; each represents progress toward developing a solution and is encapsulated as an idea. The sessions' ideas were tracked at a roughly per-sentence level of granularity and we tracked only overt statements of ideas, rather than second-order implications. All of the ideas from a given session were then used to create a large graph, organized by time, in which each node was a single idea, along with a short description and a unique identifying code. Examples of these idea codes and short descriptions include:

Intersection-M8: Only some combinations of lights in an intersection are valid.

Intersection-M8a: Two left turns is a valid combination of lights.

Intersection-M8b: All the "opposing" lanes could have the same state.

Intersection-M8c: One possible combination is a straight and a left, with the other side red.

Intersection-M9: Intersections feed into one another via shared roads.

Intersection-M9a: The flow of cars from an intersection might be blocked if a road is "full."

The second part of this process was to establish relationships between different ideas. To facilitate this process, the ideas' codes were organized hierarchically, when possible, as illustrated in the examples above. One relationship that we established was *repetition*: whenever one idea was a rephrasing or reiteration of a previous one, this relationship was established between their nodes. The other important relationship was *needs*: whenever one idea provided a context for a later idea, or otherwise established the intellectual progress needed for a later idea, this relationship was established between them. For example, in the list above, idea Intersection-M8's notion of valid combinations of lights was *needed* in order for Intersection-M8a's concept of "left turns is a valid combination" to be established.

Figure 11.1 presents three computer-generated visualizations, each of which depicts the ideas generated during one of the three design sessions. In each row in the image, the ideas from a given session are represented by small rectangular nodes. These nodes are arrayed from left to right, according to the time that they occurred in the session (the small hash marks below each row correspond to 10-minute increments of time). Finally, the *needs* and *repetition* relationships between ideas are depicted as arcs connecting the corresponding

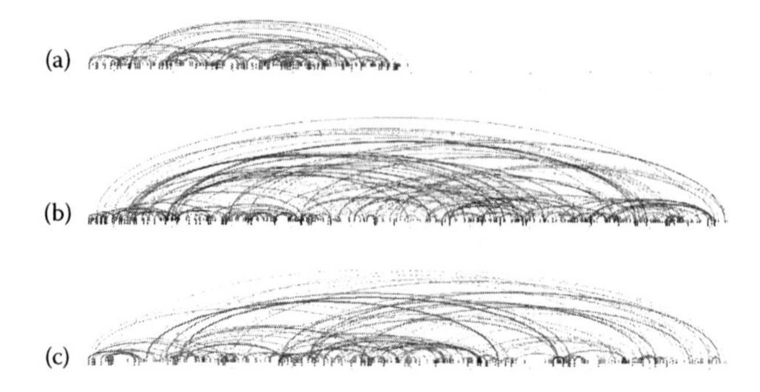

FIGURE 11.1 Visualization of all of the ideas generated by the Intuit (a), AmberPoint (b), and Adobe (c) sessions.

nodes. A full-color version of this image can be found in Baker (2010), available online at http://sdcl.ics.uci.edu/downloads/abaker.pdf (see Figure 6.1, p. 130). In the online version, the nodes and arcs are color-coded according to the subject that the designers are discussing, and noticing clusters of ideas and tracking arcs between them is greatly facilitated. The reader is strongly encouraged to examine the color version; the version presented here serves primarily to illustrate the number of ideas identified, the many relationships between them, and the relatively persistent rate at which they were generated by the designers. Also, by depicting all of the ideas, this image also provides a point of comparison for the subsets of ideas shown later in the paper in Figures 11.2 and 11.5.

Conceptually, this concept of an *idea* is important because it represents the bottom level of granularity at which we consider these design sessions, and our other ways of describing these sessions are built up based on patterns of ideas. In particular, the subjects of these ideas, as well as the ways that ideas that related to each other clustered together, lead us to the observations in the following sections.

11.2.2 Subjects

When considering the sessions, it becomes evident that the designers have an internalized notion of the subjects that they need to address. Often, after a pause, they would make statements to establish a topic of conversation; for example: "So it seems like our real thing is ... now are roads a significant thing here?" (Intuit 0:18:36), "Okay, students and, and ... each intersection, with or without an option to have sensors that detect ... so that's another thing" (AmberPoint 0:12:17), or "Okay. Maybe we should start talking about the UI" (Adobe 1:03:48). Sometimes this would simply take the form of flatly naming an aspect of the system, such as: "Intersections." (Intuit 0:09:28), "Ok, then. Cars." (AmberPoint 1:18:24), or "So, controller." (Adobe 0:09:57), in an effort to nudge the design in that direction.

In order to better classify the designers' ideas, we identified the primary subjects used by the designers in each of these sessions, according to two main criteria:

1. We collected subjects that the designers themselves identified by name. This would often be done as a way to suggest a direction of discussion, as in the examples above,

though sometimes decisions or requirements would be described by the designers as belonging to a particular subject.

2. We further categorized subjects as pertaining to modeling or user interface (UI) issues. Designers would occasionally declare they wanted to talk about how the user would interact with the system or how it would be modeled, but often this distinction was implicit, because statements were obviously directed at one aspect or the other based on their context. Therefore, we sometimes created divisions along these lines, even in cases where no overt declarations were made by the designers.

Furthermore, a broad effort was made to find a set of subjects that could be used to describe all of the designers' discussions, but within which no subject addressed a subpart of another subject; the subject list was overtly designed to be unhierarchical, with only top-level categories included. For example, this meant not including *sensors* or *car creation* as subjects, but rather using the *intersections* and *traffic* as subjects, respectively, to cover such discussions. As a result of this high-level subject approach, there was considerable overlap between the sessions' lists of subjects, but in a few cases one team focused on a part of the design that other teams chose not to. The subjects identified for each session are summarized in Table 11.1.

We utilized these subjects in two ways: each idea that we identified was classified as belonging to exactly one of the identified subjects and these subjects were used to partition the design sessions into segments called *cycles*.

11.2.3 Cycles

All three pairs of designers generated numerous ideas throughout their design process. However, their progress was not altogether steady: they would often generate a flurry of ideas, but then progress would dwindle, before picking up again.

For example, we devised an algorithm whereby, for a given set of ideas, those ideas that were not *needed* by any other ideas were removed. By applying this filter repeatedly, a core set of more-influential ideas were derived. Figure 11.2 shows the ideas remaining in each of the three sessions after applying the childless-removal rule four times. (Once again, a full-color version is available in Baker [2010] at http://sdcl.ics.uci.edu/downloads/abaker.pdf;

TABLE 11.1 Subjects Identified for Each of the Three Design Sessions

Intuit	AmberPoint	Adobe
Modeling intersections	Modeling intersections	Modeling intersections
Modeling traffic	Modeling traffic	Modeling traffic
Modeling map	Modeling map	Modeling map
Modeling roads	Modeling roads	Modeling roads
Intersection UI issues	Modeling feedback	Modeling feedback
Traffic UI issues	Intersection UI issues	Modeling controller
Map UI issues	Traffic UI issues	Intersection UI issues
	Map UI issues	Traffic UI issues
	Road UI issues	Map UI issues
	Feedback UI issues	Feedback UI issues

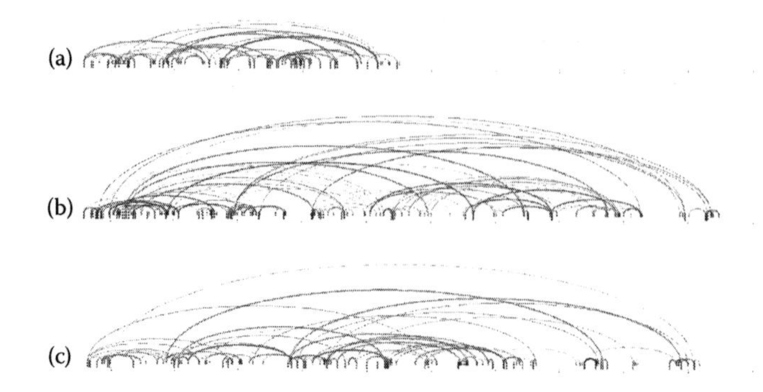

FIGURE 11.2 Visualization of the key ideas generated by the Intuit (a), AmberPoint (b), and Adobe (c) sessions.

see Figure 6.3, p. 134.) Looking at this diagram, it is clear that these central ideas are not created uniformly throughout, but rather there are bursts of influential ideas, followed by lulls, lending an ebb and flow to the design process.

When we investigated these periods of lull suggested by the diagram more closely, we found that they very often corresponded to moments of focus setting in the design sessions. The designers, of course, could not keep the entire design in mind at once. So at various points they would describe an aspect of the problem that they wanted to consider (often in terms of one or more subjects, as discussed in the previous session) and designed within those boundaries for a time. When ideas within that framework began to dwindle, they stopped working on the design itself for a moment and established a new aspect of the design to focus on. These focus-setting moments happened frequently, throughout each of the design sessions, and generally bookended periods of effective progress. We are interested in understanding what the designers focused on and how these design subsessions combined to create progress on the design at hand.

To this end, we have devised the concept of a *cycle*. A cycle is a period of time that begins with a moment of focus setting by the designers and lasts until the next focus-setting moment in that session. In this way, each moment in a session belongs to exactly one cycle. Unfortunately, finding the dividing lines between cycles is not a trivial matter; sometimes the designers discuss what they should focus on, without actually changing focus. Other times, their focus shifts without an overt declaration of this occurrence. By repeatedly considering the sessions, particularly investigating the moments suggested by our visualizations, we arrived at the following five indicators that collectively suggest when a new cycle has begun:

1. *Strategizing*: When the designers openly discussed what to do next, this suggested that their progress on the problem at hand was dwindling and that they were searching for a new subject to discuss.

2. *Abrupt subject raising*: When a subject was raised without an obvious connection to the previous statements, this often represented an implicit declaration of what the designer wanted to focus on next.

3. *Question asking*: When designers asked questions, this often indicated uncertainty and a break in the creative process. This, in turn, often suggested that a new point of focus was forthcoming.

4. *Consecutive non-idea statements*: The designers frequently made statements not directly aimed at solving the problem at hand. When both designers did this consecutively, it often indicated stalling and an overall state of being stuck, which tended to indicate that a new focus was needed.

5. *Long pauses*: Finally, when the designers remained silent for a number of seconds, this indicated a break in the creative flow and often resulted in a new design direction. A break of 4 seconds was considered the minimum threshold in our analysis.

None of these indicators is an ineffable indicator in its own right. For example, questions were asked within the flow of a session and pauses occurred for a variety of reasons. Sometimes abrupt subject changes did not cause a change of focus, but simply acted as non sequiturs, or were explicitly dismissed as a new area of focus (e.g., see the 0:17:32 point in the Adobe session). However, we found that in cases where several of these factors occurred in a small space, a shift in focus was strongly suggested. We identified places where at least three of these five factors occurred within a single designer's statement, or within the span of two changes of speaker (e.g., if Male 1 spoke, then Male 2, and then Male 1 once again, this represents the maximum span we considered). This approach was successful in identifying 46 cycle breaks, resulting in 49 cycles among the three design sessions.

Before moving to our observations, we will make a final note about our methods. The coding of ideas, as well as the breakdown into cycles, was performed by a single researcher. In the case of the breakdown, this took place over the course of three iterations. Each session was read through, and the major subjects were identified. Then, a pass was made to identify all of the ideas, categorize them according to subject, and relate them to one another, during which the subject list was adjusted as the need emerged. Finally, as the idea data were formatted for use in our automated tools, an additional "sanity check" of the ideas and their relationships was performed. The consistency of the subjects identified across these three sessions provides some degree of confidence in the subjects that we selected.

11.3 OBSERVATIONS

Having established ideas, subjects, and cycles as lenses for considering the data in this corpus, we have made a number of observations about the creative process that governs the three design sessions. In the following four sections, we discuss the following topics: the subjects discussed within design cycles, the tendency to pair subjects for consideration, the tendency for subjects to develop evenly, and the overall shape of the design process.

11.3.1 Subjects of Design Cycles

Once we established an initial set of boundaries that defined design cycles, we turned to the issue of what happened within those cycles: What subjects were discussed in each cycle and in what configurations? Most cycles contained ideas that related to a number of subjects, but we noticed that there were often one or two subjects that were especially prevalent. Specifically, three main categories of cycles emerged:

- *Single-subject cycles*: These are instances where the designers talk about a single subject in depth. Other subjects may be raised, but these mentions are short and are merely used to illustrate something about the primary subject.
- *Paired-subject cycles*: These are instances where one subject is raised, and another is considered alongside it. Both aspects are considered in depth, both inform one another, and they develop in tandem.
- *Low-depth cycles*: These are instances where the designers are talking about a number of subjects, without engaging with any one subject in any particular depth. In these cycles, three or more subjects are considered, and none is discussed for more than a sentence or two at a time.

We examined the 49 cycles closely in terms of the subjects they contained and found that 41 of them fell into one of these three categories, while 8 did not. However, on further investigation of those eight cycles, we found in each a point where the focus shifted in a manner that our previous criteria had failed to detect. Each of these eight additional cycle breaks was characterized by two criteria:

1. Each began with an idea that did not need or challenge any recently raised ideas in the session.
2. The ideas that immediately followed that idea tended to refer back to it and to one another, and tended not to refer back to the ideas that preceded the cycle break.

In short, each such cycle break was not preceded by sufficient uncertainty to be detected by our previous criteria, but clearly represented a new train of thought for the designers. Five of these supplementary breaks were found in the Adobe session, and that team's fluid approach to designing may have served to hide these breaks from our initial consideration. After further dividing the sessions according to these additional break points, all of the resulting cycles fell readily into one of the categories above. A final exception to this categorization is in Cycle 22 in the Adobe session, which involved the tightly integrated discussion of three different subjects; this occasion marked the only three-subject cycle in the three design sessions.

Table 11.2 lists the start time of each design cycle and indicates which subjects were discussed in depth during that cycle, if any ("M" indicates a modeling subject, "UI" a user interface subject).

Figure 11.3 graphs the three sessions according to the type of cycle, with time depicted along the *x* axis. The hash marks along the bottom of each grid represent a 10-minute span.

TABLE 11.2 Details of Cycles Identified in Each of the Three Sessions

	Adobe		AmberPoint		Intuit	
01	0:05:07	Low-depth	0:05:45	Low-depth	0:05:36	Intersection-M
02	0:05:54	Intersection-M	0:07:10	Low-depth	0:09:22	Intersection-M Roads-M
03	0:09:37	Controller-M Intersection-M	0:08:53	Low-depth	0:15:30	Intersection-M
04	0:23:33	Traffic-M Controller-M	0:11:03	Intersection-M	0:17:11	Traffic-M
05	0:26:59	Intersection-M Traffic-M	0:12:10	Low-depth	0:18:36	Roads-M
06	0:33:32	Controller-M Intersection-M	0:15:11	Map-UI Intersection-UI	0:19:50	Roads-M Intersection-M
07	0:40:25	Roads-M Traffic-M	0:22:46	Traffic-M Map-M	0:27:05	Traffic-M Roads-M
08	0:42:23	Map-M Intersection-M	0:25:09	Intersection-UI Intersection-M	0:29:07	Roads-M Intersection-M
09	0:46:32	Controller-M Intersection-M	0:41:13	Traffic-M Traffic-UI	0:34:21	Traffic-M Roads-M
10	0:51:20	Roads-M	0:48:47	Low-depth	0:38:10	Traffic-M Roads-M
11	0:52:21	Intersection-M	0:52:38	Feedback-UI Feedback-M	0:43:52	Intersection-M Roads-M
12	0:56:51	Traffic-M	1:02:27	Traffic-M Traffic-UI	0:50:22	Intersection-M Roads-M
13	0:58:24	Controller-M	1:05:54	Map-UI Map-M	0:56:59	Low-depth
14	1:03:48	Map-UI Map-M	1:12:26	Intersection-UI Intersection-M		
15	1:11:06	Low-depth	1:18:13	Map-UI Traffic-M		
16	1:14:19	Traffic-M Traffic-UI	1:20:06	Traffic-UI Traffic-M		
17	1:17:56	Low-depth	1:23:54	Low-depth		
18	1:22:19	Feedback-UI	1:27:55	Feedback-UI		
19	1:26:00	Roads-M Traffic-M	1:36:43	Low-depth		
20	1:29:54	Feedback-M	1:47:01	Low-depth		
21	1:35:05	Low-depth				
22	1:39:54	Map-M Roads-M Intersection-M				
23	1:46:50	Intersection-M				
24	1:50:21	Low-depth				

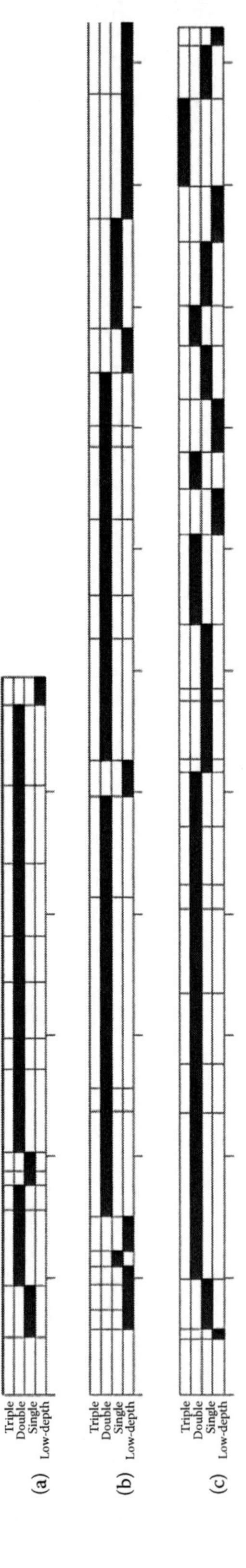

FIGURE 11.3 The types of cycles in the Intuit (a), AmberPoint (b), and Adobe (c) sessions.

Some of the patterns illustrated by these data are unsurprising. For example, the presence of low-depth cycles meshes with characterizations of software designers as seeking "balanced development" of their systems (Adelson and Soloway, 1984). These cycles often occurred at the beginning of the process, and on further investigation, they generally served to provide the designers with a sense of the solution space (Simon, 1969) before them. They were also found at the end of some of the sessions, as the designers made an effort to organize and coalesce their ideas in a unified way.

Similarly, the presence of single-subject cycles is to be expected; these parts of the design process are in line with previous observations about selective in-depth design, such as those by Ennis and Gyelzly (2005) and others, as surveyed by Stauffer and Ullman (1998). These cycles also appear primarily in the early parts of the sessions, as the designers establish the basic nature of the components that they will be working with. The major exception is the Adobe designers, who realize late in the session (1:23:55) that they need to focus on providing feedback to the user and thereafter dedicate two single-subject cycles to the subject. The two-subject cycles, though, require much more investigation, and they are the focus of the following section.

11.3.2 Paired-Subject Cycles

We arrive at this section with questions about two-subject cycles: Why were there so many of them, and what role did they serve in the design process? To address these questions, we must first consider how these two subjects relate to one another in these situations.

A common scenario was that one of the designers would declare a particular subject as a candidate for consideration, and as the discussion proceeded, a second subject would quickly be brought into the conversation as well. In these situations, the two subjects were regarded with roughly equal flexibility. It was not the case that the designers would consider one subject settled and fit the other to meet its needs; rather both halves were treated as malleable, and a fit suiting both aspects of the design was found. This can be illustrated through a series of examples from the design sessions.

Example 11.1: Adobe, Cycle 19 (1:26:44)

As this section begins, the designers have realized they want their roads to be of varying length. In order for this length attribute to be meaningful, longer roads would need to take longer to traverse, necessitating a concept of *speed* in their traffic model. This in turn demands that a car be able to be partially down a road and to make progress over time, according to its speed. The designers encounter a problem though; they have previously decided to model each road as a *queue*, which is a data structure that simply stores items in an ordered list. A program that conceptualized each road as a list of cars would not know *where* along the road a given car was, it could only know what order the cars had arrived in. At 1:27:26, one designer therefore evolves the road's model to contain a series of "spaces," which the cars would move down over time. This then causes the notion of traffic to evolve once again; the designers suggest that each space could contain only one car and describe the order in which cars move and the ways they might impede one another's progress.

Example 11.2: Intuit, Cycle 6 (0:20:21)

At this point in the session, the designers have discarded the idea of roads having lanes (0:19:33), but they have also sensed that they want to model intersections as optionally having left-hand turn lanes and are struggling to couple these parts. At first, a change to the intersections is suggested, whereby the intersection would always have exactly two lanes, even if they were both straight-through lanes, as a way to standardize this interaction (0:21:35). They then further simplify by changing the road concept to always having exactly two lanes, each of which would feed into a single intersection lane (0:24:21). They seem unsatisfied with this though, noting that the left-turn issue introduces a lot of problems (0:26:19). Eventually they realize that they simply need to track how intersections and roads connect to one another, and that a lane concept is not needed at all.

Example 11.3: AmberPoint, Cycle 9 (1:41:23)

In this section, the designers are considering how the user will control traffic flows, a subject that is tightly coupled with the underlying traffic modeling issues. They mention that the user should be able to specify cars' origins and destinations (a UI issue), leading to a comment about the cars only entering the map from the edges (modeling), suggesting that the user set the rate of cars on a per-input basis (UI). Then the female designer reiterates that the cars should have a destination (modeling, 0:42:04), prompting the male designer to suggest that each intersection could alter how cars move through it (modeling), causing the female designer to counter with a usage scenario about creating types of cars from a palette (UI). Undaunted, the male designer describes how the intersections could handle car flows (modeling) and the female designer counters that that would not allow for sufficient user control (UI). Eventually (0:44:23), this disagreement reveals an underlying misunderstanding between the designers about how traffic is being conceptualized (modeling). In the interest of saving space, we will discontinue this description here, but the discussion flips back and forth between considerations of the users' needs and the underlying models several more times hereafter.

In each case, the two issues at hand are discussed in a tightly coupled manner. It is clear that however one part is designed, it will have an effect on the other, and the designers aim to ensure that each part is designed as elegantly while working with the other part under discussion. When we consider the 28 paired-subject cycles in these sessions, we note that the relationships between the two subjects fall into three main categories:

1. *Containment*: When one part of the system serves primarily as a container and organizer of other elements, the design of that container depends on the design of the object it is containing. Example 11.1 illustrates this relationship. This category arose most commonly in road/traffic pairings and when the map was being discussed relative to intersections or roads.

2. *Communication*: When two aspects of the system need to pass information between each other, what is the protocol at hand? What interfaces does each part present to the other? Example 11.2 illustrates this relationship. This often came up in intersection/road discussions, where the designers needed to decide how cars would be tracked.

3. *Control*: When one aspect of the system needs to manipulate another, it must be determined what the target is capable of before the controller can know what to do with it. Meanwhile, the controlee's capabilities must be designed relative to what is needed. Example 11.3 illustrates this relationship. This category arose in the case of the Adobe team's controller class and in situations where the user interface of an element was being discussed alongside the way it would be modeled.

In each case, neither element can be designed in the absence of the other, and these kinds of interconnected relationships create a complex situation. For example, when deciding how to model roads, the designers might find it necessary to consider the following:

1. How roads are designed to best work with their model of traffic (containment)
2. How roads pass control of cars to intersections (communication)
3. How the map arranges roads into larger structures (containment)
4. How the controller moves cars down the road (control)
5. How the user affects the roads' attributes (control)

As Guindon (1987) found, mentally simulating the relationships between subparts of a solution is difficult, and we reason that it would be exceptionally difficult to keep all of the issues in mind at once. On the other hand, considering the road in depth, in isolation, would likely create demands that would bend the design of the other parts of the system in undesirable ways; after all, the intersections, map, and traffic all have their own pressures to accommodate. The paired-subject cycles we observed seem to represent a compromise between these pressures; perhaps this is a strategy subconsciously adopted by the designers. By considering two subjects at a time, the designer can consider aspects of the design problem in depth, without becoming overwhelmed by the problem as a whole.

11.3.3 Even Development

The next question of interest concerned the distribution of the subjects over the cycles. Given that the designers identified as many as nine different subjects to focus on in a given session, but were only able to focus on two at a time, how did they proceed? Our primary observation on this matter is that:

> The designers tend to develop all of the identified subjects in their design in parallel and avoid allowing any part of the design to "fall behind."

This phenomenon is illustrated in Figure 11.4. Here, each of the three timelines represents one of the three design sessions, with height correlating with the subject under consideration. The blocks in the bottom half of each row represent the discussion of modeling-oriented subjects, while blocks in the top half of each row designate discussion of user interface issues. The first subject mentioned in each cycle is depicted in black, while the other subjects in a cycle are depicted in dark gray. Light gray areas designate low-depth cycles, and vertical lines are used to divide cycles with identical subject designations. As in Figure 11.3, the small hash marks at the bottom indicate 10-minute increments.

Looking at the timelines above, we can see that the movement between the subjects is fluid, and rarely is a given subject left unconsidered for long. We will talk about the ways that each team exhibited this pattern in more detail here. Firstly, the Intuit designers did not consider issues such as user interaction and an overt controller mechanism. Rather, they focused on the three key parts of the system: intersections, roads, and traffic. Early in the process, they discussed each of these subjects on their own to establish the basic concepts underlying each of them. Then they moved fluidly between these three subjects, primarily focusing on the communication between roads and intersections and the containment relationship between roads and traffic. Note that once established, none of these three subjects went without discussion for more than two cycles.

The AmberPoint session, meanwhile, focused much more on user interface issues. In particular, the designers tended to focus on one aspect of the system (intersections, roads, etc.) and then aimed to find a good match between creating a cohesive underlying model and ensuring that the necessary features were available to the user. The interplay between these two goals was demonstrated in Example 11.3 in the previous section, but there are several other examples of particular interest:

- Cycle 8, where issues such as how sensors should work, whether intersection states are mirrored versions of each other, and how timing issues should be expressed, are considered in terms of how they will affect the user experience
- Cycle 11, where the experience of learning about the effectiveness of a configuration is intertwined with the underlying metrics
- Cycle 13, where they discuss how the map will be created, which is intertwined with discussion of how the map is logically constrained and how streets will be named

Compared to the Intuit session, the AmberPoint designers spent little time relating different aspects of the system to each other, instead moving rapidly back and forth between modeling and user interface issues (the main exceptions are Cycles 7 and 15). However, on

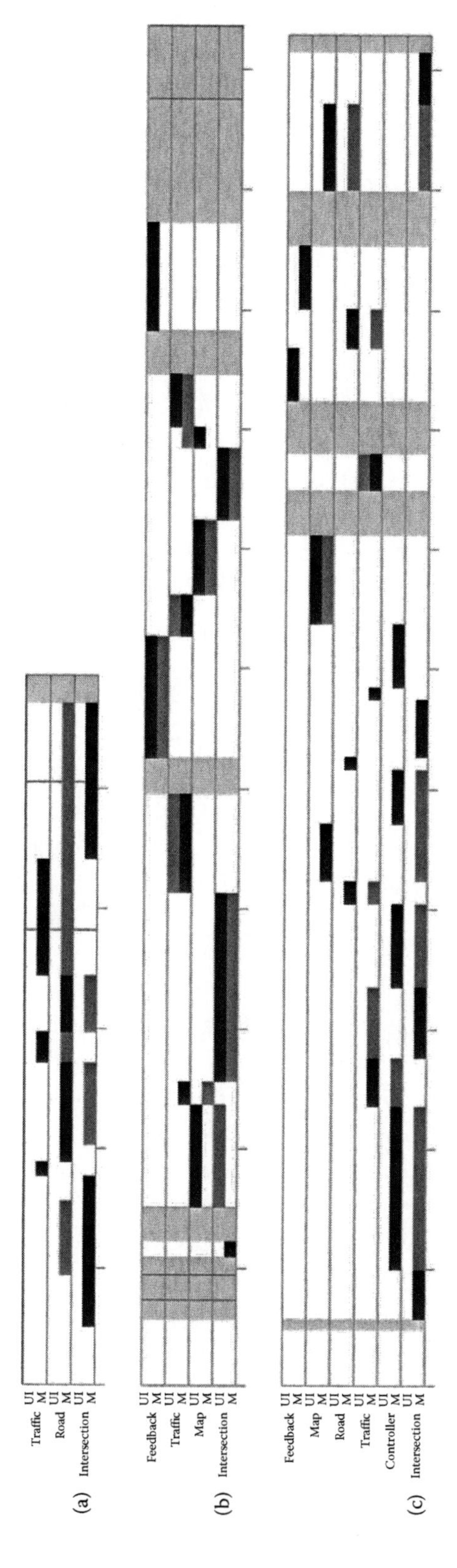

FIGURE 11.4 The subjects of each cycle in the Intuit (a), AmberPoint (b), and Adobe (c) sessions.

a higher level, they made an effort to readdress the intersection, traffic, map, and feedback aspects of the system, returning to each in turn, as is illustrated in the smooth, up-and-down curve of subjects in Figure 11.4.

Finally, the Adobe designers were the most thorough of the three teams, in terms of considering a breadth of subjects that might be important to the design. The session begins much like the Intuit session. The designers spend the first five cycles working through various combinations of three subjects, in this case intersections, traffic, and the controller. They then achieve a "creative leap" (Cross, 2007) at 0:37:48, when they realize that intersections cannot be considered in isolation from one another, and concepts of roads and an overarching map emerge. All three of these subjects are raised multiple times later in the session, and they are considered together in the session's penultimate, three-subject cycle.

Another example of even development in this session bears mention. In Cycles 10–13, the designers use references to the requirements to spur short forays into single-subject discussions; in this case roads, intersections, traffic, and the controller are each mentioned, providing excellent coverage of the subjects currently under consideration. In fact, the only subject unmentioned in this single-subject sequence is discussed in detail in the very next cycle. This cycle (14) and Cycle 16 are also noteworthy in that they follow the AmberPoint pattern of discussing an aspect of the system in terms of its modeling and its user interface issues.

In summary, we note that while they exhibited it in varying ways, all three of the design teams clearly had an impulse to keep the ideas they were working on up-to-date with one another. There were no cases where an idea was discussed in any detail early in the session and then set aside until the end. Rather, each was maintained, and attention partitioned out as evenly as possible. This pattern of returning again and again to the same ideas, in new contexts, was also observed at a more fine-grained level, as the next section discusses.

11.3.4 Repetition of Design Ideas

In each of the three design sessions, one can see the same ideas arising again and again. Our *repetition* relationship between ideas demonstrates this fact: in the three sessions, 21% (Adobe), 31% (Intuit), and 39% (AmberPoint) of the ideas were reiterations or rephrasings of previously stated ideas. Why did this happen? And what patterns governed this behavior?

Figure 11.5 depicts a subset of the complete set of ideas shown in Figure 11.1. (Note also that this is a distinct subset from the one shown in Figure 11.2.) This diagram shows all of the ideas that are repetitions of previous ideas, with arcs pointing backwards from each repetition to the moment in the session where that idea was previously stated. This diagram serves simply to illustrate the great variety in these reiterations: they happened in all parts of the process and sometimes referred back to recent ideas, while sometimes referring back to ideas stated over an hour ago. On closer investigation of the sessions themselves, it appears that the designers frequently needed to remind themselves of their previous statements, in order to understand the context in which further decisions were to be made. In particular, sometimes the designers would summarize the design in progress in terms of its major decisions, resulting in clusters of repetitions (see, e.g., at 0:31:06 in

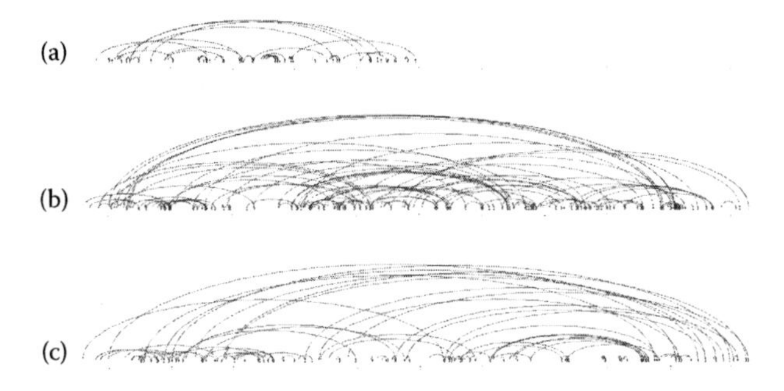

FIGURE 11.5 Reiterated ideas and their origins in the Intuit (a), AmberPoint (b), and Adobe (c) sessions.

the Intuit session, 0:49:38 and 1:40:30 of the AmberPoint session, and 1:35:28 of the Adobe session).

It is also worth noting that among such repetitions, the designers were slightly more likely to refer to an idea they themselves originally proposed, with percentages of 49% (Adobe), 56% (AmberPoint), and 60% (Intuit). The average percentage represents a very slight majority, and the fact that the designers so frequently repeated one another's ideas (as opposed to each designer fixating on their own ideas) suggests that they were effective in understanding each other's perspective. However, we note a particularly interesting exception that we have dubbed the *pet concept.*

Pet concepts are parts of the design that (a) one designer talks about far more than their partner in the session and (b) the designer brings up again and again, even in times where the flow of discussion does not demand it. Here we use the word *concept* to clarify that we are not referring to a specific *idea*, but a broader level of discourse. Examples of such pet concepts include designers' concerns with "cars controlling their own motion," "cars being created as they enter the map," and "analytics on a traffic configuration," each of which spawned a number of related, underlying ideas. The word *pet*, meanwhile, was chosen to evoke the fact that the designer has an unusual affection for an idea, but aims to avoid a negative connotation; in most cases, the designer's fixation on the idea was positive in the long term. This is related to the concept of attachment to ideas, as discussed by Cross (2007).

For example, Adobe's Male 2 adopts the concept of *analytics*, first describing the notion at 0:48:20. He later re-raises analytics and uses the concept to

- Explain analysis of the system as happening per-clock tick (0:59:01), leading to a concept of a *traverser*, which becomes a central architectural feature of the system
- Inform the design of visual feedback to the user (1:13:16)
- Inspire the idea of tracking a car through the map (1:14:19) (which later inspires the need to model individual cars)
- Motivate the overall flow of the system (1:31:01)
- Provide feedback about an intersection's performance (1:34:11)

To drive home the degree of this designer's affection for this concept, the very last transcribed comment from the session is his statement "and analytic data." The analytics concept became a lens for the designer, and it was applied to many parts of the design, often proving a useful way to develop analytics-oriented ideas, but also as a way to shed light on other, related issues.

Another example lies in the Intuit session, where Male 1 begins discussing how *cars enter the map* at 0:18:21. Later (0:34:38) he steers a conversation about traffic behavior around to the inflow of cars. Then he again raises the issue of cars entering the map to help advance the team's concept of the system's timing (0:38:12), uses it to introduce a concept of pausing the simulation (0:40:33), and finally uses it to construct a metaphor about road capacity (0:41:35).

In one last example, we see a situation where one designer did not embrace the other's pet concept, yet it served a useful purpose nonetheless. In the AmberPoint session, the female designer understood the system as involving cars as individual objects, while the male designer considered actual cars unnecessary, instead seeing traffic more abstractly. Given her perspective, the female designer felt it was necessary to have a concept of traffic *speed* (0:10:32, 0:14:57, and 0:15:11), while the male designer repeatedly dismisses the notion that speed is needed at all, or casts it as a road attribute (0:24:05).

Later in the session, the designers have developed very different views of how traffic should be directed. After a three and a half minute sequence where the designers are missing each other's points, the male designer finally asks "Are you saying there's actually a car out there?" (at 0:44:42), crystallizing the point of disagreement (this moment is further discussed in Nickerson and Yu, Chapter 19, this volume). As the session continues, there are also occasions where they outright disagree on this issue, but where the male designer presents an alternative approach to the female designer's suggestion (1:03:15, 1:23:54, and indirectly at 1:29:10). At other points, the male designer avoids engaging with the female designer's suggestions, but progress is made on other, peripheral issues that are raised (the span from 0:49:38 to 0:53:00, and at 1:44:05). While agreement was not reached on this issue, the female designer's vigorous consideration of this key concept sparked discussion and indirectly resulted in many of this prompt's trickiest issues being debated and considered.

In the form of pet concepts and basic reiteration, the overall pattern discussed here is that individual ideas recur again and again. One simple reason for this is that, as we have discussed in previous sections, the designers were constantly revisiting and recombining high-level subjects and the landscape of the software design so far (Brooks, 2010) was constantly changing. Therefore, ideas needed to be re-raised and adapted over time. In some cases, it seemed that designers happened upon an idea that they sensed was important, but were unable to integrate it into the current design. Unwilling to let the idea go as the design moved forward, the designer would keep it in mind, and would attempt to reintegrate it later, as we saw in our examples of pet concepts. Rather than representing a failure on the part of the designers, this repetition seems to be a necessary character of successful design sessions. Each time an idea is resurrected, it is placed in a new context and compared to different aspects of the system. In this way, a concept of compatible, elegant design ideas is slowly converged upon.

11.3.5 Designers' Strategies

Finally, we want to take a high-level look at the designers' overall strategies. How did each team frame the design problem? How did they go about solving the parts of the problem that they deemed important? In discussing these questions, we will further illustrate some of the observations discussed in the previous sections and provide some commentary on the strengths and weaknesses of the designers' approaches.

The Intuit team is perhaps the simplest to consider in these terms. They derived a model of the core of system, in the form of *roads*, connected to *intersections*, carrying *traffic*, and focused sharply on the relationships between these parts. For example, consider the relationship between intersections and roads, which is established initially in Cycle 2 and then undergoes three important developments.

1. In Cycle 6, the designers figure out how to reconcile the road's concept of lanes and the intersection's concept of lanes, as inspired by their increased understanding of how traffic and roads relate in Cycles 4 and 5.
2. In Cycle 8, they talk about how intersections and roads combine to create a map, as inspired by Cycle 7's discussion of how cars move around in the simulation.
3. In Cycles 11 and 12, the designers discuss how exactly intersections and roads communicate to transfer ownership of cars, as informed by the previous cycle's discussion about how exactly cars are added to and removed from roads.

In each case, the relationship between intersections and roads was illuminated by a discussion about a peripherally related subject, inspiring further development of that intersection–road relationship. By returning to this subject again and again, each time with new knowledge in hand, the designers created a mature, elegant notion of how these parts would work together. More broadly, the designers benefited from having a small number of subjects under consideration; they were able to return to them more frequently and were able to keep a larger portion of the design's needs in mind at a time. They handled important, related issues such as control flow and sensors in terse, but effective, terms.

The main problem, of course, lies in the nearly complete lack of user interface consideration. On one hand, maybe such aspects could be grafted to the model separately. On the other hand, the other two sessions give us pause, as they reveal how many tricky decisions can be found in this aspect of the design; it is possible that the Intuit team's model would not elegantly accommodate all of the features that might be desired.

The AmberPoint team can be characterized by its tendency to consider both halves of a given aspect of the system at a time, tightly pairing the discussion of its abstracted structure and the user's experience in working with it. This led to several promising interface approaches, including summary tables, graphical depictions of light timings, cycling sets of default settings, a directly manipulable map, and an overall emphasis on ease of use. This team's focus on the user interface also led to advances in their model:

- Considering light timings visually led to realizations about how these timing values should relate to one another to better simulate real-world scenarios (0:34:30).
- Their focus on providing useful feedback to the users led to an unusually well-developed sense of how a configuration might be evaluated by the system and how different analyses might be aggregated into higher-level feedback.
- Their interest in ease of use led them to discuss an effective idea: separating traffic configurations, light timing schemes, and map layouts on an architectural level.

As with the Intuit session, the AmberPoint designers rotated evenly through the aspects of the system, returning to intersections, traffic, the map, and feedback issues several times each. However, while they made great strides in developing user interface features, when they returned to a previously discussed subject, they seemed to "start over" when it came to how to model that aspect of the system. The example in the previous section illustrates how basic notions of how traffic should work were raised again and again, without achieving tangible progress, and near the end of the session, basic questions about intersections' operation were still unanswered (see 1:39:29). The problem was perhaps most strongly evidenced when they were trying to coalesce their design into a cohesive whole and found themselves needing to address the basic relationships with the parts (1:42:36). These were issues that the other two design teams considered much earlier in their processes, and such issues are difficult to resolve once the parts of the system have conceptually solidified. The male designer's comment "Yeah, that's kind of where this stuff lives. We haven't really ... this is where we find some conflict." (1:43:57) illustrates his recognition of the problem. This failure to incrementally grow a unified, central model might also help explain why this team's idea repetition rate (60%) was the highest among the three teams.

Considered together, the AmberPoint and Intuit sessions illustrate each other's weaknesses. The AmberPoint designers found ample opportunity for user interface innovation and reveal how many difficult issues the Intuit designers failed to address. Conversely, though, the Intuit session illustrates how tightly intertwined the parts of this problem are, and how difficult it might be for the AmberPoint designers to repair their solution.

The Adobe session, meanwhile, represents an attempt to handle both of these aspects of the design problem in earnest. The designers largely succeed, though at times they are spread quite thin by the effort of keeping all of the possible subjects in mind. One issue that this team faced was their failure to recognize roads as important concepts during the early parts of their process, as illustrated by this series of episodes:

1. Early on, Cycle 3 (0:09:37) shows the designers talking about how cars are added to the intersection, but at this point, they are conceptualizing each intersection in isolation, and roads are not considered.

2. In Cycle 6, roads are first acknowledged as being first-class entities, spurred by the acknowledgment that "the simulation is not supposed to be just one [intersection]" (0:37:48). But at this point, over 30 minutes' worth of design has taken place; the intersections, traffic, and controller have been discussed in some detail, and decisions surrounding them have begun to solidify. Roads, rather than being brought into the broad, coevolving rotation, are still considered secondary.

3. In the following cycle (0:40:25), some progress is made in repairing this situation, as a need for separate car-generating "outer roads" is recognized. But the discussion of traffic flows is stunted. Intersections are seen as needing cars that are destined to go left, right, and straight, and the responsibility for providing these is assigned to the burgeoning road concept.

4. In Cycle 15 (0:51:20), roads are given an overt focus, but are described as needing to be created in pairs, one pointing in each direction. This inelegant solution is an artifact of a previous decision that roads are owned, in sets of four, by intersections, and symptomatic of the lack of focus on roads.

5. By Cycle 19 (1:26:44), they have solidified their idea of roads as movers of traffic between intersections, as well as the idea that cars need to take time to move down the road, such that they are able to come to an appropriate way to model roads for their system (roads have a series of "spaces" that cars move down).

Space constraints prevent a complete description, but a close viewing of the Adobe session reveals another blind spot in the form of their notion of "car types," which does not mesh well with other parts of the design but inspires stubborn assumptions that the designers struggle to free themselves from. As we discussed in Section 11.3.3, the Adobe designers demonstrated patterns found in the Intuit and AmberPoint sessions; their discussions sometimes paired aspects of the system to be modeled and at other times paired modeling and UI issues of the same aspect. They generally succeeded, and the way that they were able to truly discuss three subjects in tandem in Cycle 22 is a testament to this success. But it also meant that some aspects of the system were comparatively underdeveloped: their model was not as cohesive as the Intuit designers', and their interface ideas were not as inspired as the AmberPoint designers'.

11.4 IMPLICATIONS

Together, the summaries in the previous section illustrate why the designers tended toward a pattern of rotating through their major identified subjects in turn, and why so much repetition was present. When designing software, the question of what makes a part or decision good or bad largely defies any universal standard of quality or aesthetics. What makes a design of a particular subpart of a software design good is, oftentimes, its fit with the needs of the other parts of the design, and decisions must be made according to the mental concept of the model as it exists so far. In some sense, no part of a model can be effectively evaluated until at least some concept of the whole model has already been established. Therefore, the designers built the design up incrementally.

Similarly, we note the need for ideas to be integrated into some whole. Perhaps because of the invisibility of software (Brooks, 1975) and the broader lack of ways to quickly visually represent software design ideas, it is difficult for designers to keep a variety of design ideas in mind. Those ideas that were not cognitively connected by both designers to the overarching mental model of the design were often disregarded or needed to be raised again and again in order to be adopted. In this way, incremental, two-subject development can be seen as a mechanism for taking new ideas and relating them to those that the designers had already reinforced in their minds.

The main implications of this work fall into three categories:

- Further studies must be performed to verify and expand upon these results. The observations we have presented here are based on a relatively small, controlled data set; they would be strengthened by comparison to observations of other designers, perhaps working in real-world settings. In addition, open questions remain about whether the designers' observed actions have parallels in other design fields, or if they arise from the nature of the problem before them.
- This work presents suggestions for programs supporting software designers. For example, these findings suggest that designers might benefit from being able to view two parts of a design-in-progress side by side, or that tools might otherwise be designed to allow for focus on a relationship between two parts. Alternatively, new, lightweight methods of recording, displaying, and comparing design ideas might help designers maintain a context for their design decisions.
- Finally, this work has implications for software design education. While further validation of our work is needed, lessons emphasizing incremental work and on considering pairs of subjects and their relationships could provide much-needed guidance for novice software designers.

11.5 RELATED WORK

The work presented here, of course, builds on the considerable body of research into studying designers in a variety of fields. Two studies in particular, though, bear special mention. Goldschmidt's (1990) work on linkography focused on visualizing the connections between *design moves*, which are analogous to our concept of design ideas. The major difference is that our work puts a strong focus on the subjects that the designers discussed as a high-level categorization, and we have derived a number of conclusions that were based on subject changes. Ennis and Gyeszly (1991), meanwhile, report a pattern in designers' behavior whereby they engage in an initial analysis and then move into a mix of opportunistic and deliberate behavior, which includes periods of focus on a single subject. This mirrors our own observations about the early parts of these designers' processes and is consistent with the presence of single-subject cycles in these sessions. On the other hand, their work does not recognize the two-subject pattern that was so prevalent in our studies.

Our study is also related to a small, but growing, body of work focusing on software design. Visser (2006) and Cherubini et al. (2007) provide similarly detailed analyses of software design, but largely focus on the artifacts that the designers used. Lakshminarayanan et al. (2006) share our interest in how software designers structure their design work, but their observations were based on interviews rather than direct observations of designers in action.

Several other studies bear more immediate connections to our work. Our concept of rotating through subjects largely matches Dekel and Herbsleb's (2007) characterization of the software design process as opportunistic, in that the designers proceed in an apparently ad hoc manner. However, our work details that such opportunism is not necessarily steeped in reaction to immediate needs, but rather stems from a broad (if loose) strategy for managing progress on all parts of the system. In this way, our work also reinforces and expands upon the description of design as being a mix of structured and opportunistic approaches by Guindon (1990). Our characterizations are also in line with Zannier et al.'s (2007) description, based on interviews with software designers, of the design process as rationalistic and naturalistic. The sessions observed in our work tended toward naturalistic decision-making, perhaps due to the open-ended nature of the problem presented and the inability to supplement the problem description with external sources of information.

Finally, our observations bear some similarities to the observations of those who describe software designers as seeking "balanced development" of their systems. Adelson and Soloway (1984) focused on the level of detail at which designers developed parts of a solution; they posited that an even level of detail among the system's parts was needed to allow the system to be effectively mentally simulated. Our work, instead of focusing on how detailed the designers' work within a subject was, focused on the patterns governing the designers' high-level selection of which subjects to consider. In addition, we suggest an orthogonal explanation for this behavior: that it is driven by a need for the parts to be evaluated relative to one another, rather than collectively evaluated against an external set of criteria. Our work also builds on the earlier work of Guindon (1987), who broadly observed the difficulty of combining the subparts of a solution, especially when they had been developed separately for a long of time.

11.6 CONCLUSION

In this chapter, we have considered the three design sessions from the 2010 Studying Professional Software Design Workshop and evaluated them in terms of the subjects they discussed, the ideas they generated, and the design subsessions, called cycles, that resulted. Our analysis shows that the sessions are highly incremental and that the designers repeatedly return to high-level subjects, as well as to specific, previously stated ideas. Furthermore, we observe that the designers often considered exactly two subjects at a time. Together, these observations allow us to characterize the design process in terms of building up relationships between the parts of a system, in an effort to generate a cohesive, elegant whole.

ACKNOWLEDGMENTS

The authors would like to thank the designers who participated in the study, as well as the workshop participants whose advice helped to improve this chapter. The results described are based upon videos and transcripts initially distributed for the 2010 international Studying Professional Software Design Workshop, as partially supported by NSF grant CCF-0845840 (http://www.ics.uci.edu/design-workshop).

REFERENCES

Adelson, B. and Soloway, E. (1984). A model of software design. Department of Computer Science Technical Report Number 342. Yale University.

Baker, A. (2010). Theoretical and empirical studies of software's role as a design discipline. PhD dissertation, SDCL, University of California, Irvine.

Brooks, F. (1975). *The Mythical Man-Month: Essays on Software Engineering.* Addison-Wesley, Boston, MA.

Brooks, F. (2010). *The Design of Design.* Addison-Wesley, Upper Saddle River, NJ.

Cherubini, M., Venolia, G., DeLine, R., and Ko, A.J. (2007). Let's go to the whiteboard: How and why software developers use drawings. In Rosson, M.B. and Gilmore, D.J. (eds), *Proceedings of the 2007 Conference on Human Factors in Computing Systems, CHI 2007,* San Jose, CA, April 28–May 3, pp. 557–566. ACM.

Cross, N. (2007). *Designerly Ways of Knowing.* Birkhäuser Architecture, Basel, Switzerland.

Dekel, U. and Herbsleb, J. (2007). Notation and representation in collaborative object-oriented design: An observational study. *SIGPLAN Notices 42,* 261–280.

Ennis, C. and Gyeszly, S. (1991). Protocol analysis of the engineering systems design process. *Research in Engineering Design 3,* 15–22.

Goldschmidt, G. (1990). Linkography: Assessing design productivity. In Trappl, R. (ed.), *Cybernetics and System 1990,* pp. 291–298. World Scientific, Singapore.

Guindon, R. (1990). Designing the design process: Exploiting opportunistic thoughts. *Human-Computer Interaction 5,* 305–344.

Guindon, R., Krasner, H., and Curtis, B. (1987). Breakdowns and processes during the early activities of software design by professionals. In Olson, G.M., Sheppard, S., and Soloway, E. (eds), *Empirical Studies of Programmers: Second Workshop,* pp. 65–82. Ablex, Norwood, NJ.

Jones, J. (1970). *Design Methods.* Wiley, New York.

Lakshminarayanan, V., Liu, W., Chen, C., Easterbrook, S., and Perry, D. (2006). Software architects in practice: Handling requirements. In *Proceedings of the 2006 Conference of the Center for Advanced Studies on Collaborative Research,* Toronto, October 16–19, Article 25. ACM.

Lawson, B. (2005). *How Designers Think.* Architectural Press, Oxford.

Simon, H.A. (1969). *The Sciences of the Artificial.* MIT Press, Cambridge, MA.

Stauffer, L. and Ullman, D. (1998). A comparison of the results of empirical studies into the mechanical design process. *Design Studies 9,* 107–144.

Visser, W. (2006). *The Cognitive Artifacts of Designing.* Lawrence Erlbaum, Hillsdale, NJ.

Zannier, C., Chiasson, M., and Maurer, F. (2007). A model of design decision making based on empirical results of interviews with software designers. *Information and Software Technology 49,* 637–653.

CHAPTER 12

The Cobbler's Children

Why Do Software Design Environments Not Support Design Practices?

David Budgen

University of Durham

CONTENTS

12.1 INTRODUCTION

The cobbler's children have no shoes.

The building of tools forms a key element of engineering practice, whatever the form of engineering involved, and indeed is one of the characteristics that distinguishes software engineering from other branches of computing. Engineers (of all forms) build tools to support their own activities and to aid them with creating larger artifacts.

In terms of supporting software development activities, tools for supporting the software design process are possibly one of the most challenging forms of application to create. Indeed, it can be argued that tool developers have often preferred to provide support for highly constrained models (such as plan-driven development) simply because supporting real design activities is so difficult. Software design itself involves manipulating and evolving ideas about invisible elements that have both static and dynamic properties [1]; the process is generally opportunistic in nature [2–5]; and the notations used to record the outcomes lack sound rationale for their form [6]. Indeed, although the unified modeling language (UML) is widely regarded as a "standard," a systematic literature review that we recently conducted to examine empirical studies of the UML indicates that there are many issues and questions about its notational forms and their usefulness [7].

The aim of this chapter is to draw upon a set of three exploratory case studies of the process of design, provided by the National Science Foundation (NSF) study data, in order to identify some principles that can be used to help design future generations of computer-assisted software engineering (CASE) tools and environments that will be able to support realistic design processes instead of constraining them. As such, it also draws upon some earlier ideas for its framework [8,9].

A question that should be asked here is whether tool support for designing is even necessary. Activities in the preliminary stages of design (and beyond) are often performed with a whiteboard, allowing freedom of expression using informal notations where appropriate and avoiding the inevitably intrusive effect of any form of user interface. (And of course, this approach is precisely what has been employed for the three design episodes provided by this study.) However, there are some good reasons for seeking to explore how we can adapt and deploy technology to support the design process. These include the following:

1. A design team may well be physically distributed and will need to perform shared design activities by means such as videoconferencing. In such a situation, employing suitable forms of technology in support, including options such as multitouch tables, becomes an attractive idea to explore.
2. The documentation stage of design whereby the records and ideas are transferred from whiteboard to a CASE tool forms a potential discontinuity that can be a source of errors, especially if it takes place after the main design activities. Among other things, the *rationale* for design choices is easily lost in such a process.

3. Using computer-supported forms gives better scope to track the evolution of a design and even to backtrack. While not of immediate relevance to this study, this can be particularly relevant where component-based design is concerned.

The aim of the analysis presented here was to identify principles for designing such systems, and as the case study approach that has been employed as the empirical framework offers little or no scope for control of experimental variables, we have no specific research question other than the one posed in the title. However, there is clearly a related question about the feasibility of supporting the software design process through software tools and how this might be done.

In the rest of this chapter, we outline the *viewpoints* framework that was used to analyze the NSF data, briefly review ideas about software design notations and support tools, describe the way that the case studies were conducted, examine the observed patterns of behavior, and consider how these might be met by design support tools in the future. As stated above, the aim is to generate some guidelines that can be used to create a new, more usable, generation of design support tools.

12.2 BACKGROUND

In this section, we address a number of related issues that contribute to the overall goal of design support.

12.2.1 Viewpoints Model

Describing ideas about software systems is complex, due to their multifaceted nature. In Ref. [10], we describe a relatively simple four-viewpoint framework that can be used to classify many forms of design representation, especially those that use "box and line" forms, as is almost invariably the case in software engineering. (For a fuller description of a wide range of such notations based upon an analysis of plan-driven design practices, see Wieringa [11].) The viewpoint framework employs the following categories:

- *Constructional* forms, in which the notation is concerned with describing the static elements of the system (such as classes, methods, subprograms, and processes) and the way that these are combined (in terms of the forms of interaction) as well as how the system is assembled. A good example of this would be a UML *class diagram.*
- *Behavioral* forms, which describe the causal links between events and system responses during execution. Such notations tend to be centered upon the idea of *state*, and two good examples would be the UML *state diagram* and the "traditional" forms of *state transition diagram* (STD).
- *Functional* forms that seek to describe the operations of a system, that is, the tasks that it performs. A good example of this would be a *data flow diagram* (DFD).
- *Data-modeling* forms that are concerned with the data (or knowledge) elements of a system and any relationships between these. The classic example of this form would be an *entity-relationship diagram* (ERD).

The framework this provides is essentially independent of architectural style (an aspect that is mainly addressed through the detailed forms used for the constructional viewpoint) and can be used to categorize abstract modeling ideas as well as notations. For the purposes of this chapter, we will use it to classify the utterances of the designers in the three case studies rather than to describe specific notations.

Of course, in practice, designing software usually involves continual interaction between these viewpoints as a designer seeks to assess the consequences of particular choices. Indeed, it is this that creates one of the major mismatches between the practices of software design and the assumptions that are apt to be embedded within both design support tools and the procedures adopted by plan-driven design methods.

12.2.2 Design Notations

The review paper by Wieringa [11] provides a valuable overview of the evolution of software design notations and associated design models. More recently, Moody [6] has examined the rationale (or more accurately, the lack of it) behind existing "box and line" notations used in software engineering and, particularly, software design. He comments about design rationale that "this is conspicuously absent in design of SE visual notations," and specifically that "the UML does not provide design rationale for any of its graphical conventions". Indeed, the UML provides a good example of the way that software engineers have managed to compound the problems when faced with software design issues: it is overcomplicated, uses excessive redundancy of notational elements, and few, if any, empirical studies of its forms and use are other than critical of its structures [7].

So, part of the problem with recording design ideas about software systems, as well as recording their evolution and rationale, is that no one would appear to have really given the design of notational forms the attention that it merits. However, for the purposes of this chapter, the issue we are seeking to address is much more one of how the use of combinations of such notations (representing the different viewpoints) can be supported, regardless of the qualities of the notations themselves.

12.2.3 Design Support Tools

Design support tools have tended to focus on the syntactic aspects of design notations. This is perhaps not surprising, given that the abstract nature of design models means that they are unlikely to have well-defined semantics. Such tools therefore mainly provide support for drawing separate, largely disconnected, forms of diagram. In earlier work, we did explore the idea of tools that provided "joined-up" viewpoints in order to try to provide a more holistic view of a conceptual design model that would also support opportunistic design development [9], but with limited success [12]. Our problems were due in part to the complexity of the user interface required to manage the model, which itself was further constrained by the technology available at the time.

If we look at some of the current tools and restrict ourselves to the UML (one of the benefits of the UML is that it is at least a "standard", albeit not a well-justified one), we can informally observe that such tools tend to fall into two categories:

1. *Drawing* tools (e.g., Visio and Dia) that are concerned with layout and symbols, but that make little syntactic interpretation of these (thus making it possible to mix symbols from different viewpoints or include symbols that are nonstandard for a given notation)
2. *Recording* tools (e.g., Rational Rose and argoUML) where a diagram has a type and the tool uses this to impose related syntactic rules, such as constraining the user to join elements together only in the prescribed way

The first group provide greater freedom to the user in terms of creativity, and hence (within the constraints of their user interface) they can be deployed in a way that is more like designing on a whiteboard, whereby some form of "blob" can gradually evolve into something more formal (such as a class or a package). However, such tools still do not empower the users by allowing them to switch between viewpoints, while thinking about a particular design element, in a manner that reflects observed practices. We examine this further in Section 12.6.2.

In many ways, the second group have changed little since the criticisms made in Ref. [8] to the effect that designing with such tools effectively constrains the designer to adopt the sort of diagram-driven practices that are usually associated with a plan-driven philosophy and hence impedes any use of opportunistic thinking. Such tools are essentially intended to record the outcomes of a design process, but have forms that do not readily integrate with that process.

12.3 RESEARCH METHOD

For the study reported here, the aim was to use the data provided by the Studying Professional Software Design Workshop to identify the major characteristics of the design process used by each pair of designers, expressed in terms of the viewpoints model. The purpose was to see if it was possible to identify how the features of design support tools could be better attuned to opportunistic development than those provided in current tools.

The research method adopted was that of the *case study*, employing the positivist approach described by Robert Yin [13] and conducting this case study using a multiple-case structure, with the three design teams providing the individual cases. Because the aim of the study was *exploratory* in nature, there were no *propositions* (the case study equivalent of a hypothesis). The first step was therefore to write a short *protocol* (research plan) that described the detailed form of the case study.*

12.3.1 Data Collection

The *unit of analysis* adopted for this study was the individual design model as produced by each of the three teams (these are referred to as Episodes 1–3). The data collection process involved an adaptation of the practices of *protocol analysis* [14], performed on the transcripts provided. This consisted of using an expert analyst (the author) to interpret and

* Templates for a range of empirical study forms can be found at www.ebse.org.uk.

encode the utterances of the designers in terms of the design viewpoints model and to identify those points where a *transition* occurs in the viewpoint being used. A *test–retest* strategy was employed for this purpose, whereby the transcripts were encoded twice, with an interval of a few weeks between the encodings, and with the two codings then being compared and assessed for consistency.

12.3.2 Data Analysis

For the viewpoints analysis, this was intended to address two aspects:

1. Identifying where the transitions between viewpoints occurred, in order to identify if there were particular patterns to these (such as ordering, focusing upon a design element, and interactions occurring between elements). The purpose of this was to see whether there were patterns that could potentially be supported by preplanned "shortcut" mechanisms in a design support tool.
2. Seeking to identify where any *breakdowns* occurred in the design process [15], whereby the designers were unable to elaborate their model further and needed to go back to an earlier stage or use a different strategy. The interest here was to identify what forms of backtracking might be needed in a design support tool. (However, we should note that, as no significant breakdowns were observed, this aspect of the analysis was not pursued.)

In addition, after analyzing the transcripts, the video recordings were examined to see how far they could provide further interpretation of the viewpoint analysis. In particular, we were interested in examining how the whiteboard was used to capture ideas about design.

12.4 RESULTS

We first examine the way that the data-coding process was organized and then consider the outcomes from this.

12.4.1 Coding the Data

As indicated above, the design episodes were coded twice, with an intervening period of several weeks, in order to provide a consistency check on the interpretation of the utterances. For all three episodes, the coding was undertaken using the transcript only.

Coding involved identifying the time of the utterance where a change of viewpoint occurred, along with the new viewpoint. Where viewpoints were entwined, this was also noted in the margin of the coding sheet.

During coding, the utterances were examined for key words (such as "object") as it was expected that this would provide an initial indicator of the viewpoint involved. However, although useful, because the discussions between designers tended to be informal in terms of vocabulary, a more subjective approach had to be employed to determine which viewpoint was involved at any point. So, as general guidelines, the following interpretations were used to help identify each viewpoint, based upon the concepts and forms used in related notations.

- For *constructional* aspects, we looked for relevant key words and also for any discussion that was related to the structure of the eventual system or elements of that system.
- *Functional* modeling was interpreted as being any discussion of what the system itself was to do—discussion of user interfaces was a good example of this (such as, "if you click on this intersection").
- The *behavioral* viewpoint is widely associated with concepts of state and possibly of transitions between these. So again, the utterances were examined for terms relating to how the state of the eventual system would change.
- For *data modeling*, any utterances referring to information use were interpreted as being data related. These were apt to relate to rather general concepts such as lists and tables rather than specific variables, but the key aspect was that they related to how the information was to be managed.

One way to help identify the viewpoint was to consider how the utterance could be modeled as a diagram (e.g., a statechart or DFD).

Of course, not all utterances fitted exactly into one category, and there were also some that fell outside of this (mainly concerned with assertions about the specification). Function and behavior can be entwined in a discussion of some aspect (essentially in terms of cause and effect), and equally data models can be related to the constructs containing them, such as objects and classes. In such cases, we tried to identify the "primary" viewpoint involved based upon the context of the utterance.

12.4.2 Consistency of Coding

The first episode to be coded twice was Episode 3 (AmberPoint). When the two codings were compared, they appeared to have very little overlap in terms of identifying transitions at exactly the same points (and with that, the same transition viewpoints).

To clarify the situation, a graphical model was employed showing viewpoint coding against time. For this, the viewpoints were assigned the following values: 1, functional; 2, behavioral; 3, data modeling; and 4, constructional. These values were chosen so that viewpoints that interact (such as functional and behavioral) were close in value.

Figure 12.1 shows the outcome of this process, with the initial coding plotted as a solid line and the second coding as a dashed line. (In addition, a "pseudo-viewpoint" with a value of "0" was used to indicate where the designers were consulting the specification.)

Overall, while there were many instances of agreement between the codings (although not always starting at exactly the same time), it is clear that there were also many inconsistencies. What was consistent, however, was the frequency with which the designers changed viewpoint (i.e., the emphasis of their thinking).

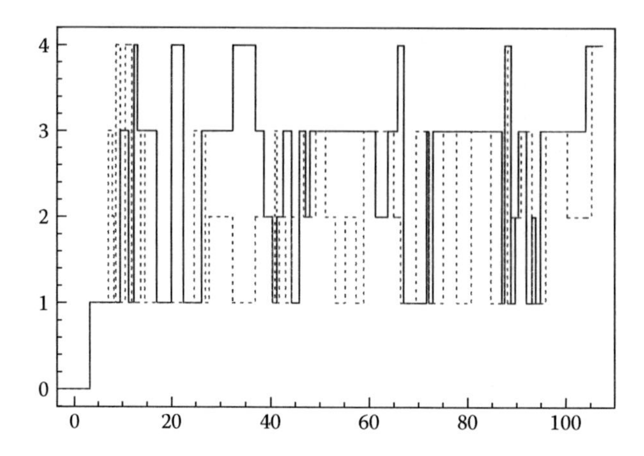

FIGURE 12.1 Comparison between codings for Episode 3 (AmberPoint) where 1 = function, 2 = behavior, 3 = data model, and 4 = construction.

However, closer inspection indicated that two factors were contributing to the differences:

- The timing recorded for the transition was not always consistent. Generally, the difference was a matter of whether the transition was considered to occur at the end of one utterance or at the beginning of the next one (particularly where some intervening and uncoded utterance occurred).
- In both cases, some relatively short-term transitions back and forth between two viewpoints were encoded differently. Looking at the longer-term envelopes for these showed rather greater consistency.

Neither are really major factors since our interest is in the broader patterns of change, rather than exact timing.

A second coding was then performed on Episodes 1 and 2 (Adobe and Intuit) to see what level of consistency was obtained for these. For Episode 2, the outcomes were much more evidently consistent—indeed, much of the limited variation that did arise occurred because in different codings, some sequences were coded in more detail (e.g., one coding might code a sequence of utterances as data modeling, while another coding might alternate between this and short periods of behavioral). A similar result was observed for Episode 1.

Finally, the two codings were compared for each of the three episodes in order to resolve these differences. It proved to be relatively straightforward to resolve all of them by inspecting the individual utterances. Inconsistencies arising from timing differences were resolved by adopting the start of the next utterance as the transition time; and those concerning the choice of viewpoint were resolved by examining the context of the utterance. At this stage the videos were also used to help resolve such issues as context. The outcomes from this process are discussed in the next section.

12.5 CASE STUDY ANALYSIS

The original plan was to look for patterns of viewpoint transitions (relevant to the design of a support tool interface) as well as to identify where any breakdowns occurred—and hence where a tool would need to provide some form of "wind-back" to an earlier design state. As no significant examples of breakdowns were observed, the analysis focused solely up on identifying larger-scale patterns and looking at how each team had addressed the viewpoints using the whiteboard.

12.5.1 Case 1: Adobe

Resolving differences in the codings on the transcripts produced the pattern of events shown in Figure 12.2.

As indicated above, the process of resolution was relatively simple and, apart from some small adjustments (chiefly in interpreting timings), produced the list of resolutions shown in Table 12.1.

Examination of the transcripts and the codings led to the identification of the following characteristics for this design episode:

- A strongly cyclical pattern of data modeling interspersed with behavioral exploration and some functional clarification.
- Very little discussion that relates to constructional ideas (but see next point).
- Three mentions of patterns: one concerned with architectural form (model-view-controller or MVC) and two others with behavioral aspects (use of the *Visitor* pattern). Where used, patterns formed a shared vocabulary and effectively acted as "labels for plans" [2].
- The functional viewpoint was chiefly used for interpreting the specification, particularly in terms of the resulting user interface needs.

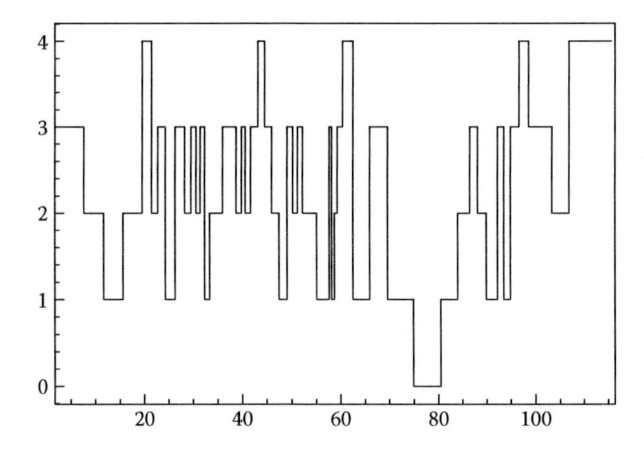

FIGURE 12.2 Resolved coding for Episode 1 (Adobe), where 1 = function, 2 = behavior, 3 = data model, and 4 = construction.

TABLE 12.1 Resolutions for Episode 1

Original Codes		Resolved Code	Count
B	F	B	4
B	F	F	1
DM	C	DM	2
DM	C	C	1
C	F	C	1

The codings were based only on the transcript, and only after the second coding was the video examined. In general, the activities shown on this matched the codings fairly well, although some interpretation of the utterances could be changed toward the end when the team were producing a UML diagram. (However, for a class diagram, there is not always a distinction between a data model and a construction model.)

The video recording was then analyzed in terms of whiteboard use (by further annotating the transcript). This showed that the development of the final design fell into roughly four stages.

1. The initial phase, about 30 minutes in duration, involved considerable interaction between the *behavioral* and *data modeling* viewpoints (this is also fairly evident in Figure 12.2). During this, the designers did the following:

    a. Built an abstract and stylized model of a road junction and used this to explore the resulting models

    b. Created a working list of candidate objects

    c. Walked through various mini-scenarios and used these to expand on object characteristics and interactions

    However, we should note that although they used behavioral ideas to develop and exercise their design model, they did not actually record many of them.

2. The second (relatively short) stage was when they attempted to capture their ideas about behavior in terms of a set of rules. However, this led to only a few items on the whiteboard. During this phase, they did however add some new objects to their list and extend the characteristics of the existing ones. So, in many ways this was an extended version of the first phase, although for this one, stimulation of ideas was based upon considering networks of road junctions more than the junctions themselves.

3. The third phase was concerned with the design of the user interface and involved rather more interplay between the *functional* and *data modeling* viewpoints, as is evidenced in Figure 12.2. In terms of whiteboard use, the designers employed an abstract model of the interface, involving “click” and “drag and drop” actions, to help them form a set of use cases (which were used to form a list).

4. In the fourth and final phase, they created a basic UML class diagram—at least one object was discarded in this process, and consideration of interactions between objects did lead to extended descriptions of the objects themselves.

So the main visual elements of this process could be summarized as being the creation and management of

- Lists of objects, with some characteristics and operations
- Abstract descriptions of junctions and networks of junctions, used to create use cases and explore the behavior of the system
- A list of use cases for the user interface
- A rather short list of behavior rules
- A UML class diagram

Few of these, apart from the last item, would be readily supported by most of the current generation of UML design tools.

In general, the decisions made in coding the transcripts were supported by the video interpretation.

12.5.2 Case 2: Intuit

Resolution of the coding for this was easier than for Case 1, and there were only three relevant changes. Figure 12.3 shows the pattern of viewpoint use. Again, examination of transcripts and codings identified the following main characteristics.

- A cyclical pattern of data modeling, explored behaviorally, with some use of the functional viewpoint
- Little discussion of constructional aspects until later in the process

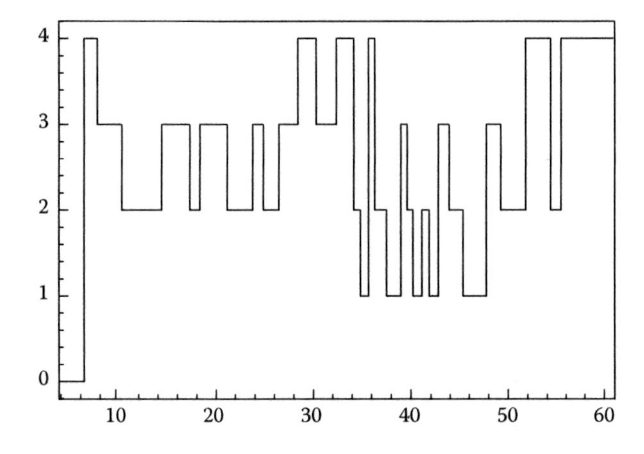

FIGURE 12.3 Resolved coding for Episode 2 (Intuit), where 1 = function, 2 = behavior, 3 = data model, and 4 = construction.

- No specific discussion of design or architectural patterns, although there was an early identification of an overall form similar to MVC

The use of the whiteboard had some similarities with Case 1, as well as some significant differences. For this case, the design activity fell into two clear phases.

1. An initial phase was very similar to the initial phase of Case 1, consisting of building up a list of "objects" with the aid of a physical abstraction of a road junction. However, there was less use of mini-scenarios and more consideration of the characteristics and dependencies of the objects.
2. For the second phase, the team concentrated on building up a "graph" showing object dependencies and interactions. Again, this was explored and developed with the aid of the physical abstraction of (two) road junctions. The resulting diagram was relatively informal, but at points the designers were effectively using it as the basis for short bursts of "mental execution"

So for Case 2, there were three main visual elements:

- Lists of objects (both physical and abstract)
- An abstract description of the physical characteristics of a junction, used to create and explore the software models throughout the episode
- A "graph" showing interactions and dependencies—clearly this would then have formed the basis of a UML class diagram had the team progressed on to this

Again, none of these forms really fit with the models and facilities provided by "traditional" design environments.

12.5.3 Case 3: AmberPoint

Resolution of this proved slightly more problematic than for Case 2. While the two codings did not disagree significantly (there were only three distinct points needing resolution), the team used the functional viewpoint much more than others and also had frequent utterances that involved a mix of viewpoints (especially functional and data modeling). The resolved sequence of viewpoints is shown in Figure 12.4. So, characteristics of this case were:

- A cyclical pattern once again, but now largely between functional and data modeling viewpoints
- A number of references back to the specification (and questions for this), largely arising from their consideration of functional aspects
- Relatively little discussion of constructional issues
- No obvious discussion of design or architectural patterns
- Considerable entwining of viewpoints within any one utterance

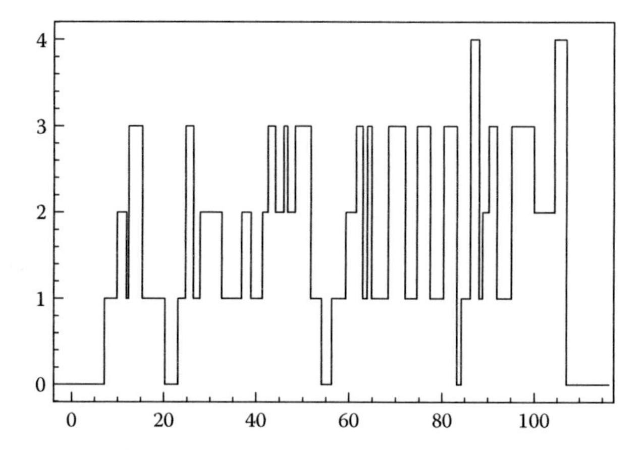

FIGURE 12.4 Resolved coding for Episode 3 (AmberPoint), where 1 = function, 2 = behavior, 3 = data model, and 4 = construction.

Turning to the use of the whiteboard, this was sampled at approximately 5-minute intervals in order to map out the development of the models being recorded in this manner. From this, it was then possible to identify four phases of diagram development as follows:

1. An initial phase drawing out a model of the user interface/controls (including a more detailed model of a junction) and using this to explore ideas and produce an initial list of "objects"
2. A phase in which the functions of the objects are listed, based on considering how the system would be used
3. The development of a more detailed user interface model
4. Creating a "dependency" diagram showing how the objects are related (again, something of an outline UML class diagram)

So, for Case 3, the main visual elements were as follows:

- Lists of objects, extended to describe their functions
- A physical abstraction of the user interface for the system
- A dependency diagram linking the objects

The team were particularly motivated by consideration of the user interface and the functions that the system was expected to provide. Their lists lacked the links and annotations that tended to be used by the other two teams. However, once again, their use of the whiteboard was very different from the models assumed by design tools.

12.5.4 Summary of Transitions

Table 12.2 shows the count for each viewpoint transition pair for each case (omitting any consultation of the specification).

TABLE 12.2 Summary of Transitions between Viewpoints

	From Viewpoint 1 Functional			From Viewpoint 2 Behavioral			From Viewpoint 3 Data Model			From Viewpoint 4 Construction		
To Viewpoint	2	3	4	1	3	4	1	2	4	1	2	3
Case 1	4	6	0	4	9	2	5	10	3	1	1	2
Case 2	1	3	1	5	4	2	0	7	2	0	3	2
Case 3	6	7	1	3	5	1	9	3	0	1	0	0
Total	11	16	2	12	18	5	14	20	5	2	4	4

When extracting the counts, a noticeably dominant pairing between behavioral and data modeling was evidenced for Case 1 and to some extent in the others too. There are also very low figures for the transitions between the functional and constructional viewpoints and vice versa.

12.6 DISCUSSION

For this section, we first consider some of the possible threats to validity involved in this case study and then go on to consider how the outcomes might be interpreted in terms of how software design could be better supported by CASE tools.

12.6.1 Threats to Validity

Case studies provide scope to probe much deeper into the underpinning rationale for design actions than is generally possible in laboratory experiments, as well as provide scope to use "field data" (which the data provided by this study closely approximates). However, they are also more vulnerable to bias and provide less scope for generalization. Here we examine some possible sources of bias.

The first of these is obviously the use of a single analyst to make the interpretations of the viewpoints (and of the outcomes). For the former task, this was addressed by performing the coding twice with a suitable interval between codings (test–retest) and then examining the two annotated transcripts in order to resolve any differences of interpretation. The resolution process tended to favor the second coding where differences occurred (although this was not always the case), which probably reflected the emergence of clearer interpretation rules with repeated reading and coding of transcripts. Indeed, an associated issue was the lack of clear rules for interpretation. Calculation of Cohen's kappa values for agreement between the two coding phases indicated that agreements for Cases 1 and 2 were "fair" (0.26 and 0.27), but that for Case 3 was "poor" (0.16). Even though the differences were easily resolved, this does demonstrate the difficulty in encoding such qualitative data and the need for clearer rules in the protocol than was the case.

The second is the nature of the problem tackled. While in itself a quite reasonable problem, it did offer scope for relatively easy visualization of the problem space (junctions, networks of junctions, traffic queues, etc.). This in turn could have influenced the way that the designers worked, since they were readily able to come to shared views of the problem space and its needs. Certainly this aspect does need to be considered when suggesting ways to support the design process.

The third point is not in itself a source of bias, but it does raise a question about the design task posed to the designers. The problem as set provided little scope for *reuse*, yet this is now a major feature of design, even for undergraduate students. The provision of Java APIs and frameworks makes it possible to design much more ambitious software artifacts than would have been possible a few years ago—but this was not reflected here. Again, from a design support perspective, this might lead to some bias in our ideas.

12.6.2 Consequences for Design Support Tools

Here we try to draw together the lessons from the analyses of the three cases provided in the previous sections and then to interpret these in terms of the facilities that might usefully be provided by design support tools.

That design in practice is opportunistic and differs markedly from plan-based forms has been recognized for many years [2,4,5]. The main benefit provided by this study was the combination of the verbal utterances with recordings of how the whiteboard was used.

Looking at the three cases, although all teams used different approaches, we can identify a number of common features:

- The use of informal lists, often annotated to show links or to enlarge upon characteristics.
- The use of informal diagrams and annotations.
- The role played by a "physical" abstraction of the problem (road junctions and networks of junctions)—all teams seemed to use these to help develop their ideas and for animation of their models.
- The difficulty of capturing reasoning (or "rules"), even when it was intended. Almost all of what were captured were outcomes.
- Frequent switching between viewpoints and between sections of the whiteboard.
- A point not really noted about the videos, but which was evident when considered as a whole was that for all three teams, the whiteboard annotation was chiefly performed by one person (only in Episode 2 did the second designer use a pen to any extent).

Viewed from a tool-building aspect, all of these provide challenges if we are to support teams of designers who are not even colocated. Some of these have already been addressed to some degree in other areas such as technology-enhanced learning, using technologies such as multitouch screens. Others such as capturing rationale could possibly be achieved by using audio recording and linking "audio blocks" to the sections of a diagram being annotated at the same point in time.

So, what recommendations might we be able to make for future design support tools? Looking at the viewpoint summary plots and the summary of transitions in Table 12.2 suggests three issues that might well merit further exploration.

1. Viewpoint transitions occur frequently and opportunistically and so need to be made as simple and quick to achieve as possible.

2. Informal diagrams and lists need to be integrated with other more formal notations. Ideally, a tool should support the transition from informal notations to more formal ones, but there is little in these case studies that can give any pointers to this.

3. As noted above, the capture of *reasoning* is needed. Again, the case studies indicate the need (as failure to record this occurred very frequently), but provide few points as to how best to achieve it.

Finally, we might also note that all of the solutions were essentially object oriented. Emerging technologies such as web services as yet seem to have little in the way of visual support from diagrams, and the design processes (which introduce some distinctive new features) are still emerging. So, in considering how design support tools might evolve, it would be useful to also consider other forms of architectural model, not just the use of objects.

12.7 CONCLUSIONS

The three episodes of design forming the cases of the case study demonstrate a number of common characteristics but also great variation, presenting challenges to anyone who is trying to build a design support environment that goes beyond simple diagram creation.

In terms of the question raised in the title of this chapter, it is probably still the case that tool builders use diagram-centric models rather than process-centric ones. None of the three cases produced other than fairly rudimentary UML class diagrams, and none produced anything that resembled the other UML notations. Equally, the lack of syntax and semantics implicit in the information forms that they did use suggests that unless transformed into a more formal description at an early stage, much of the meaning and rationale involved would quickly be lost.

The challenge for tool building is clearly one of moving beyond the straitjacket of the "diagram." Software is now being used to support a range of educational and developmental activities, many involving relatively free-form activities, and we need to look outside of the CASE tool tradition for new ideas about using software to design software.

ACKNOWLEDGMENTS

The author would like to thank the organizers of the design workshop for the opportunity to use the data sets described in this chapter.

REFERENCES

1. FP Brooks Jr. No silver bullet: Essences and accidents of software engineering. *IEEE Computer*, 20(4):10–19, 1987.
2. B Adelson and E Soloway. The role of domain experience in software design. *IEEE Transactions on Software Engineering*, 11(11):1351–1360, 1985.
3. B Curtis, H Krasner, and N Iscoe. A field study of the software design process for large systems. *Communications of the ACM*, 31(11):1268–1287, 1988.

4. R Guindon. Designing the design process: Exploiting opportunistic thoughts. *Human-Computer Interaction*, 5:305–344, 1990.
5. W Visser and J-M Hoc. Expert software design strategies. In J-M Hoc, TRG Green, R Samurçay, and DJ Gilmore, eds, *Psychology of Programming*, pp. 235–249. Academic Press, London, 1990.
6. DL Moody. The "physics" of notations: Toward a scientific basis for constructing visual notations in software engineering. *IEEE Transactions on Software Engineering*, 35(6):756–779, 2009.
7. D Budgen, A Burn, P Brereton, B Kitchenham, and R Pretorius. Empirical evidence about the UML: A systematic literature review. *Software—Practice and Experience*, 41(4):363–392, 2011.
8. D Budgen, M Marashi, and A Reeves. CASE tools: Masters or servants? In *Proceedings of 1993 Software Engineering Environments Conference*, pp. 156–165. IEEE Computer Society Press, Los Alamitos, CA, 1993.
9. A Reeves, M Marashi, and D Budgen. A software design framework or how to support real designers. *Software Engineering Journal*, 10(4):141–155, 1995.
10. D Budgen. *Software Design*, 2nd edn. Addison-Wesley, Harlow, England, 2003.
11. R Wieringa. A survey of structured and object-oriented software specification methods and techniques. *ACM Computing Surveys*, 30(4):459–527, 1998.
12. D Budgen and M Thomson. CASE tool evaluation: Experiences from an empirical study. *Journal of Systems and Software*, 67(2):55–75, 2003.
13. R Yin. *Case Study Research: Design and Methods*, 3rd edn. Sage Books, Thousand Oaks, CA, 2003.
14. KA Ericsson and H Simon. *Protocol Analysis: Verbal Reports as Data*. MIT Press, Cambridge, MA, 1993.
15. I Vessey and S Conger. Requirements specification: Learning object, process and data methodologies. *Communications of the ACM*, 37:102–113, 1994.

III

Elements of Design

CHAPTER 13

The Blind Men and the Elephant, or the Race of the Hobbyhorses

Mark D. Gross

Carnegie Mellon University

CONTENTS

13.1 INTRODUCTION

Modeled on the Delft Protocol Workshop of 1994, the Studying Professional Software Design (SPSD) Workshop at Irvine asked us each to examine the SPSD design protocols using a specific method of analysis or perspective. The chapters in this book that result from the workshop remind me of the tale of the blind men and the elephant—one feels the leg and says an elephant is like a pillar; another feels the trunk and says an elephant is like a rope, and so on. Some authors showcase the felicities of their method or their theoretical framework. Others argue, on the basis of their analysis, for their design support tool. In short, like all the authors in this book, the authors in this section ride their hobbyhorses.

And that is all right. That is what we were asked to do. Our project, recall, has two purposes: on the one hand, to understand designing; on the other hand, to understand what the various methods and tools of analysis have to offer design research. In that spirit, we begin with an overview of the questions, methods, and findings of each of the four chapters in this section, followed by a few reflections on what we might (or might not) learn from them about design.

The chapter by Ball and colleagues (Chapter 14, this volume) analyzes the relationship between complexity and uncertainty of design requirements and the strategies that designers employ to work the problem. Previous design research work has generally found that, in contrast to beginning designers, experts tend to work breadth first. Yet, sometimes they do not. Which do they do, when, and why?

Ball et al. (Chapter 14, this volume) posit that, faced with requirements that they find complex or uncertain, designers choose to explore depth first. Ball et al. state, "… the preferred strategy of expert designers is a top-down, breadth-first one, but they will switch to depth-first design to deal strategically with situations where their knowledge is stretched." They also posit that when designers find requirements to be complex or uncertain, they engage in mental simulation. They describe mental simulation as "A mental model 'run' in the context of software design typically entails the designer engaging in detailed, reflective reasoning about functional, behavioral or end-user aspects of the developing system."

To investigate these hypotheses, Ball's team analyzed the written transcripts of the three SPSD study teams, and coded them for the three things that they are interested in: requirements, simulation, and uncertainty. Key words (such as "complex," "control," and "crash") identified 22 requirements from the brief, with 3 levels of complexity for each requirement (high, intermediate, and low). They used similar key-word methods to code for uncertainty and mental simulation in the transcripts.

So when did the designers engage with which category of requirements? Early in each episode, the designers brought up all the requirements categories, supporting the proposition that designers begin working breadth first. The designers addressed high-complexity categories in the first half of each episode, but left them relatively alone in the second half. This supports the proposition that high-complexity requirements promote a depth-first investigation. Also, the teams engaged more in mental simulation when discussing high-complexity requirements than intermediate- and low-complexity ones. In short, faced with gnarly problems with uncertain requirements, designers dig deep and simulate.

Matthews's chapter (Chapter 15, this volume), "Designing Assumptions," examines the roles that assumptions play in designing, using a textual analysis. "The purpose of this particular investigation is to document a set of designers' 'languaging' practices—social actions done through talk—in the course of designing." Matthews focuses on assumptions: "the real time emergence of assumptions in design activity and the work such assumptions do."

He looks at the play between two kinds of work that the transcripts of the design conversations reveal. The first kind of work is "social/interactional" and the second kind is "design." Matthews argues that these two kinds of work are inextricably intertwined and a proper analysis must take both into account. Designers' talk cannot be abstracted or extracted from its social/interactional context. As Matthews states, "… to focus exclusively on the content—of what ideas are introduced, of what considerations are taken into account, of what problems are anticipated—is to overlook the details of how design work actually gets done."

Here, we see the different functions that assumptions serve in the discourse. Through a rigorous analysis of the conversation, not only searching for key words but also marking significant pauses and nonverbal utterances, Matthews finds the various ways that the partners in the design conversation use assumptions. In one example, he shows that what participants label as a simplifying assumption serves at once the purpose of distinguishing between the system and the world, "(gently) offering it up for criticism" and allowing designers to "inspect the consequences for the model they are building of instantiating this 'simplification.'" But in another case, "a simplifying assumption is not a label that criticises an unrealistic proposal, but a suggested strategy that enables [the designers] to 'conserve ambiguity' … so as not to be waylaid by too many real world issues and constraints."

The chapter by Ossher and colleagues (Chapter 17, this volume) on "flexible modeling" argues for a particular kind of design support tool, which they call "flexible modeling tools." They argue through a language analysis of the transcripts of these case studies that flexible modeling tools are appropriate for software design. The authors draw a distinction between modeling tools that are informal and "flexible" (such as their BITkit software) and more traditional modeling tools that are formal and strictly structured. The disadvantage of the more traditional tools is that they require users to adhere to a previously determined system of domain objects and relationships.

The authors arrived, iteratively, at an analysis of the transcripts that identified concerns and concepts spoken out loud and drawing and pointing activities at the whiteboard. (They note that their analysis is not inter-rater reliable, and should therefore be considered "suggestive.") They also created several visualizations to help them analyze the transcripts. For example, they created a time line that highlighted specific concepts and when the designers discussed them. From their analysis and its associated visualizations, the authors observe the following:

- Concerns evolve; they do not only originate from requirements documents.
- Concern concepts are revisited throughout the design process.
- Concern domains are different for teams with different backgrounds.
- Multiple representations are used.
- Representations evolve, becoming increasingly formal and detailed.

These observations hardly challenge conventional wisdom in design, but they do support the authors' argument for their BITkit tool. The authors conclude that flexible modeling tools are appropriate for (at least these three cases of) software design.

Petre's chapter (Chapter 16, this volume) focuses on the ways that each of the three software design teams uses the whiteboard, and more generally what a whiteboard affords a design team. To compare and contrast the teams' performances and to discuss the characteristics of the whiteboard as a medium or platform for designing, Petre uses the cognitive dimensions framework initially developed by Thomas Green and subsequently refined by others (Green and Petre 1996). As its name suggests, cognitive dimensions analysis looks at the use of notations and external representations ("information artifacts") in terms of several key dimensions: abstraction, closeness of mapping, consistency, diffuseness, hard mental operations, hidden dependencies, premature commitment, progressive evaluation, provisionality, role-expressiveness, secondary notation, viscosity, and visibility.

Notable about the cognitive dimensions analysis of these three teams' design episodes is how little difference it reveals across the teams. One might expect the various teams to use the whiteboard differently, but that does not appear to have happened to any significant degree. This observation lends support to the cognitive dimensions analysis: that it is accurately gauging the characteristics of the whiteboard as a representational medium. Petre's chapter also points to the social properties of this particular representation: the role that gesture plays, if not precisely as part of the whiteboard itself, then as a social activity around it. Likewise, in discussing persistence, Petre alludes to the embedding of the whiteboard in the social context of the workplace. These and similar properties of the representation, or perhaps more accurately the representational medium, are important, but they are often overlooked.

13.2 DISCUSSION

Werner Heisenberg (1958) is quoted as saying, "We have to remember that what we observe is not nature herself, but nature exposed to our method of questioning." The methods we use to examine design processes have inherent viewpoints about the nature of design, and the authors of these four chapters have strong and divergent (yet not conflicting) viewpoints about the nature of design. My colleague Susan Finger tells the story from when she was a program director at the U.S. National Science Foundation (NSF) in the early 1980s, administering an initiative then called "design theory and methodology." Every day, she says, someone would call up to pitch a great idea. The pitches always started like this: "design is X and I am an expert in X, therefore ...." By the end of her tenure at NSF, she had a long and interesting list of sentences all beginning with "design is X." Reading the chapters in this section reminds me of Susan's story.

So on the one hand, we are interested in what each approach can reveal about the design process—in this case, the design processes followed by the three teams of professional software designers in our workshop. On the other hand, each method of analysis brings to bear its own biased viewpoint, and will lead us to a different view of what design really is.

In the late 1960s, during the first wave of research on design and design methods, the dominant dialectic was between Herbert Simon's view expressed in his Karl Taylor Compton Lecture "The Science of Design" and Horst Rittel's view of design later articulated in "The Reasoning of Designers." Simon (1969) saw design as a process of searching for solutions, whereas Rittel (1987) saw designing as deliberative argument and negotiation among stakeholder viewpoints. Since then the scope of the dialectic has broadened considerably, but it echoes still in research on design, and specifically in these four chapters.

For Ball et al., looking at the behavior of expert designers, there is no question: *design is search*, pure and simple. The only question is how designers proceed in that search. How do they traverse the decision tree of design? This perspective is a most clear and direct descendant of Simon's view of design as the search of a multidimensional space within constraints.

Matthews's view of *design as social negotiation* is reminiscent of Rittel's view, but he makes an important point, familiar to us all from personal experience, that an argument among people is more than logic. The subtleties of verbal, visual, and gestural language play powerful roles in how we resolve arguments. If design is, as Rittel would have it, argument, then we had better attend not only to the content of the argument but also to its encompassing social attributes.

Ossher et al. (Chapter 17, this volume) present not so much a model of designing as an argument for a style of tool to support designing. They see *design as modeling*:

making a model of a something that has yet to be built—in this case, a piece of software for simulating traffic. Model making as a central activity in designing has a long and illustrious history. Maurice K. Smith, an extraordinary designer who taught architecture at MIT for four decades, was fond of saying, "design is surrogate building." He was referring to the direct correspondence between selecting and assembling elements in a design and the process of constructing the completed building. The purpose of creating a model is to examine it and test it against desiderata. It is much easier to revise a design than to reconfigure a building or a complex piece of software.

It is a truism that the modeling tools cannot but affect the models made. Consider a spectrum of tools, from highly structured at one extreme to entirely open-ended at the other. A structured tool presupposes the elements of the solution and requires the designer to assemble them in syntactically correct ways, which ensures that the resulting designs are buildable. The cost is that the designer follows a fairly narrow prescribed sequence of steps, which (unless the designer is schooled in the method) tends to work against free exploration of the design space. At the other end of the structured-unstructured tool spectrum is the blank sheet of paper or whiteboard. Whereas a structured tool ensures that the design is at least syntactically correct, you can draw anything you like on a whiteboard—and later find that it makes no sense, or cannot be built. The unstructured *tabula rasa* requires designers to do the work, in their heads, of keeping the design constraints. "Flexible modeling" falls in the middle of this spectrum, striking a balance between the invitation to madness that the blank whiteboard presents and the straitjacket of a structured tool.

In the end, so what? Each analysis here engages solidly with the specific domain of software design; yet, they all make more general arguments about the design process. None of these chapters seem likely to be proven wrong; neither do they conflict with one another. Yet, each presents a starkly different perspective on the design vignette videos we analyzed. In this, the four chapters in this section exemplify the workshop as a whole: they contain many good ideas that have yet to come together to form a unified understanding of design.

REFERENCES

Green, T.R.G. and Petre, M. (1996) Usability analysis of visual programming environments: A 'cognitive dimensions' framework. *Journal of Visual Languages and Computing* (special issue on HCI in visual programming), 7 (2), 131–174.

Heisenberg, W. (1958) *Physics and Philosophy: The Revolution in Modern Science.* Lectures delivered at University of St. Andrews, Scotland, Winter 1955–1956.

Rittel, H. (1987) The reasoning of designers. In: *International Congress of Planning and Design Theory.* Boston, MA: ASME.

Simon, H. (1969) *Sciences of the Artificial.* Cambridge, MA: MIT Press.

CHAPTER 14

Design Requirements, Epistemic Uncertainty, and Solution Development Strategies in Software Design*

Linden J. Ball

Lancaster University

Balder Onarheim and Bo T. Christensen

Copenhagen Business School

CONTENTS

* Reprinted from *Design Studies* 31 (6), Ball, L.J., Onarheim, B., Christensen, B.T., Design requirements, epistemic uncertainty and solution development strategies in software design, 567–589, Copyright 2010, with permission from Elsevier.

14.1 INTRODUCTION

The findings reported in this chapter are based on an analysis of videos and transcripts initially distributed for use in the 2010 international workshop, Studying Professional Software Design (SPSD; see http://www.ics.uci.edu/design-workshop). Full details of the design task and the data capture methods used in this study are presented in Appendix I. Suffice it to say here that the transcripts were obtained from three pairs of professional designers tackling an identical design brief, with each design dyad being from a different company: Adobe, AmberPoint, and Intuit. The duration of the design work for the former two teams was nearly identical (i.e., 1 hour 53 minutes vs. 1 hour 54 minutes, respectively) and differed from the Intuit team, which spent just a single hour on the design task. The analysis that we present in this chapter adopted an information-processing perspective (e.g., Goel and Pirolli, 1989, 1992), whereby design is viewed as a form of complex problem solving that involves the search for a product specification that satisfies an overarching design goal. In the present case, the goal, as stated in the given design brief, was for the designers to produce a conceptual design for a traffic flow simulation program that was requested by a client, Professor E, who was a civil engineering lecturer. Professor E's expectation was for the conceptual design ultimately to be realized as a program that students could use to "play" with different traffic signal timing schemes in different scenarios.

By adopting an information-processing standpoint on expert software design, our aim was to examine how aspects of an overarching design goal motivate particular solution development strategies. Our contention, based on our previous research in various design domains, including electrical and electronic engineering (Ball et al., 1994), microelectronic engineering (Ball et al., 1997), and software engineering (Ormerod and Ball, 1993), was that the organization of solution development processes would be seen to be determined in important ways by the inherent *complexity* of the design requirements associated with the design goal. We further aimed to examine the possibility that complex design requirements impact on designers' feelings of *epistemic uncertainty* as they attempt to deal with these requirements by generating and evaluating putative solution ideas. Our working hypothesis was, therefore, that designers' perceptions concerning the complexity of requirements would be observed to fuel strategic shifts in the solution-search process, with such shifts potentially being mediated by feelings of uncertainty.

14.2 ROLE OF REQUIREMENTS IN DESIGN

The SPSD design brief unpacked the overarching design goal in terms of a set of *design requirements*. In many design tasks, such requirements relate to aspects of an artifact's functionality, structure, and behavior, but they can also extend to issues that include verifiability, testability, maintainability, usability, and the like (Simon, 1973). In the present case, functional and behavioral requirements predominated and focused primarily on allowing students to use the traffic flow simulation software in the following ways:

- To create visual maps of an area, laying out roads of varying lengths in patterns involving arrangements of at least six four-way intersections
- To describe the behavior of traffic lights at intersections, including a variety of sequences and timing schemes such as left-hand turns protected by green arrow lights, with such schemes potentially being driven by sensors that detect the presence of cars within lanes.
- To simulate traffic flow on the map according to the chosen sequences, timing schemes, and traffic densities in a manner that is conveyed visually in real time.

The brief also encompassed requirements associated with ease and flexibility of use. For example, it was stated that the design "should encourage students to explore multiple alternative approaches. Students should be able to observe any problems with their map's timing scheme, alter it, and see the results of their changes on the traffic patterns".

Closely linked to the notion of design requirements is the concept of *design constraints*, which are factors that limit the search for solution ideas (Simon, 1973). Constraints are typically external in origin (i.e., imposed by clients), with the dominant ones being the time and cost available for product development. Some authors (e.g., Savage et al., 1998) suggest that constraints can also be internal, relating to the designer's domain knowledge, while others may be inherent to particular design tasks (e.g., deriving from a product's physical characteristics such as its size). Constraints—whether external, internal, or inherent—are generally used by designers in both a generative manner (to inform and limit the space of design possibilities; Ball et al., 1994), as well as in an evaluative manner (as *criteria* to determine the best solution from a range of options; Ball et al., 2001). Many design researchers avoid being drawn on the distinction between requirements and constraints since it is difficult to demarcate the boundaries between the two concepts. In this chapter, we use the terms interchangeably.

14.3 ILL-DEFINED VERSUS WELL-DEFINED DESIGN TASKS

One feature of conceptual design tasks that renders solution development particularly challenging is that they are typically *ill-defined* (Simon, 1973), with only partial or imprecise requirement information being given up front. This lack of definition means that designers have to uncover a deeper understanding of the requirements during the solution development process itself. Studies of real-world design have shown that the quality of the methods employed to identify and clarify requirements has a strong impact on the quality of the emerging design solution (Chakrabarti et al., 2004; Walz et al., 1993). Intriguingly, these studies have also revealed that many requirements are left unsatisfied in commercial design contexts, especially when designers spend insufficient time studying the given brief.

The SPSD design prompt seemed representative of real-world design briefs in terms of encompassing a range of ill-defined requirements. For example, the requirement associated with the visual, real-time presentation of the simulation was associated with the comment that, "It is up to you how to present this information to the students using your program … you may choose to depict individual cars, or to use a more abstract representation." Numerous statements throughout the brief reinforced the notion that requirements could be tackled in an "open-ended" way. Some requirements were, nevertheless, fairly well defined in their level of prescriptive detail. Although researchers have noted that a prespecified list of requirements—especially well-defined ones—may hinder requirements exploration (Chevalier and Ivory, 2003), we are also mindful of other evidence that designers will tend to treat even well-defined design requirements *as if* they are ill-defined (Malhotra et al., 1980).

One aspect of the SPSD brief that may give grounds for caution in generalizing any findings is that the brief only superficially determined the client's *rationale* for the stipulated requirements. For example, why was it that intersections could only be four-way and not involve other common types such as T-junctions? All that the designers were told was that this requirement related to "theories of traffic patterns," but it was left unspecified what Professor E wished to highlight in this respect. The lack of any rationale for design requirements may encourage designers to assume that requirements have been carefully screened and selected by the client beforehand. As such, they might believe that they do not need to challenge or redefine such requirements by gathering information or restructuring them—as would normally be the case in a typical iterative design process where a degree of client liaison would take place (Herbsleb and Kuwana, 1993; Malhotra et al., 1980). Notwithstanding this caveat, we nevertheless believed that an analysis of the transcripts would produce a range of interesting observations concerning the way in which expert designers handle a realistic design task.

14.4 INTERPLAY BETWEEN DESIGN REQUIREMENTS AND SOLUTION DEVELOPMENT STRATEGIES

The concept of design requirements is central to our analysis, which aimed to inform an understanding of the interplay between requirements and the solution development strategies that designers use to organize their search through the design space. The topic of solution development strategies in design has long been of considerable interest, with research revealing some intriguing developments in our understanding, which we now turn to (for detailed reviews, see Cross, 2001; Visser, 2006). Pioneering studies in the 1980s presented a view of solution development in design as involving a systematic, top-down process that is structured and hierarchical. For example, Jeffries et al.'s (1981) software design study observed experts and novices tackling design tasks using a top-down, problem reduction approach that entailed the *modular* development of the program from abstract design levels through to levels of increasing detail. Jeffries et al. noted, however, that the experts' schedule for tackling problems was predominantly *breadth-first* in nature, whereas the novices' was predominantly *depth-first*.

A breadth-first strategy has three characteristics (Ball and Ormerod, 1995). First, a top-level design goal is reduced into a number of subgoals, which typically reflect design submodules that meet particular design requirements. Second, these subgoals are decomposed a full level at a time into sub-subgoals (and so on). That is, the design approach is "balanced" (Adelson and Soloway, 1986) in that the whole solution to the overarching design goal develops in an integrated manner through each level of abstraction. Third, the process of sequential, modular development proceeds until a requisite level of design detail is reached. In contrast, the depth-first strategy entails taking one top-level subgoal at a time and developing it in considerable detail before then moving on to do the same with the remaining top-level subgoals (Ball and Ormerod, 1995). This strategy seems less effective than a breadth-first one because it requires costly backtracking if the current solution ideas turn out to be incompatible with earlier ones. Nevertheless, the depth-first strategy has the advantage of facilitating an early resolution of whole branches of the goal tree.

One noteworthy observation from Jeffries et al.'s (1981) study was that some experts periodically deviated from a predominantly top-down, breadth-first strategy. Such deviations typically involved the designer focusing on the in-depth development of a solution to meet a high-level subgoal—especially when this subgoal was perceived to be complex, unfamiliar, or important. Thus, while these experts were still operating in a top-down manner, they were apparently mixing breadth-first and depth-first design strategies in a context-driven manner that was flexible yet effective.

In the early 1990s, a new wave of research made contrasting claims to the view of design as being a predominantly structured, top-down activity, with several authors instead proposing that both novice and expert designers are highly unstructured. For example, Guindon (1990) noted that the software engineers she studied exhibited 53% of activities that deviated from top-down, breadth-first design. Visser (1990) claimed that while the expert designer she studied described himself as "intending" to pursue a top-down, depth-first strategy (rather than a supposedly expert top-down, breadth-first one), his actual behavior revealed considerable deviation from a structured approach. Accordingly, solution development processes in design started to be described as primarily *opportunistic* in nature, with a key emphasis being placed on the way in which designers take immediate advantage of solution opportunities that become available, even if their pursuit leads to deviation from a structured design approach.

Not all studies in the 1990s supported the extensiveness of opportunistic processing in design. For example, Ball and Ormerod (1995), Ball et al. (1994, 1997), and Davies (1991) all presented evidence showing that design might best be viewed as predominantly top-down and structured in nature but with opportunistic episodes arising to circumvent design impasses or knowledge deficits, to capitalize on emerging opportunities, or to deal with working memory failures. Most design researchers nowadays seem in general agreement that design involves an adaptive combination of both structured *and* opportunistic processing (cf. Visser, 1994). Nevertheless, one important aspect of the theoretical account of design expertise espoused by Ball et al. (1997) that is of particular relevance to the present analysis was that experts will often tend to *mix* breadth-first and depth-first design (with some opportunism present as well). According to this account, the preferred strategy of expert designers is a top-down, breadth-first one, but they will switch to depth-first design to deal strategically with situations where their knowledge is stretched. Thus, depth-first design is a response to factors such as problem complexity and design uncertainty, with in-depth exploration of solution ideas allowing designers to assess the viability of uncertain concepts and gain confidence in their potential applicability.

14.5 AIMS OF PRESENT ANALYSIS

The SPSD transcripts afforded a unique opportunity to examine the involvement of mixed-mode solution development in software design. We were especially interested in gaining new insights into the causal determinants of switches from breadth-first to depth-first activity, in particular the role of requirement complexity in driving designers toward depth-first solution exploration. We were also interested in examining whether complex design requirements impact on designers' feelings of epistemic uncertainty as they attempt to deal with them by means of the generation and evaluation of putative solution ideas. It is possible, for example, that feelings of uncertainty serve to *mediate* between complex requirements and the strategic design shifts that result in depth-first solution development.

One further issue that we wished to explore related to the role of *mental simulation* in design (Christensen and Schunn, 2008). Mental simulation is a cognitive strategy whereby a sequence of interdependent events is consciously enacted or "run" in a dynamic mental model to determine cause–effect relationships and to predict likely outcomes (Nersessian, 2002). In design contexts, this simulation strategy appears to be primarily evaluative in nature, where the designer's imagination is used to test ideas and validate solution concepts (Ball and Christensen, 2009). The mental models that underpin mental simulation are assumed to involve qualitative rather than quantitative reasoning, relying on ordinal relationships and relative judgments (Forbus, 1997). Thus, when running mental models people neither estimate precise values and quantities nor carry out mathematical computations. In essence, mental simulation provides a quick and economical way for a designer to test out an aspect of the artifact being developed, including how a system might function under changed circumstances or with altered features. Mental simulation can also extend to imagining the way in which an end user might interact with a system.

Recent research has identified instances of mental simulation in design protocols and has clarified its nature. For example, Christensen and Schunn (2008) tested three assumptions of mental simulation theories: that mental simulations are run under situations associated with subjective uncertainty; that mental simulations inform reality through inferences that reduce uncertainty; and that the role of mental simulations is approximate and inexact. Christensen and Schunn demonstrated support for all three assumptions: initial representations in simulations had higher than baseline levels of uncertainty;

uncertainty was reduced after the simulation run; and resulting representations contained more approximate references than either baseline data or initial representations. The SPSD data set provided an opportunity to examine further the links between mental simulation and uncertainty as part of the process whereby designers move from requirements toward solution concepts that address those requirements.

14.6 METHOD

14.6.1 Transcript Segmentation

We made no attempt to check or correct the SPSD transcripts prior to segmenting and coding them on the assumption that any transcription errors would be minor, at worst adding a small degree of noise to our analyses. To divide the three transcripts into discrete units of spoken discourse we used a turn-taking approach, with segmentation arising at points where the speaker switched from one designer to the other, or where the interviewer interjected with comments or answers to queries. This approach revealed 913 segments. Of these, 20 involved interviewer verbalizations, which were subsequently removed, leaving 893 segments.

14.6.2 Categorization of Design Requirements

An examination of the design brief revealed 22 design requirements that were explicitly mentioned and that could be isolated from one another. We grouped these 22 requirements according to two categorization schemes: one that related to the "focus" of each requirement and the other that related to the "level of complexity" of each requirement (Table 14.1). In terms of focus, five requirements were of a general type (e.g., concerning ease of use), five dealt with the nature of interactions between the user and the simulation, four covered the design of intersections, five concerned traffic-light behavior, and three dealt with the roads within the simulation. In terms of level of complexity, the requirements were categorized according to whether they were at one of three levels: high, intermediate, or low. The complexity metric reflected an a priori judgment of how easy the requirement would be to add to the simple "first-pass" solution attempt that each design team was observed to make. We note that variants of a fundamentally equivalent first-pass solution were mentioned in all three transcripts and revolved around a grid-based road system and a two-state (red–green) traffic-light concept, where each grid meeting point was a four-way intersection. These first-pass solutions met some (but not all) of the stated requirements.

Table 14.1 provides clear examples of the difference between intermediate and high complexity requirements in relation to requirement 16 ("Optional sensor at all intersections"; intermediate complexity) and requirement 14 ("Allow for left hand turns"; high complexity). Both requirements focus on the way that traffic lights operate, and both add complexity to the initial first-pass design solution. The optional sensor provides another trigger for traffic-light changes in addition to mere timing, but is otherwise not particularly complicated. Allowing left-hand turns, however, adds multiple states in relation to how traffic lights operate and also fails to specify whether there are multiple lanes approaching intersections, including one for going left.

TABLE 14.1 Coding Scheme for Requirements

Code	Design Requirement	Focus	Level of Complexity
1	Maps need not be **complex**	General	—
2	**Easy** to use	General	—
3	Encourage students to explore multiple **alternative** approaches	General	—
4	Design of **interaction** (student/software), with basic appearance of interface	General	—
5	Design basic **structure** of code	General	—
6	Show **visual map** of an area	Interaction/simulation	Low
7	Students should be able to **observe** problems with ***timing*** scheme, make changes, and observe again	Interaction/simulation	Low
8	Different **arrangements** of intersections	Intersections	Low
9	Accommodate **six** intersections (at least)	Intersections	Low
10	No intersection without traffic lights	Intersections	Low
11	Only **four**-way intersections	Intersections	Low
12	**Control** of behavior of traffic lights at each of the intersections	Lights	Intermediate
13	Allow for a variety of **sequences** and **timing** schemes for lights	Lights	Intermediate
14	Allow for ***left***-hand turns protected by green arrow	Lights	High
15	No **crash**-combinations allowed	Lights	Intermediate
16	Optional **sensor** at all intersections	Lights	Intermediate
17	Free choice of road **pattern** by students	Roads	Low
18	Roads of varying **lengths**	Roads	Intermediate
19	Visual simulation of traffic **flow** (dynamic)	Interaction/simulation	Low
20	Should be **real-time**	Interaction/simulation	Low
21	Visual simulation of **light state** (dynamic)	Interaction/simulation	Low
22	Allow for choice of **traffic density** on any given road	Roads	High

Note: Words in bold were used as search terms to facilitate identification of references to requirements in transcripts.

In categorizing requirements in terms of complexity, we identified two as being of high complexity: "Allow for left-hand turns" (already mentioned) and "Allow for choice of traffic density on any given road." These require the addition of multiple states to the traffic-light codes and add multiple controls to the road system. Adding the density requirement makes the simulation more complex than a "length = space" model since car trajectories become important. Five requirements were of intermediate complexity: "Optional sensor at all intersections," "Allow for a variety of sequences and timing schemes for the lights," "Roads of varying lengths," "No crash-combinations allowed," and "Control of behavior of traffic lights at each of the intersections." With the exception of general requirements, all remaining requirements were categorized as low complexity, being relatively easy to add to the first-pass solution; indeed, in most cases they were largely resolved by the initial solution.

The general requirements (e.g., "Easy to use") were excluded from the complexity categorization system (as well as from subsequent analyses) since it was evident that these requirements were hardly dealt with by any of the teams, the exception being the Adobe team, which did pay some limited attention to elements of interaction design and code structure late in the design process. We note that the failure of the teams to deal with general requirements is curious since such requirements are often particularly important and are likely to have a pervasive influence on the emerging design. Indeed, such requirements are notoriously difficult for designers to deal with late in the design process, and good designers should really be addressing these requirements early on. For example, Golden et al. (2005) provide evidence that when general usability requirements are not dealt with early in the software design process, they can add considerable complexity to redesign because they typically necessitate specific architectural choices.

14.6.3 Coding of Design Requirements

Each segment was coded for whether at least one requirement was mentioned. This involved a three-step process. At Step 1, we conducted word searches using the terms depicted in bold in Table 14.1 to identify segments where requirements were potentially being referred to. At Step 2, all segments were examined to determine whether they contained some mention of requirements. Step 2 also enabled confirmation that words identified within segments during the word search genuinely referred to requirements. At Step 2, if a segment contained at least one requirement it was coded with a "1"; a "0" was used to denote the absence of a requirement. At Step 3, each segment coded as "1" was recoded in terms of the specific requirements mentioned using the codes from Table 14.1.

14.6.4 Coding of Mental Simulation

The codes relating to the presence of mental simulation within segments were based on those used by Ball and Christensen (2009), as adapted from Trickett and Trafton (2002). A mental model "run" in the context of software design typically entails the designer engaging in detailed, reflective reasoning about functional, behavioral, or end-user aspects of the developing system. As such, mental simulation enables software designers to reason about new ways to develop the software's functions, features, or attributes, but without any need for physical construction or manipulation of program code. An example of a simulation run is presented in Extract 14.1, which was produced by the Adobe team when tackling Requirement 14 ("Allow for left-hand turns protected by green arrow"; high complexity).

EXTRACT 14.1 Example of a Complex Requirement (Requirement 14) Leading to a Simulation

Bold font indicates possible requirement-related words identified during the word search (see Section 14.6.3). Underlining indicates hedge words expressing uncertainty as described in Section 14.6.5.

[0:27:36.0] *Male 1:* A couple more advanced details is there's the **left**-hand turn, right? There's a, you can copy **left**-hand turns protected by green arrow **lights** so if there's a protected **left** turn then these need to be broken up almost.

[0:28:02.9] *Male 2:* Two queues.
[0:28:03.0] *Male 1:* Two queues: one going straight, one going **left**.
[0:28:06.0] *Male 2:* If it's a single lane of traffic, could those **left**-turners and straight-throughers be queued up in the same queue?
[0:28:15.6] *Male 1:* Yeah there could be two. So the—
[0:28:23.2] *Male 2:* So that kind of comes into car, right? We now have two types of car. Straight-through guy and **left**-turn, right turn, **left**-turn.
[0:28:41.5] *Male 1:* Now **left**-turn is a little more special because it can't turn if the road in the opposite has cars in it.

14.6.5 Coding of Epistemic Uncertainty

Epistemic uncertainty was coded using a purely syntactic approach, adapted from that employed by Trickett et al. (2005), which involved using *hedge words* to locate segments displaying uncertainty. In the present analysis, these hedge words included terms such as "probably," "sort of," "maybe," "possibly," and "don't know" (see Extract 14.1). Segments containing these words were coded as *uncertainty present*, but only if it was also clear that the hedge words were not simply being used as politeness markers. All segments that were not coded as uncertainty present were coded as *uncertainty absent.*

14.6.6 Intercoder Reliability Checks

The primary coding of the transcripts was undertaken by an unbiased coder who was not associated with the research but who was experienced in coding design transcripts. To assess coding reliability, one-third of the data was coded by the second author (for design requirements) and the third author (for uncertainty and simulation). Kappa reliability coefficients are reported in Table 14.2. All coding categories reached acceptable levels of reliability (kappa coefficients > 0.70), with near perfect reliability for the uncertainty code.

14.7 RESULTS

14.7.1 General Descriptive Aspects of the Data

Across the three transcripts, 40 references to requirements were made. The amount of transcript spent dealing with these requirements was 447 segments—or 50% of the data set. A total of 176 complete simulations (covering one or more segments) were identified in the three transcripts (mean = 37 simulations per hour of verbal data). The range of complete simulations across transcripts was between 51 and 64. The syntactic code for epistemic uncertainty revealed that 34% of the segments contained words or phrases indi-

TABLE 14.2 Kappa Coefficients for Intercoder Reliability

Coding Category	Kappa Coefficient
Design requirements	0.74
Mental simulation	0.85
Epistemic uncertainty	0.97

cating uncertainty. The range of segments containing epistemic uncertainty was between 28% and 46% across the three transcripts.

14.7.2 Quantitative Findings

14.7.2.1 First Mentions of Requirement Focus Categories

Our initial quantitative analysis focused on the time at which each requirement focus category (intersection, lights, roads, and simulation/interaction) was first mentioned by each team (Table 14.3). Within the first 10 minutes, both the AmberPoint and Intuit teams had introduced all requirement focus categories that would subsequently be developed, a finding we view as indicative of the initial deployment of a highly breadth-first solution development strategy. The Adobe team was slower to bring all requirement focus categories into play, with two categories being introduced within the first 10 minutes and with subsequent categories arising at 22 minutes and 45 minutes. Nevertheless, the Adobe team still evidenced aspects of breadth-first solution development since all requirement focus categories had been introduced before 40% of the entire duration of the design process had elapsed.

14.7.2.2 Temporal Locus of the Addition of Low-, Intermediate-, and High-Complexity Requirements

Having established that the initial aspects of solution development were largely breadth-first, we next examined the point within the transcripts (early vs. late) at which low-, intermediate-, and high-complexity requirements were dealt with. As specified earlier, the intermediate- and high-complexity requirements are those that should be conceptually more difficult to add to the first-pass solution attempt, although it remains unclear at what stage in the design process these requirements might be tackled. To pursue this analysis we calculated the number of segments appearing within the first half of each transcript that dealt with requirements at each respective complexity level (Table 14.4). We then calculated the equivalent scores for requirements arising in the second half of each transcript. Only segments where a *single* requirement category was being addressed were included in this scoring approach to ensure independence of data points in subsequent statistical analyses.

We next conducted a 3 × 2 chi-square test on the overall data (i.e., collapsing across teams; refer to scores in bold in the "overall" data columns in Table 14.4) in which we

TABLE 14.3 Time (Hours:Minutes) at Which Each Requirement Focus Category Was First Mentioned by Each Team

Requirement Focus Category	**Team**		
	AmberPoint	**Intuit**	**Adobe**
Intersection	0:09	0:05	0:45
Lights	0:09	0:05	0:09
Roads	0:07	0:05	0:22
Interaction/simulation	0:06	n/a	0:06

TABLE 14.4 Total Number of Segments Dealing with Requirements at Each Complexity Level in the First Half versus the Second Half of the Transcripts[a]

	Position in Transcript							
	First Half				Second Half			
Complexity Level	**AmberPoint**	**Intuit**	**Adobe**	**Overall**	**AmberPoint**	**Intuit**	**Adobe**	**Overall**
Low	24	5	28	**57**	28	0	18	**46**
Intermediate	19	0	7	**26**	9	14	11	**34**
High	29	51	48	**128**	6	0	8	**14**

[a] Broken down by design team as well as collapsed across teams. See scores in bold.

contrasted the three levels of requirement complexity with whether the requirements appeared in the first half or the second half of the transcripts. This analysis revealed a difference between complexity levels in terms of whether the requirements were dealt with predominantly early or late, $\chi^2(2, N = 305) = 57.32, p < .001$. Individual 2 × 2 chi-square tests showed that there were significant differences between high- and intermediate-complexity requirements, $\chi^2(1, N = 202) = 51.01, p < .001$, and between high- and low-complexity ones, $\chi^2(1, N = 245) = 39.10, p < .001$, but not between intermediate- and low-complexity requirements, $\chi^2(1, N = 163) = 2.19, p = .14$. The latter null effect is unsurprising since these two levels both relate to relatively straightforward requirements that are fairly easy to integrate in a breadth-first manner—as indicated by their roughly equal distributions across both halves of the transcripts. What is more interesting is the evidence that high-complexity requirements were dealt with predominantly in the first half of the transcripts, with little subsequent development of these requirements in the second half.

The same pattern of data—whereby only high-complexity requirements are prioritized early—was evident across all three teams (Table 14.4), with chi-square analyses at the team level providing good support for the aggregate findings reported above (note that the Intuit team had too few expected instances to allow for statistical analysis). In descriptive terms, Table 13.4 indicates that the Intuit team dealt with 100% of high-complexity requirements in the first half of the session, with the equivalent scores for the AmberPoint team and the Adobe team being 83% and 86%, respectively.

These previous findings are highly suggestive of two important facets of the present design process. First, it appears that the designers prioritized addressing the complex requirements immediately after the first-pass, breadth-first design solution had been produced. Second, it seems that these complex requirements were addressed in a primarily depth-first manner, with solution concepts being developed in the first half of the transcripts but not the second half. A balanced, breadth-first approach for dealing with these requirements would have been revealed by their iterative development over both transcript halves.

One way to substantiate the argument for the depth-first handling of complex design requirements is to examine the transcripts for "first mentions" of each requirement (Table 14.5). A chi-square test showed a nonsignificant trend toward high-complexity

TABLE 14.5 First Mention of Specific Requirements at Each Complexity Level in the First Half versus the Second Half of the Transcripts[a]

	Position in Transcript	
Complexity Level	First Half	Second Half
Low	7	11
Intermediate	3	6
High	5	1

[a] Collapsed across teams.

requirements being more likely to be mentioned in the first half of transcripts than the second half, with a different pattern evident for other requirement levels, $\chi^2(2, N = 33) = 4.32$, $p = .12$. The chance of finding a significant effect with this data set was compromised by the low number of observations ($N = 33$), which also rendered a by-teams analysis impossible. Nevertheless, the descriptive trend in the aggregate data points to the tendency for these designers to deal in an early and apparently depth-first manner with complex requirements.

14.7.2.3 Number of Segments Devoted to Addressing Low-, Intermediate-, and High-Complexity Requirements

The next issue we examined was whether complex requirements were dealt with over more segments on average, which would indicate that they took longer to resolve. For each requirement, we computed how many segments had been spent dealing with it (Table 14.6). A one-way ANOVA on the aggregate data that collapsed across the three teams revealed a significant difference between the three requirement complexity levels, $F(2, 16) = 20.17$, $p < .001$. Follow-up analyses using Tukey tests showed more segments were associated with addressing high-complexity requirements than intermediate- or low-complexity ones ($p < .001$ for both). There was no difference between low- and intermediate-complexity requirements ($p = .36$).

We also examined the data at the team level and found that the main effect of complexity observed at the aggregate level was replicated for the AmberPoint team, $F(2, 16) = 5.83$, $p = .014$, and for the Intuit team, $F(2, 16) = 7.86$, $p = .005$, but not for the Adobe team. The follow-up analyses for the AmberPoint and Adobe teams also supported the aggregate

TABLE 14.6 Mean Number of Segments Devoted to Dealing with Requirements at Each Complexity Level[a]

Complexity Level	Number of Requirements	Mean Segments	Standard Deviation
Low	11	9.36	8.04
Intermediate	4	19.25	19.93
High	2	67.50	16.23

[a] Collapsed across teams.

findings that differences typically resided in the number of segments spent dealing with the high-complexity requirements relative to the intermediate- or low-complexity ones.

In sum, these analyses indicate that complex requirements took longer to deal with than the easier ones. Although this evidence is perhaps unsurprising, when viewed in conjunction with the finding that complex requirements were dealt with primarily in the first half of the transcripts, it is clear that these results again attest to the depth-first manner in which complex requirements were addressed.

14.7.2.4 Length of Individual Episodes Spent Dealing with Low-, Intermediate-, and High-Complexity Requirements

In this analysis we counted how many *continuous* segments were spent dealing with a requirement once it had been introduced into the solution. This *episode* measure reflects a more refined estimate than the previous analysis of how much time designers spent "in one go" in addressing a specific requirement at each level of complexity (Table 14.7). Our working assumption was that longer episodes are indicative of a greater depth of development. To ensure independence of data points, all segments dealing with more than one requirement were eliminated from the analysis.

A one-way ANOVA on the aggregate data revealed significant differences between requirements at the three complexity levels, $F(2, 44) = 4.53$, $p = .017$. Follow-up analyses using Tukey tests indicated that more continuous segments were associated with addressing the high-complexity requirements than the low-complexity requirements ($p = .015$), but neither the difference between high- and intermediate-complexity requirements nor that between the low- and intermediate-complexity requirements was reliable ($p = .12$ and $p = .83$, respectively). We note that the relatively low number of requirements associated with the intermediate-complexity category may well have hindered the emergence of a significant effect when comparing this category with the high-complexity requirements, since the trend in the data is as expected.

Limitations with the data set prevented a team-based analysis for the Inuit team and also potentially contributed to the absence of a significant complexity effect for the AmberPoint team. However, the complexity effect was highly reliable for the Adobe team, $F(2, 18) = 15.13$, $p < .001$, with follow-up analyses indicating that more continuous segments were associated with addressing the high-complexity requirements relative to both the intermediate- and low-complexity requirements ($p < .001$ for both).

TABLE 14.7 Mean Number of Continuous Segments Making Up Episodes That Were Devoted to Dealing with Requirements at Each Complexity Level[a]

Complexity Level	Number of Requirements	Mean Segments	Standard Deviation
Low	19	3.68	3.98
Intermediate	11	4.91	4.22
High	15	9.47	5.84

[a] Collapsed across teams.

Our overall conclusion from these analyses is that when complex requirements were being mentioned the designers spent rather longer dealing with them compared to easier requirements, with this finding being particularly evident in relation to the design work of the Adobe team (see Extract 14.2 for an example of an in-depth simulation deriving from this team in relation to Requirement 22: "Allow for choice of traffic density on any given road"; high complexity). Again, these findings provide further support for the role of depth-first design in dealing with high-complexity requirements.

EXTRACT 14.2 Example of a High-Complexity Requirement (Requirement 22) Leading to a Long, Detailed Design Episode in the Adobe Team That Spanned Eight Transcript Segments

Bold font indicates possible requirement-related words and underlining indicates hedge words expressing uncertainty.

[0:17:02.8] *Male 1:* Yeah, exactly. So, it's, the cop is looking at well **state** of the signal and the some, probably also needs to define some sort of great [inaudible] cuz that's something that user should be able to control rate of the traffic. How heavy the traffic is—

[0:17:32.4] *Male 2:* Right, maybe we can peg that for the moment and introduce that as we flush this out.

[0:17:40.2] *Male 1:* Right, so the cop, yeah, per tick at this point the cop just changes the **state** of the model consistently.

[0:17:51.1] *Male 2:* Right, so tick happens and changes time, cop is watching time, for each tick cop has some set of rules that it knows and based on those rules changes the queue of cars, well <u>maybe</u> those rules are really what's going on in the intersection and, you know, how fast the car is traveling say, but then there's also the notion of **state** of traffic so, you know, what is that a road, some number of roads that have, and just wonder if <u>maybe</u> the important thing is since we are trying to simulate traffic, traffic happens, so this road is <u>kind of</u> like queue, queue is the road say cuz that's what were trying to determine is, trying to minimize the average size of the queue by optimizing the **timing** on the **light**s. So, the road has a queue, blue boxes on the road and that's a [inaudible] **structure**. So, let's say, so talking just about how you empty the queue not how you fill up the queue. So, like pop event would have, the clock would tick, the cop would say a tick has happened and the cop looks at, maybe there's a **light** at the intersection, **light** with some green, red—

[0:19:59.0] *Male 1:* Right, so somehow represents the **state** of each traffic signals.

[0:20:03.8] *Male 2:* Right, so the cop looks at the intersections says okay it's a green **light** on these roads. This road has some number of cars in it. If the **lights** green one tick means we pop and we do that queue and second picture. So, from T zero to T one that's how we've gone through two. So, one thing we haven't talked about is how we in queue a car.

[0:20:36.2] *Male 1:* Yeah and also, so that's one thing, and also like how long these signals stay green. Is that also something the cop looks at?

[0:20:45.3] *Male 2:* Well, that could be, maybe the cop—

14.7.2.5 Relation between Simulation and Low-, Intermediate-, and High-Complexity Requirements

Our next analyses explored the relation between mental simulation and requirement complexity. For each requirement at the various levels of complexity, we calculated how many separate simulations arose (Table 14.8). A one-way ANOVA on the aggregate data comparing across the three levels of complexity revealed a reliable effect, $F(2, 16) = 15.42$, $p < .001$. Follow-up analyses using Tukey tests indicated that more simulations arose for high-complexity requirements compared to both intermediate- and low-complexity requirements ($p = .002$ and $p < .001$, respectively), but there was no difference between intermediate- and low-complexity requirements.

Examining the data at the team level revealed that the main effect of complexity observed at the aggregate level was evident for the AmberPoint team, $F(2, 16) = 4.60$, $p = .03$, and for the Intuit team, $F(2, 16) = 9.21$, $p = .003$, but not for the Adobe team. The follow-up analyses for the AmberPoint and Adobe teams supported the aggregate observations that differences primarily resided in the number of simulations devoted to dealing with the high-complexity requirements relative to the intermediate- or low-complexity requirements.

Overall, these findings again support our claims that the designers were not only tackling the complex requirements straight away but were also exploring them more intensively and more deeply than the easier requirements—in the present case as reflected in an increased quantity of simulation runs used to explore design ideas at greater levels of detail.

14.7.2.6 Role of Uncertainty in Mediating between Requirements and Solution Development Strategies

Our final analyses examined the associations between uncertainty, requirements, simulations, and solution development processes in order to explore the potential mediating role of uncertainty in determining design strategies. Our first examination of these

TABLE 14.8 Mean Number of Simulations Devoted to Dealing with Requirements at Each Complexity Level[a]

Complexity Level	Number of Requirements	Mean Simulations	Standard Deviation
Low	11	3.00	2.28
Intermediate	4	5.50	4.80
High	2	21.00	11.31

[a] Collapsed across teams.

mediation issues focused on assessing whether segments associated with handling requirements embodied more uncertainty than segments that were not associated with handling requirements. A *t*-test on the aggregate data revealed a borderline significant effect whereby segments with requirements present showed more uncertainty ($M = .37$; $SD = .48$) than segments with requirements absent ($M = .31$; $SD = .46$), $t(891) = 1.95$, $p = .051$. It therefore appears that requirements handling is associated with uncertainty, at least when the data are analyses in aggregate. At the team level this effect showed up most markedly in the AmberPoint team, with segments with requirements present revealing more uncertainty ($M = .54$; $SD = .50$) than segments with requirements absent ($M = .37$; $SD = .49$), $t(303) = 2.89$, $p = .004$. The effect was not reliable for the other two teams.

Our second analysis examined whether segments associated with mental simulations embodied more uncertainty than segments that were not associated with mental simulations. A *t*-test on the aggregate data revealed a highly significant effect whereby segments with simulations present showed more uncertainty ($M = .40$; $SD = .49$) than segments with simulations absent ($M = .24$; $SD = .49$), $t(885) = 4.94$, $p = .001$. This effect was replicated across all three design teams in a by-teams analysis ($t > 2.2$ and $p < .05$ for all). It thus seems clear that simulations are also associated with uncertainty.

Our final analysis explored whether the complexity of requirements was associated with uncertain simulations. For each simulation segment we recorded whether it referred to a requirement at low, intermediate, or high complexity. We excluded segments involving more than one complexity level to ensure data independence. We also limited the data set to segments only associated with simulations. Table 14.9 shows the resulting data collapsed across design teams. We conducted a one-way ANOVA on the data set comparing across complexity levels. The analysis revealed a significant effect, $F(2, 184) = 6.31$, $p = .002$. Follow-up analyses using Tukey tests indicated that more simulations arose for high-complexity requirements compared to intermediate-complexity requirements ($p = .002$), and although the trend was in the right direction for the comparison between high-complexity requirements compared to low-complexity requirements, this difference failed to reach significance ($p = .099$).

Overall, the analyses reported in this section indicate that simulations are strongly associated with uncertainty across all three design teams and, moreover, that this uncertainty increases further as a result of the designers dealing with complex requirements. Thus, we see compelling evidence here to support our novel prediction that high-complexity requirements are linked to uncertain simulations. Extract 14.3 shows an example of a

TABLE 14.9 Mean Segment Uncertainty of Simulations Devoted to Dealing with Requirements at Each Complexity Level[a]

Complexity Level	Number of Segments	Mean Uncertainty	Standard Deviation
Low	67	.37	.49
Intermediate	38	.21	.41
High	80	.54	.50

[a] Collapsed across teams.

high-complexity design requirement leading to a simulation run that is also associated with expressions of uncertainty.

EXTRACT 14.3 Example of a High-Complexity Requirement (Requirement 14) Leading to a Simulation in the Intuit Team

Bold font indicates possible requirement-related words and underlining indicates hedge words expressing uncertainty.

[0:20:21:3] *Male 1:* I'm still wondering about the **left**-turn lane.

[0:20:32.4] *Male 2:* What about it?

[0:20:38.3] *Male 1:* What I'm wondering is that, is it one of those—how does that work? One of those where it's kind of one lane and then you have the option of going in there to you have a protected **left** and this has a certain distance, or is that—

[0:20:58.5] *Male 2:* The problem is you can chop up lanes, I mean with that they have two intersections coming to **four** turn lanes, I mean that's the problem

[0:21:03:8] *Male 1:* Yeah, so which ones do we want to have because that is really going to affect the complexity of the problem and so, we need a protected left. So let's go with the minimum so

[0:21:19.0] *Male 2:* Yeah, I'm saying this is probably the best assumption.

[0:21:21.5] *Male 1:* So there's this protected left. But then this could have a pile-up effect. So we want simple.

14.8 GENERAL DISCUSSION

Our aim in examining the SPSD data set was to improve our current understanding of the relation between requirements and solution development strategies in software design. A particular interest concerned the extent to which the organization of solution development might be determined by the complexity of the requirements associated with the overarching design goal. Our motivation for exploring this issue derives from our previous empirical research that has shown how expert design strategies in various domains involve a sophisticated combination of top-down, breadth-first and top-down, depth-first solution development (for a review see Ball and Ormerod, 1995). We anticipated finding further evidence to validate the existence of a mixed-mode solution development approach, while also predicting that high-complexity requirements would be causally important in driving expert designers' shifts from breadth-first to depth-first design. The rationale for this prediction is that high-complexity requirements will be difficult to address in terms of the degree of change that would need to be made to an initial design concept. As such, they should necessitate detailed exploration of possible solution ideas to enable their effective integration with the emerging design.

We were also interested in determining whether complex requirements impact on designers' feelings of epistemic uncertainty. A positive association between high-complexity requirements and epistemic uncertainty in the absence of an equivalent association between lower-complexity requirements and epistemic uncertainty would support the mediating role of uncertainty in driving strategic switches from breadth-first to depth-first design. We note that our previous design research has indicated that epistemic uncertainty may be related to strategic recourse to a "satisficing" heuristic in design evaluation (Ball et al., 1998), to strategic transitions from structural to functional modes of representation during design sketching (Kavakli et al., 1998; Scrivener et al., 2000), and to strategic deployment of analogizing and mental simulation processes aimed at uncertainty resolution (Ball and Christensen, 2009; Christensen and Schunn, 2007, 2008). There are, therefore, a priori grounds for predicting that uncertainty might also be associated with the deployment of a depth-first design approach aimed at dealing with requirement complexity.

Our data analyses gave rise to several important observations relevant to our predictions, as follows:

- All design teams rapidly produced an initial "first-pass" solution that involved a grid-based road layout. This early-stage design activity is indicative of breadth-first solution development, since initial solution concepts incorporated many requirements stated in the design brief.
- High-complexity requirements were subsequently dealt with much earlier in the transcripts (almost exclusively within the first half) in comparison to intermediate- and low-complexity requirements, which were tackled to fairly equal degrees across both transcript halves. This finding was generalized across all three design teams and suggests the use of a depth-first strategy to handle high-complexity requirements and a breadth-first strategy to deal with low- and intermediate-complexity requirements.
- The deployment of a depth-first strategy in the case of high-complexity requirements is further supported by the observations that these requirements were dealt with over more segments compared to other requirements and led to more simulations that were also more extensive. These observations were generalized across two of the three design teams, the one exception being the Adobe team.
- Not only were simulations found to be associated with epistemic uncertainty (an effect that was generalized across all teams), but there was also evidence of a unique association between high-complexity requirements and uncertain simulations.

Overall, these findings point to a sophisticated interplay between structured breadth-first and depth-first development in software design and support previous claims for a mixed-mode approach in expert design (see Ball and Ormerod, 1995). It would appear that the present experts were carefully attuned to optimizing the development of a conceptual design solution by capitalizing upon the benefits of both breadth-first and depth-first strategies, dependent on the types of requirements being tackled. For example, the early attainment of conceptual solutions for high-complexity requirements is likely to determine the eventual success of a piece of software, since solutions for such requirements will form a framework onto which easier requirements can be added.

Not only did we support previous findings that requirement complexity is causally related to depth-first design (e.g., Ball et al., 1997; Jeffries et al., 1981), we additionally found evidence to support our novel prediction that epistemic uncertainty (indexed by designers' use of hedge words) may mediate between complex requirements and a depth-first design strategy. This effect supports other evidence noted previously that has demonstrated how epistemic uncertainty may be critical in triggering strategic shifts in relation, for example, to the deployment of analogizing and simulation processes (e.g., Ball and Christensen, 2009). Future studies across a range of software design contexts would be valuable to clarify the mechanism by which uncertainty mediates between requirements and expert solution development and the generality of this uncertainty effect. One issue that would be particularly interesting to examine concerns the way in which requirements that have multiple dependencies are tackled during the design process. Given that such requirements are likely to be both complex and important, we would predict that they would be dealt with early rather than late by experienced designers and would also be associated with considerable uncertainty.

It also remains important to validate the present observations in studies of real, ongoing software engineering projects using *in vivo*, ethnographic methodologies (Ball and Ormerod, 2000a,b; Christensen and Schunn, 2007, 2008). Clearly the design discussions that have been analyzed here are not instances of genuine company-based design work, instead reflecting activities associated with artificial tasks staged specifically for the purpose of the study where the available time was limited and unrealistically focused and where designers were on camera and knew that they were being recorded for others to watch. Real software design scenarios studied over longer time periods via ethnographic methods might well provide evidence for a range of more nuanced phenomena surrounding requirements handling, including the possibility of preliminary solutions collapsing or being restructured when new requirements arise during client–designer dialogs.

Notwithstanding provisos regarding the generalizability of the present findings, we suggest that our analysis still provides an original and useful contribution to the design research literature by clarifying the roles played by requirement complexity and epistemic uncertainty in triggering strategic shifts to depth-first software development. We acknowledge that our interpretation of "depth-first" design in this chapter equates with metrics such as the time spent addressing a requirement and the design phase during which a requirement is predominantly tackled. We also admit that we have not pursued an analysis of design subgoaling (e.g., Ball et al., 1997), which would be another way to substantiate depth-first processing in the present transcripts. Nevertheless, the cumulative evidence that we have presented across several different types of analyses provides good support for the claim that handling complex requirements takes place in a predominantly depth-first manner that is mediated by feelings of uncertainty, whereas the handling of easier requirements is more breadth-first in orientation and is not mediated by uncertainty.

ACKNOWLEDGMENTS

Results described in this chapter are based upon videos and transcripts initially distributed for the 2010 international workshop Studying Professional Software Design and partially supported by NSF grant CCF-0845840 [http://www.ics.uci.edu/design-workshop]. We thank the workshop organizers, André van der Hoek, Marian Petre, and Alex Baker, for convening this event and for providing highly constructive feedback on our work. We are also grateful for the valuable critical feedback provided by three anonymous reviewers who read a previous version of this chapter. Finally, we thank Kristoffer Riis Pedersen for his assistance in primary transcript coding. This writing of this chapter was partly supported by the Initial Training Network "Marie Curie Actions," funded by the FP7 People Programme (PITN-GA-2008-215446) entitled "DESIRE: Creative Design for Innovation in Science and Technology."

REFERENCES

Adelson, B and Soloway, E (1986) A model of software design. *International Journal of Intelligent Systems*, 1(3), 195–213.

Ball, L J and Christensen, B T (2009) Analogical reasoning and mental simulation in design: Two strategies linked to uncertainty resolution. *Design Studies*, 30(2), 169–186.

Ball, L J and Ormerod, T C (1995) Structured and opportunistic processing in design: A critical discussion. *International Journal of Human–Computer Studies*, 43(1), 131–151.

Ball, L J and Ormerod, T C (2000a) Applying ethnography in the analysis and support of expertise in engineering design. *Design Studies*, 21(4), 403–421.

Ball, L J and Ormerod, T C (2000b) Putting ethnography to work: The case for a cognitive ethnography of design. *International Journal of Human–Computer Studies*, 53(1), 147–168.

Ball, L J, Evans, J St B T, and Dennis, I (1994) Cognitive processes in engineering design: A longitudinal study. *Ergonomics*, 37(11), 1753–1786.

Ball, L J, Evans, J St B T, Dennis, I, and Ormerod, T C (1997) Problem-solving strategies and expertise in engineering design. *Thinking and Reasoning*, 3(4), 247–270.

Ball, L J, Maskill, L, and Ormerod, T C (1998) Satisficing in engineering design: Causes, consequences and implications for design support. *Automation in Construction*, 7(2/3), 213–227.

Ball, L J, Lambell, N J, Ormerod, T C, Slavin, S, and Mariani, J A (2001) Representing design rationale to support innovative design re-use: A minimalist approach. *Automation in Construction*, 10(6), 663–674.

Chakrabarti, A, Morgenstern, S, and Knaab, H (2004) Identification and application of requirements and their impact on the design process: A protocol study. *Research in Engineering Design*, 15(1), 22–39.

Chevalier, A and Ivory, M Y (2003) Web site designs: Influences of designer's expertise and design constraints. *International Journal of Human–Computer Studies*, 58(1), 57–87.

Christensen, B T and Schunn, C D (2007) The relationship of analogical distance to analogical function and pre-inventive structure: The case of engineering design. *Memory and Cognition*, 35(1), 29–38.

Christensen, B T and Schunn, C D (2008) The role and impact of mental simulation in design. *Applied Cognitive Psychology*, 22, 1–18.

Cross, N (2001) Design cognition: Results from protocol and other empirical studies of design activity. In C M Eastman, W M McCracken, and W C Newstetter (eds), *Design Knowing and Learning: Cognition in Design Education*. Elsevier, Oxford, pp. 79–103.

Davies, S P (1991) Characterising the program design activity: Neither strictly top-down nor globally opportunistic. *Behavior and Information Technology*, 10(3), 173–190.

Forbus, K D (1997) Qualitative reasoning. In A B Tucker (ed.), *CRC Handbook of Computer Science and Engineering*. CRC Press, Boca Raton, FL, pp. 715–733.

Goel, V and Pirolli, P (1989) Motivating the notion of generic design within information-processing theory: The design problem space. *AI Magazine*, 10(1), 18–35.

Goel, V and Pirolli, P (1992) The structure of design problem spaces. *Cognitive Science*, 16(3), 395–429.

Golden, E, John, B E, and Bass, L (2005) The value of a usability-supporting architectural pattern in software architecture design: A controlled experiment. In *Proceedings of ICSE 2005—27th International Conference on Software Engineering*. ACM Press, New York, pp. 460–469.

Guindon, R (1990) Designing the design process: Exploiting opportunistic thoughts *Human-Computer Interaction*, 5(2/3), 305–344.

Herbsleb, J D and Kuwana, E (1993) Preserving knowledge in design projects: What designers need to know. In *Proceedings of the INTERACT'93 and CHI'93 Conference on Human Factors in Computing Systems*. ACM Press, New York, pp. 7–14.

Jeffries, R, Turner, A A, Polson, P G, and Atwood, M (1981) The processes involved in designing software. In J R Anderson (ed.), *Cognitive Skills and their Acquisition*. Erlbaum, Hillsdale, NJ, pp. 255–283.

Kavakli, M, Scrivener, S A R, and Ball, L J (1998) Structure in idea-sketching behaviour. *Design Studies*, 19(4), 485–518.

Malhotra, A, Thomas, J C, Carroll, J M, and Miller, L A (1980) Cognitive processes in design. *International Journal of Man–Machine Studies*, 12(2), 119–140.

Nersessian, N J (2002) The cognitive basis of model-based reasoning in science. In P Carruthers and S Stich (eds), *Cognitive Basis of Science*. Cambridge University Press, New York, pp. 133–153.

Ormerod, T C and Ball, L J (1993) Does programming knowledge or design strategy determine shifts of focus in Prolog programming? In C R Cook, J C Scholtz, and J C Spohrer (eds), *Empirical Studies of Programmers: Fifth Workshop—ESP 5*. Ablex, Norwood, NJ, pp. 162–186.

Savage, J C D, Miles, C, Moore, C J, and Miles, J C (1998) The interaction of time and cost constraints on the design process. *Design Studies*, 19(2), 217–233.

Scrivener, S A R, Ball, L J, and Tseng, W S-W (2000) Uncertainty and sketching behaviour. *Design Studies*, 21(5), 465–481.

Simon, H A (1973) The structure of ill structured problems. *Artificial Intelligence*, 4(3/4), 181–201.

Trickett, S B and Trafton, J G (2002) The instantiation and use of conceptual simulations in evaluating hypotheses: Movies-in-the-mind in scientific reasoning. In *Proceedings of the Twenty-Fourth Annual Conference of the Cognitive Science Society*. Erlbaum, Mahwah, NJ, pp. 878–883.

Trickett, S B, Trafton, J G, Saner, L, and Schunn, C D (2005) 'I don't know what's going on there': The use of spatial transformations to deal with and resolve uncertainty in complex visualizations. In M C Lovett and P Shah (eds), *Thinking with Data*. Erlbaum, Mahwah, NJ, pp. 65–86.

Visser, W (1990) More or less following a plan during design: Opportunistic deviations in specification. *International Journal of Man–Machine Studies*, 33(3), 247–278.

Visser, W (1994) Organisation of design activities: Opportunistic, with hierarchical episodes. *Interacting with Computers*, 6(3), 239–274.

Visser, W (2006) *The Cognitive Artifacts of Designing*. Erlbaum, Mahwah, NJ.

Walz, D B, Elan, J J, and Curtis, B (1993) Inside a software design team: Knowledge acquisition, sharing, and interaction. *Communications of the ACM*, 36(10), 63–77.

CHAPTER 15

Designing Assumptions

Ben Matthews

The University of Queensland

CONTENTS

15.1 INTRODUCTION

Design theorists have differed widely on whether design is, in essence, a form of some other activity. Candidates for the "essence" of design have included problem solving, problem defining, creating fit, negotiation, selection, generation, and decision making, among others. "Foundations" for design practice have been proposed from the arts and crafts, as well as the natural, technical, and social sciences (Buchanan 1992). There are also those who reject the idea that there is, or needs to be, an "essence" to design at all (e.g., Jones 1992; Walker 1989). But whatever one's preferred notion of design, the vernacular idea of design relates to bringing something new into being. And this is certainly an aspect of design that makes it a compelling topic of study: the object of designers' thought, work, and activity is not (yet) something that exists. This chapter aims to show how such creative work in design—imagining a nonexistent future—is facilitated by designers' use of assumptions.

There is a historical track of design studies that have explored some of the characteristic ways in which designers work with the as-yet nonexistent. Designers necessarily sketch out possible futures (e.g., of form, configuration, function, interaction, behavior, etc.) in the course of designing. To date, one of the most visible strands of inquiry in this regard has focused on the various design representations that are used in the process, such as sketches, diagrams, prototypes, and CAD models. Recently however, important attention has also been paid to some of the more mundane ways in which this work of design is done, particularly through talk (e.g., Dong 2007; Lloyd and Busby 2001; McDonnell and Lloyd 2009). The contention that creative design work is done through talking is perhaps not as sexy to the design research field as the idea that the ability to design essentially rests in other competencies that are (more or less) unique to the discipline and that are honed in studio-based educational environments; for instance, visualization or prototyping. Nevertheless, more often than not design activity is a collaborative endeavor and is dominated by talk even when other media (pens, papers, computers) are available to designers and other competencies are evident in what they do.

The purpose of this particular investigation is to document a set of designers' "languaging" practices—social actions done through talk—in the course of designing. The particular practices I am interested in here are those associated with the concept of assumptions. Assumptions are interesting in a design context for a number of reasons, not least of which is their constructive role in sketching out (in language) the future. Although this role of assumptions may seem obvious, the attention that has been paid to designers' assumptions in the literature is predominantly concerned with designers' *incorrect* assumptions—their mistakes—that are embedded in resultant designs. Discussions often center on designers' mistaken conceptions of users, use context, and the "natural" intelligibility of the features they inscribe into products and systems (e.g., Frishberg 2006; Strohecker 2001). The clear prescription in such cases is the recommendation to replace such assumptions with (as close as possible to) firsthand knowledge of users, users' settings, and their endogenous sense-making practices (e.g., Crabtree 2003). While these kinds of designers' assumptions are undoubtedly the source of many design problems and are worth remedying in the ways prescribed, the real-time emergence of assumptions in design activity and the work such assumptions do remains an unexamined topic in design studies. Video recordings of collaborative design activity, such as those that have been made available as data for this volume, provide a valuable opportunity to scrutinize designers' assumptions in action.

Additionally, the task these designers have been assigned lends itself well to this investigation, as it is essentially a software modeling task for a real-world phenomenon—traffic. As such, discussion content in each of these sessions explicitly concerns itself with attempts to determine which aspects of the world (e.g., time, cars, roads, lanes, signals, distances, speeds, traffic density, etc.) need to be included in the software (and how) for it to do what it is supposed to do: model traffic flow. In any case, assumptions figure explicitly in each of the three sessions. Initially, then, it was fairly straightforward to find some cases where designers specifically deploy an "assumption" in the course of what they are doing.

15.2 ANALYTIC APPROACH

The particular approach to analysis that I have taken draws on discursive psychology (Edwards 1997), which is a form of discourse analysis informed by ethnomethodology. Discursive psychology specializes in treating psychological themes as they arise in ordinary conversation. One of the strands of this program of research has been to analyze a range of vernacular uses of a particular concept or lexical item (Edwards and Potter 2005) in ordinary conversation, in order to identify the pragmatic work that the concept/word is employed to do in everyday life. Discursive psychology has typically treated concepts of central importance to psychology: for example, thought, knowledge, intention, attention, emotion, memory, and the like. My aim in this chapter is to use this particular analytic lens to look at the vernacular notion of "assumption" as it appears in these sessions. This will have several upshots. Firstly, it will detail the range of deployments of the notion of assumptions in design conversations as a means of elucidating the use and usefulness of the notion for both social/interactional and design task ends (Sections 15.3 and 15.4); it will enable a comparative discussion of the differences between theoretical/conceptual analyses of concepts such as assumptions and the empirical/pragmatic analysis undertaken here (Section 15.6); and it provides for several important methodological points to be made regarding the difficulties and complications arising from analyses that attempt to abstract the content of designers' talk away from the social and interactional context within which the talk does work in the setting (Sections 15.4 and 15.5).

15.3 ANALYZING ASSUMPTIONS

The first example is drawn from the Adobe transcript. This is toward the end of their session, while M1 is writing on the board, populating a set of variables within an "intersection class" at this point ("current [light] state," "horizontal flow," "vertical flow," etc.). The transcripts throughout have been rendered in a simplified version of Jefferson's (2004) transcription notation. An explanation of the notation system used here has been included as an appendix.

EXCERPT 15.1 (Adobe) It's a Simplifying Assumption

[1:47:01.8]
1 *M2:* oh I guess then current (.) current state ri:ght
2 (1.6)
3 *M1:* o:h
4 *M2:* which would (.) be a horizontal flow or (0.2)
5 *M1:* vertical flow
6 *M2:* or a vertical flow and that really dictates (1.4) well
7 there's horizontal flow, vertical flow, but then there's
8 horizontal left (.) horizontal (.) vertical left.
9 *M1:* yea::h
10 (1.3)
11 *M2:* and that's (1.7) can:: (0.8) could you have a case
12 where (1.2) you can: (.) make an advanced left here
13 but not (.) an advanced left there?
14 (0.7)
15 *M2:* or are they- (.) or are- (.) all the opposing (0.3) like lights
16 the same state?

17 (1.0)
18 *M2:* °probably not°
19 (0.5)
20 *M1:* it's a simplifying assumption.
21 (2.8)
22 *M1:* c'z you could also I mean (0.3) .hhh theoretically you could
23 also have an intersection where (1.2) this- this goes left a:nd
24 straight and this is just (.) kinda red.
25 *M2:* yeah so maybe we need, uhm, (1.4) we can say horizontal and
26 vertical but then we can say then take two lights,

At Line 15, M2 asks, "Are all the opposing like lights the same state?" There is an accountable absence of an immediate response. In ordinary conversation, agreement generally comes quickly, that is, within a few tenths of a second; disagreement is prefaced by delays such as this (Pomerantz 1984). The ensuing silence of 1 second is a fair indication that the two designers are misaligned on this point. M2 appears to orient to this, as he comes in at Line 18 with an answer to his own question in the negative, "Probably not." After another pause, M1 offers, "It's a simplifying assumption." As an answer to the original question, it almost works as an effective "no," but not quite. As M2's question at Line 15 was phrased, it is not explicit whether the question is in reference to "the world" (real traffic lights) or "the model." By labeling the opposing-lights-always-the-same-state condition as a "simplifying assumption," M1 indexes a distinction between world and model. That this is what M1 is actually doing here becomes clear in how he continues: he creates a "theoretical" scenario of how traffic lights might actually be implemented, a scenario that exposes the "simplifying assumption" as unrealistic. M1's turn here is met with immediate agreement from M2 (line 25), who proceeds to introduce (now) two different possible light states for opposing traffic into the intersection class.

The half second of silence following M1's assumption ascription at Line 20 is also notable. This is a transition-relevant place: M1 has finished a complete thought, and the floor is M2's if he wants to take it. This is a space in which the other participant might agree with the label "simplifying assumption," infer for himself the consequences of the simplification, take up, reject, or challenge the label, provide a minimal response token to at least register that he has heard it (e.g., "hmm" or "yeh"), or a number of other relevant actions. However, M2 withholds a turn here, and so M1 proceeds to argue, through a theoretical scenario, what precisely is being simplified by this assumption (and also demonstrates to M2's apparent satisfaction that it is indeed an assumption, rather than, e.g., a fact of the matter).

So M1's ascription of "assumption" to this proposal in this case highlights, and clarifies, the distinction between system and world. But we can also see that the action of labeling a particular proposal as "a simplifying assumption" was a means of (gently) offering it up for criticism, and it here allows the designers to inspect the consequences for the model they are building of instantiating such a simplification. But as we will see in the following sequence, this is not always the case.

In the Intuit data, for example, we see the following exchange.

EXCERPT 15.2 (Intuit) What Simplifying Assumptions Could We Make?

[0:23:42.9]
1 *An2:* I guess the issues that's gonna to come into play is if (.)
2 some car here (.) is wanting to go straight and this is
3 backed up, they cannot. they can't get into the left. I mean
4 tha- that's the real scenario its- right
5 (3.0)
6 *An1:* I hate to (0.3) bog down in- (0.2) in that de:tail.
7 *An2:* yeah- yeah let's (0.2) true. very true.
8 (1.4)
9 *An1:* so to keep it moving, (1.2) wha- what simplifying assumptions
10 could we make initially? (0.4) a:::nd find those so that (0.3) if
11 this has a len:gth? (2.2) it never overflows?
12 (0.3)
13 *An1:* you know if ten- five cars can fit then that's all:
14 *An2:* other option is to make a full (0.2) t- second lane.
15 (0.2)
16 *An2:* don't worry about (0.3) turning in, just have a full second lane
17 here so that we have
18 (0.3)
19 *An1:* yea:h
20 *An2:* that way we don't have that whole trying to figure out how-
21 what that length considered to the other part of the lane i:s and
22 *An1:* right
23 *An2:* if you have a full lane (0.2) everywhere in the intersection so
24 everything fits they're two lanes at that point.
25 *An1:* it's two lanes, one of them
26 *An2:* yeah (0.2) and then (.) you just have dictate people waiting for
27 that tur:n are going to pile u:p an-
28 *An1:* uh [huh
29 *An2:* [as if it was straight
29 (0.6)
30 *An1:* right
31 (1.1)
32 *An2:* I think it's a simplification
33 *An1:* okay.

Here we have a rather different use of "simplifying assumptions." Beginning in Line 1, An2 identifies an issue in reference to a "real scenario" of traffic flow. The issue he introduces relates to vehicles that are turning into a road with heavy traffic and must cross a busy lane to get into a turning lane on the left side. An1's response, "I hate to bog down in that detail," treats as *problematic* such details of traffic conditions, and is a remark that garners immediate assent, without delay, from An2. An1 proceeds to construct a turn that, in the interests of keeping "it moving," proposes, "what simplifying assumptions could we make initially," to obviate the issue. In this case, as in the first, an assumption marks a contrast with the world, that is, An2's "real scenario" (Line 4). But an importantly different action is being performed with the idea. An1's simplifying assumption is employed as a component of a proposal for a course of action to pursue, as a provisional remedy (note the "initially") to getting bogged down in details. Within that turn he appears to suggest the simulator sport a lane that never overflows (something clearly unrealistic for a traffic simulator!). In contrast, An2 offers an "other option" (marking his suggestion as a counterproposal to the "never overflows" idea on the table), to make a "full second lane" for the turning lane. He assesses even this option of his as "a simplification" (Line 32), but such a simplified proposal is met with agreement ("okay") from An1, and they proceed to develop the program along these lines.

In this case, a simplifying assumption is not a label that criticizes an unrealistic proposal, but a suggested strategy that enables them to "conserve ambiguity" (this is Minneman's [1991] felicitous phrase) so as not to be waylaid by too many real-world issues and constraints. This move enables the designers to find a temporary means of getting around the issue without having to immediately specify every aspect of real traffic conditions in order to perfectly model the protected left turn. The generation of simplifying assumptions is proposed and collaboratively pursued as a strategy for managing the complexity of the problem.

Where the two uses of assumptions we have seen so far have been involved with distinguishing between model and world, the following example is a bit different. Early in the AmberPoint transcript, we have the following:

EXCERPT 15.3 (AmberPoint) We'd Have to Assume

[0:06:27.4]
1 *F:* bu- so it's like both a- a draw:ing tool
2 *M:* yeah
3 *F:* that can allow them to lay down these intersections
4 (0.2)
5 *F:* and also some mechanism by: (.) specifying these timing switch
6 (0.4)
7 *F:* for the lights
8 *M:* yeh
9 *F:* distances between intersections, and (.) car speeds
10 I suppo:se, or I don't know. does it say anything about n-
11 (0.3)
12 *M:* yeah I didn't see any[thing about speeds right
13 *F:* [I guess tha-

14 *F:* yeah I suppose there's some traffic (.) speeds
15 at which these cars travel and they probably vary
16 (0.2)
17 you know so we'd have to assume that .hhh from
18 one intersection to another it can be different
19 (0.8)
20 *F:* is tha- so if they go from here to here at 25 max
21 and here 35 max is [that how is that
22 *M:* [yeah, that's probably enough yeah
23 (6.0)
24 *M:* yeah I don't know if they can set the speeds.
25 (0.2)
26 *M:* they can set the (.) density.
27 (1.4)
28 *F:* °yeah.°
29 (5.5)
30 *M:* want to draw something?
31 (1.2)
32 *F:* oka::y:?

Here, what "we'd have to assume" at Line 17 also concerns an issue with respect to the distinction between the system being designed and something else. But in this case, the something else is not the real world. What is being negotiated here, what is to be assumed, is about what the necessary inclusions in the system will be. She begins by listing elements: the drawing tool to draw the intersections, the timing settings for the lights, and the distances between intersections. When she shifts to suggest car speeds, though, she softens this as a suggestion ("car speeds I suppose, or, I don't know," Lines 9–10). Then she phrases a question "does it say anything about ...." Thus the issue for the moment becomes what the brief requires (not, e.g., what a real traffic scenario might be). When he says he "didn't see anything about speeds," he reinforces this reference to the brief as an appropriate arbiter, with the implication being that they may not need to consider speed. However, she continues to propose car speeds as an inclusion in the system, and in the absence of an explicit requirement to consider speed in the brief, she instead "supposes" and "assumes" what they would have to take into account: "I suppose there's some traffic speeds ... so we'd have to assume from one intersection to another it can be different." Although this ends her turn (Line 18), this proposal is not met with any audible response from him (compare his first turn in this excerpt, where he comes in quickly with "yeah" after she has suggested it should be a drawing tool). Without any uptake or agreement from him here, she proceeds to introduce an example (Line 20), but he cuts her off, "yeah that's probably enough yeah," which is produced in overlap with her speech (the onset of overlap is indicated in the transcript by the open square brackets). The long 6-second silence at this point prefigures the ensuing disagreement that he produces: "I don't know if they can set the speeds, they can set the density." It is relevant that traffic density (as opposed to speed) features quite prominently in the brief. Interactionally, there is a good deal of evidence of misalignment between the participants in this sequence. The positions of silence after her turns where he could come in with agreement or understanding tokens (Lines 4, 6, 16, and especially 19) and the long silences toward the end following both participants' turns (Lines 23, 27, 29, and 31) are indications of "dispreferred" structures of interaction. (In American English, for example, a silence in conversation longer than a second is, all things being equal, an accountable delay that likely prefigures interactional "trouble" of some kind.)

Returning to her assumption in this case, we can see it is not a "simplifying" assumption; it is a complicating one. And as it turns out, precisely what "we'd have to assume" is itself a contestable matter, and one that is momentarily denied further consideration by reference to the brief's silence on the issue of speed.

There is another way in which the notion of assumption comes to be of use to designers in these sessions, and that is in order to do what I will call "hedging work." The following sequence appears early on in the Intuit transcript, where An1 is annotating a diagram of a four-way intersection on the whiteboard.

EXCERPT 15.4 (Intuit) I'm Assuming Probably

[0:12:12.9]
1 *An1:* so do we have to model the behavior where (.)
2 we have these cars turning left, thes::e
3 stopped, (.) these cars going straight, and then when
4 this (.) stops these cars can then go
5 *An2:* I'm assuming probably we'd have to do that
6 *An1:* and so:::, I mean
7 *An2:* (if a lane would help there?)
8 *An1:* I'm not a civil engi↑nee:r, but (.) yeah we can
9 do, you know, the lefts can go at the same time and then

An1 asks a question about the necessity of including a particular traffic scenario in the behavior the system should allow for. An2 responds generally positively at Line 5, with a form of agreement that employs the word "assuming": "I'm assuming probably we'd have to do that." Here it becomes quite easy to see what the concept of assumption does in such a sequential position (i.e., as a component to an answer to a question of what to include in the system). This instance sheds some light on its other uses documented above. One can see what semantic/pragmatic effect "assuming" has in An2's reply if we compare it to two other replies that use only some of the same phraseology he employs here. If An2 had simply said, "we'd have to do that," he insists upon the necessity of this behavior requirement, something "I'm assuming probably we'd have to do that" does not do. Alternatively, the reply "probably we'd have to do that," would clearly soften this; but this too is an evaluative statement. Saying "I'm assuming probably we'd have to do that" commits An2 to a cautiously endorsing position. Phrased like this, he produces a form of agreement as his own opinion, what he is personally assuming. Furthermore, both the "assuming" and the "probably" work to hedge An2's commitment against the possibility that this position might turn out to be incorrect. In this grammatical position, "I'm assuming" is somewhere between "I'm sure we'd have to do that" and "I haven't the slightest idea if we'd have to do that." This is what I am calling "hedging work." Here it is a form of conditional agreement for the time being, which in this particular formulation already hedges against the possibility that future considerations might show this to be an unwarranted, unnecessary, or unfounded position.

15.4 SUMMARY AND ELABORATION

To this point, I have sought to identify a range of different employments of the concept of assumption in these design data. However, I want to first distinguish between the social or interactional work in which assumptions are implicated and the design work and/or consequences of their use. Taking each of these examples in turn, we find the following assumptions:

- In Excerpt 15.1, a component of a gentle criticism implying that the prior suggestion was unrealistic (social/interactional work); a means of introducing and articulating a distinction between system and world (design work)
- In Excerpt 15.2, a component of a proposal or invitation for a course of action to pursue (social/interactional work); a strategy for simplifying and/or managing the complexity of considerations to take into account (design work)
- In Excerpt 15.3, a component of an incremental elaboration of an assertion of what the system should take into account (social/interactional work); a means of filling in system requirements left out of the brief, proposing additional necessary considerations (design work)
- In Excerpt 15.4, a component of an expression of guarded agreement that does "hedging work" in interaction (social/interactional work); a means of publicly marking a design option as imperfect, possibly wrong, or unrealistic (design work)*

I have only attempted to separate "social" from "design" work above for the sake of clarity and as a means of drawing into relief the utility of adopting a discursive/pragmatic approach to the analysis of design interactions as I have done here. Generally speaking, analysts of conversation are typically interested in the social/interactional work performed—that a particular utterance is a proposal, invitation, criticism, explanation, etc. They will analyze how the turn at talk is produced, its sequential placement, lexical choice, and so on to reveal precisely how it does the interactional work that it does in conversation. In contrast, design researchers are, generally speaking, only interested in the design implications of what is said—utterances are scrutinized for their relevance to the design being developed, the ideas expressed, the content of the talk, and the strategies it might contain for organizing the

* I am indebted to an anonymous reviewer for this observation and its subsequent elaboration in Section 15.6.

process or tackling the design task. Before returning to the notion of assumptions, then, it is worth noticing what is missing in either the typical pragmatic or the typical design research approach to the analysis of conversation. The former frequently does not progress to the point of being able to make connections to the topics of interest to design research, which include, among other things, strategies for doing design, managing complexity, and how ideas emerge in dialog. The second approach, by abstracting away the interactional details in favor of pursuing its particular interests, "loses" them as phenomena (cf. Garfinkel 2002). That is, if we see the question in Excerpt 15.2 "what simplifying assumptions can we make" exclusively as a strategy for managing complexity in design, we are prone to conceive of it as an ever-present interactional device that, when employed in design conversations, can always do the work of managing the complexity of the task. To presume so would be a mistake, however, since it gains its local sense *as a strategy* in this instance from its relations to what has just transpired and how it is subsequently responded to in interaction. In other words, it is its *sequential placement* that allows us (as analysts) to see it as a strategy with particular design consequences. If we reify it by abstracting it from its endogenous sequence, we lose its occasioned character. If, in contrast, we focus solely on its sequential placement and consequences, we run the risk of mistaking its design character—its relevance to the business at hand. Clearly, attention must be paid to both these aspects.

With this in mind, the features extracted and listed in bullet points above with respect to assumptions in design should not be seen as independent or mutually exclusive. Similarly, they cannot be reduced to a single formulation or definition of the concept of "assumption." They are best seen as partially overlapping, partially unique, shades of meaning that are deployable in the performance of a range of actions in design (and other) settings. If we continue to examine some of the other examples of the appearance of assuming or assumptions in these data, we can recognize in each case their relation to this family of employments (social and design) specified above.

EXCERPT 15.5 (Intuit) Do We Want to Assume One Lane

[0:11:56.3]
1 *An1:* do we want to assume that it's (0.4) one lane of
2 traffic coming i::n:, an::d
2 (0.6)
3 *An2:* u-u- (.) I don't think we need to make an assumption of lanes,
4 period. I think we just (.) go off the number of signals at the
5 (.) [each intersection. um at-
6 *An1:* [okay
7 (0.5)
7 *An2:* the only thing the lanes really going to
8 affect (.) i:::s (1.5) how (do) the lights affect us.

In this case, for instance, a proposal is (gently) made through a question about what the designers ("we") might "want to assume." The question is answered in the negative, after a short delay prefiguring the ensuing dispreferred response. Interestingly, the suggestion in the question appears to be an attempt to simplify the model; the negative answer is a recommendation to avoid having to consider the complication of lanes altogether. In terms of the design task, assumptions are again employed here as a means of distinguishing between world and system, and managing the complexity (or conserving the simplicity) of the program being developed.

The use of "assumption" as border police of the world/system divide is also clear in the following example from the AmberPoint data.

EXCERPT 15.6 (AmberPoint) This Would All Assume

[0:17:20.3]
1 *F:* again, this would all assume very <regularly placed,
2 (.) roads and (.) intersections,> right? so
3 (1.0)
4 *M:* hmm::? [inaudible overlap]
5 *F:* [In the real world] you wouldn- you know,
6 A can be curvy an:d
7 *M:* right
8 *F:* intersect 1 three times and (.).hhh.tch do an
9 S [and dead-end].
10 *M:* [inaudible] different arrangements of intersections
11 (9.0)
12 *M:* Okay, well I think we've got the visual map of the area huh?
13 *F:* Mm Hmm

Here, "this would all assume" is a reference to the system that is currently depicted on the whiteboard (see Figure 15.1). And precisely what "this would all assume" is explicitly contrasted to the case "in the real world." Produced here, in reference to the sketch of the interface on the whiteboard, it is a critical realization of a limitation of the system model (exposed by the way that intersections are being named within it), similar to the way "it's

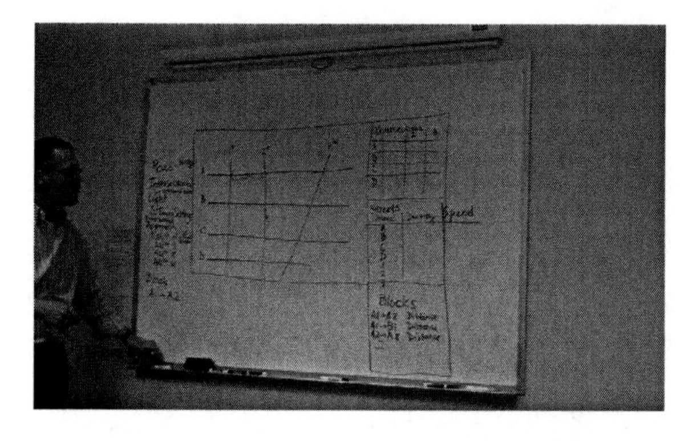

FIGURE 15.1 AmberPoint whiteboard corresponding to Excerpt 15.6.

a simplifying assumption" was used in Excerpt 15.1—as a suggestion that this particular feature or option is unrealistic. But this criticism is effectively deflected for the moment, as her colleague responds by reading a line from the brief, where it only says the system should accommodate "different arrangements of intersections." After his quote of the brief here, there is a very long, 9-second silence. Then he attempts to change topic, moving on with an "okay, well," and a summary of where they have got to so far. The critical realization that she is making by explicitly identifying an assumption in the map on the board in this excerpt is deflated by his reference to the explicit requirements of the brief.

Assumptions and filling in missing material from the brief are connected again in the following example:

EXCERPT 15.7 (AmberPoint) Do We Assume

[0:09:31.1]
1 *F:* So I guess roads, intersections, (0.4) right?
2 they need to place roads, intersections, .hhhhh
3 uh, what else are we gonna need to draw it?
4 do we assume- did it say that the intersection has
5 definitely traffic lights?
6 *M:* yeah everyone has a light

In this case, the designer self-corrects in Line 4, having started a question "do we assume," rephrasing to "did it say." M answers the question in the affirmative, implying that the brief specifies this requirement of the modeled world, and that there is (therefore) no assumption to make.

A final example is drawn from the Intuit transcript, where the designers have been discussing different types of real intersections.

EXCERPT 15.8 (Intuit) This is Probably the Best Assumption

[0:21:03.8]
1 *An1:* yeah, so which ones do we want to pick out because that
2 is really going to
3 *An2:* yeah
4 *An1:* affect the complexity (.) of the problem and so, (0.8)
5 *An2:* I would sa:y
6 *An1:* to ge- (0.4) we need a: protected left. (0.8) And so (.)
7 think (0.2) let's go with the minimu:m:? [and ha:ve
8 *An2:* [Yeah, I w'd say
9 this (.) this is probably the best assumption
10 *An1:* °the minimum° so there's this (1.8) °protected le:ft°.
11 but then this can be (0.4) have a pile up effect.
12 (2.0)
13 *An1:* so we want simple, (1.6) u::hm:

14 *An2:* I mean does it say anything about (.)
15 *An1:* yeah [(inaudible)
16 *An2:* [number of lanes? heheheh because we can
17 make the assumption of just two lanes everywhere
18 and that would be
19 *An1:* yeah [exactly
20 *An2:* [make life a lot easier.

Here the designers themselves make several explicit links between assumptions and the complexity/simplicity of the issues. When An1 proposes they "go with the minimum" (Line 7), An2 comes in immediately with agreement: "yeah I w'd say this this is probably the best assumption." An2 here relabels An1's "the minimum" proposal as an assumption. But note here that "probably" makes another appearance along with "assumption" (as it did earlier in Excerpt 15.4), and that this works, as before, to hedge his agreement against possible future misfortune. Furthermore, the management of complexity aspect of assumptions also resurfaces a couple turns later, where An2 ties "the assumption of just two lanes everywhere" (Line 17) to making "life a lot easier."

15.5 THE TERM "ASSUMPTION" VERSUS ITS ACTIONS

Having made these illustrations about assumptions in the analysis, it is important to emphasize that there is not necessarily anything essential about the explicit use of the term "assumption" in these cases. In other words, many closely related, very similar actions can be performed without the use of the word. When we identify the kinds of actions that are performed with the concept, we can see that it (the concept/word) is better understood as one lexical option for designers in the course of doing something, and it is rarely the only one available to them. For example, in the Adobe transcript, some different phrases are used that similarly negotiate the boundary between system and world, without any reference to assuming or assumptions.

EXCERPT 15.9 (Adobe) In Some Ways for Our Purposes

[0:51:35.4]
1 *M2:* so this is interesting so now (.) should allow for roads of
2 any length to be placed so length, in some ways, for our
3 purposes, is just the number of cars that it will hold right?
4 *M1:* the maximum yeah, so that's like a property of the road

EXCERPT 15.10 (Adobe) As Far as We're Concerned

[0:52:15.7]
1 *M2:* Yeah, right. right because roads kind of come in
2 pairs as far as we're concerned
3 *M1:* Yeah, in-road. There's a corresponding in-road in every
4 intersection. That holds the same for maximum cars.

In both of these excerpts, we find elements of the shades of uses identified for assumptions. There is an implied distinction between model and world when one defines the length of a road "in some ways, for our purposes" in numbers of cars that the queue can hold. The "in some ways, for our purposes" identifies the model as the object of conversation, an identification clearly taken up by M1 in his response. Similarly, "as far as we're concerned" highlights the specifically selective interest in roads that the designers have for their modeling purposes here. It is not difficult to imagine paraphrases for either of these actions that employ the word "assume" or "assumption." In the first case, perhaps "so we can assume that length is just the number of cars that the road will hold, right?" would effectively do the same job. For the second case, "right, because we're assuming roads kind of come in pairs" might work in the same way. In each of these four cases (the two originals and the two paraphrases), we can see that they single out the system (as opposed to world, brief, scenario, or whatever) as being under consideration; we can see that they hedge against possible error (e.g., "in some ways," "for our purposes," and "as far as we're concerned" also do hedging work); and in each case they simplify or delimit the considerations designers are dealing with. Of course, it is important to note that the two paraphrases I have invented do not do an identical job to the original formulations, but they are clearly close in kind.

An important methodological point can be reinforced here. Firstly, to the extent that paraphrasing (as I have above) legitimately works as a demonstration that the same actions in design discourse can be performed in different ways, with different words, this signals a methodological difficulty for forms of analysis that specifically focus on words or phrases used in design discourse or that code data into categories (either predefined or "grounded") based on transcripts. That is, very different things can be done with the same words if they occur at different times in different places, in response to different actions. And much the same things can be done with different words or even, at times, without words at all. Analysts run the risk of missing, or mistaking, the phenomena being investigated if these are identified without reference to *where* (i.e., the sequence in which) they appear and what actions they perform there. This is a point that bears close relation to a methodological argument made elsewhere (Matthews 2007). There is a strong temptation in many analyses of design activity to look solely at the content of the talk, the drawings, and other productions of design work. In a sense, this is natural, since what design researchers are frequently most interested in is precisely the "something from nothing" that comes into being, that is, the object or system being designed. But to focus exclusively on the content—what ideas are introduced, what considerations are taken into account, what problems are anticipated—is to overlook the details of how design work actually gets done. These details are important for several reasons. Firstly, they are a means of making sure the phenomena of interest are not distorted beyond recognition through some subsequent analytic procedure or manipulation. Secondly, by preserving the "social work" in the analysis, they show how design work is actually done and how the "social" and "design" aspects of discourse are not separable except for post hoc analytic purposes. For example, the "hedging" done by an assumption (an interactional move that distances a proposal from the designer making it), creates a landscape of immediate interactional possibilities that are absent (or much more difficult to bring off) without it. If you mark your idea as possibly wrong when you announce it, it makes it much easier for me to point out its shortcomings or to offer a different opinion, because I can do so without baldly contradicting you. Furthermore, in hedging with an assumption you have publicly marked a design direction as possibly requiring future revision, something that may need to be returned to later and amended. It is not just that these interactional moves are made within design activity; they are worth paying attention to on account of their consequences for the possibilities they afford an ongoing design process.

This is why the approach I am taking here (as elsewhere, e.g., Matthews [2009] or Matthews and Heinemann [2012]) establishes an important supplement to content-based analyses of talk. I have attempted to start with a consideration of how certain actions are performed in design discourse—criticisms, assessments, suggestions, proposals, and the rest—in order to first identify what is being done with the concept (assumptions) under investigation.

15.6 DISCUSSION

A theoretical (as opposed to empirical) consideration of "assumptions" might begin with consideration of a dictionary definition: for instance, to take something to be true; or to take for granted (Merriam-Webster 2012). Were we to permit ourselves idle philosophical reflection on the "nature of assumptions," we might initially be tempted to conceive of them as being much like beliefs. Like beliefs, assumptions clearly have a lesser epistemic status than knowledge—indeed an assumption would be epistemically distinguishable from knowledge in almost every conceivable instance. Like beliefs, assumptions can be ill conceived, erroneous, untested. Like beliefs, they can also be accurate, useful, true, and reliable. Like beliefs, it is possible to hold assumptions that we are not presently aware we hold (although in these cases there would be certain circumstances that make the beliefs/assumptions we tacitly hold evident to us, in which case we would then be able to recognize and admit to them).

Further reflection would reveal the comparison to beliefs to be inapt in certain respects. As opposed to beliefs, assumptions do not need to be taken to be true by those who propose them (in spite of what the dictionary may tell us). Whereas "I believe that P, even though I know P is untrue" is not a logical possibility for the concept of belief (Collins 1987), I can assume that P, even though I know P is untrue. This could be a valid conjecture. We are entitled to assume P just for the sake of an argument; but we cannot *believe* P for the sake of an argument. And we could go on in this manner, in order to identify further salient features of the conceptual landscape of "assumptions."

The consideration of such similarities and differences between assumptions, beliefs, and knowledge certainly helps us trace some of the "logical geography" (Ryle 2004) of our vernacular concepts. It would also enable us to speculate (but perhaps no more than that) on the value and perils of assumption making in design work. And yet this kind of philosophizing about the concept of assumption does not bring us to the same kind of understanding of the employment of the concept or its usefulness to design that an empirical analysis can demonstrate.

For instance, talk of an assumption in these data is received (by the participants themselves) as something quite different to knowledge claims or propositions. To talk of an assumption does something different—it is clearly weaker in certain respects and can therefore be used in different ways. We saw it work as an invitation for revision (cf. Excerpts 15.3 and 15.5). The suggestion that a proposal is an assumption can itself be a criticism (e.g., Excerpts 15.1 and 15.6). When featured as components of proposals for action, the making of assumptions can become a strategy for managing complexity or conserving the simplicity of the solution (Excerpts 15.2, 15.4, 15.5, 15.8 through 15.10); they can also be employed in the course of introducing complications or additional considerations (Excerpts 15.3, 15.5, and 15.6).

Assumptions are often only identified (or articulated) after a particular problem or issue has been encountered, and then they are, in a sense, "retrofitted" to the current circumstance under consideration (as in Excerpt 15.6). It would be misleading to presume that assumptions are typically things we walk around with in our heads. As we can see from Excerpt 15.6, they are an occasional discursive means of identifying what we had failed to take into account after that failure becomes noticeable. In such cases, they are articulations of what we have *not* been thinking about (c.f., Suchman 1987).

Furthermore, "assumption talk" is frequently employed to distinguish between model or brief and the real world—that is, making clear a discrepancy (whether desirable or not) between the system or its requirements and the way the world is or should be. Thus, assumptions are a way of clarifying that the topic of conversation is not, for example, a real intersection but an intersection-according-to-our-model. In their very concept, assumptions preserve the possibility of error, inappropriateness, or unrealism. And as we can see, they are defeasible (i.e., open to potential refutation): sometimes by reference to the real world, sometimes by reference to the design brief, sometimes by reference to a theoretical scenario. And this surely makes them very useful in a design context, where work is directed at a future that is presently both uncertain and yet malleable. They are employed in a range of practices as a means of working along with error, complexity, simplicity, and/or unreality rather than being stalled by such things. This is undoubtedly an aspect of "satisficing" (Simon 1981) work in action, that is, one of the ways in which designers temporarily tame the intractable problems they face, enabling them to make further moves without knowledge, certainty, excessive details, or realism for the time being. These are among the possible demonstrations of an empirical investigation of the concept of assumption in design work.

REFERENCES

Buchanan, R. 1992. Wicked problems in design thinking. *Design Issues* 8 (2): 5–21.

Collins, A. W. 1987. *The Nature of Mental Things*. Notre Dame, IN: University of Notre Dame Press.

Crabtree, A. 2003. *Designing Collaborative Systems: A Practical Guide to Ethnography*. London: Springer.

Dong, A. 2007. The enactment of design through language. *Design Studies* 28 (1): 5–21.

Edwards, D. 1997. *Discourse and Cognition*. Thousand Oaks, CA: Sage.

Edwards, D. and Potter, J. 2005. Discursive psychology, mental states and descriptions. In H. te Molder and J. Potter (Eds), *Conversation and Cognition*, pp. 241–259. Cambridge: Cambridge University Press.

Frishberg, L. 2006. Presumptive design, or cutting the looking-glass cake. *Interactions* 13 (1): 20.

Garfinkel, H. 2002. *Ethnomethodology's Program: Working Out Durkheim's Aphorism*. Lanham, MD: Rowman & Littlefield.

Jefferson, G. 2004. Glossary of transcript symbols with an introduction. In G. Lerner (Ed.), *Conversation Analysis: Studies from the First Generation*, pp. 13–31. Amsterdam: John Benjamins.

Jones, J. C. 1992. *Design Methods*. New York: Wiley.

Lloyd, P. A. and Busby, J. A. 2001. Softening up the facts: Engineers in design meetings. *Design Issues* 17 (3): 67–82.

Matthews, B. 2007. Locating design phenomena: A methodological excursion. *Design Studies* 28 (4): 369–385.

Matthews, B. 2009. Intersections of brainstorming rules and social order. *CoDesign: International Journal of CoCreation in Design and the Arts* 5 (1): 65–76.

Matthews, B. and Heinemann, T. 2012. Analysing design conversations: Studying design as social action. *Design Studies* 33 (6): 649–672.

McDonnell, J. and Lloyd, P. 2009. Editorial. *CoDesign: International Journal of CoCreation in Design and the Arts* 5 (1): 1–4.

Merriam-Webster. 2012. Merriam-Webster Online Dictionary. www.Merriam-Webster.com/dictionary/assume.

Minneman, S. 1991. The social construction of a technical reality. PhD Dissertation, Stanford University, Palo Alto, CA.

Pomerantz, A. 1984. Agreeing and disagreeing with assessments: Some features of preferred/dispreferred turn shapes. In J. M. Atkinson and J. Heritage (Eds), *Structures of Social Action*, pp. 57–101. Cambridge: University of Cambridge Press.

Ryle, G. 2004. *The Concept of Mind*. London: Penguin.

Simon, H. A. 1981. *The Sciences of the Artificial*. Vol. 2, rev. and enl. Cambridge, MA: MIT Press.

Strohecker, C. 2001. M. Kyng and L. Mathiassen (Eds), Computers and Design in Context [Book review]. *User Modeling and User-Adapted Interaction* 11 (3): 261–266.

Suchman, L. A. 1987. *Plans and Situated Actions: The Problem of Human–Machine Communication*. Cambridge: Cambridge University Press.

Walker, J. A. 1989. *Design History and the History of Design*. London: Pluto Press.

APPENDIX

Transcription Conventions Used in This Chapter

Symbol	Explanation
(0.4)	Numbers in parentheses indicate silences timed in tenths of seconds
(.)	A Period in parentheses indicates an untimed micropause, usually less than (0.2)
oh [I see [okay	Square brackets indicate onset of overlapping speech, transcribed on successive lines
(for sure)	Words in parentheses indicate unclear speech and/or uncertainties in transcription
goo::d	Colons indicate an elongation in pronunciation of the immediately prior sound (e.g., the "oo" is elongated in goo::d)
.hhh	Audible in-breath, often prior to speech
really?	Question mark indicates rising intonation
anyway,	Comma indicates continuing intonation
we-	Dash indicates a sudden cut-off of the word
I know	Underline indicates prosodic emphasis
KEEP this	Capital letters indicate noticeably louder speech
°oh my°	Degree signs indicate noticeably quieter speech
<so what>	Angled brackets out indicate noticeably slower talk
>or not<	Angled brackets in indicate noticeably faster talk
h↑e↓llo	Arrows up and down indicate rising and falling pitch

CHAPTER 16

Reflections on Representations

Cognitive Dimensions Analysis of Whiteboard Design Notations

Marian Petre

The Open University

CONTENTS

16.1 INTRODUCTION

Whiteboards are familiar features of software design discussions. Whiteboards provide a focus for developmental discussions, offering a flexible, shared medium that can be used in a way that keeps pace with the discussion. Whiteboard use is dynamic; what is portrayed on boards often changes form and content quickly during use. Designers' use of whiteboards in *early design* is an example of how they represent design ideas when they are largely unconstrained; that is, they are constrained mainly by their own rendering ability and imagination, by the dynamics of the conversation, and by the physical constraints of the board—rather than by a tool or by a formalism.

Whiteboards are of interest because of the way they facilitate design discussions—and hence provide a glimpse into the design process. Design discussions at the whiteboard typically include explicit expressions of design decisions and rationale. What is drawn on the board typically feeds into later stages of design, and so the decisions made and the designs captured at the board are carried into subsequent representation and development. Some of the decisions that are made in collaboration at the whiteboard are very significant: key requirements, architectures, and main user interfaces are often sketched at this point.

This study considers what representations software designers use on the whiteboard during early design. It draws on three videos of pairs of professional designers designing a traffic simulation program, captured for the Studying Professional Software Design (SPSD) Workshop (Petre et al., 2010). The study is concerned with early, formative representations generated during design discussions. These representations can provide an insight into early design thinking, and analyzing them allows us to consider how the use of particular "materials" (in this case, informal representations on a whiteboard) can influence that early design process. It allows us to see the precursors to the more formal representations that will be used to capture the design downstream. Watching the evolution of those emerging representations during design discussions provides some insight into the designers' developing mental models of the problem and the solution. It enables us to track transitions and gaps in what is represented and discussed, and it has the potential to expose discontinuities in thinking—or discontinuities between thinking and representation—and hence to expose assumptions, misconceptions, or misrepresentations.

This analysis uses the cognitive dimensions (CDs) framework (Green, 1989, 1991; Green and Petre, 1996), which provides a broad-brush assessment of how the representation (what is drawn on the whiteboard) serves the task (software design) as conducted by the users (professional software designers). The CDs framework provides a vocabulary (see Table 16.1) for considering the ability of a representation—a "cognitive artifact"—to support the task as conducted by the user, in this case, the ability of the designers' renderings on a whiteboard to support software design. "Cognitive artifacts" are "those artificial devices that maintain, display, or operate upon information in order to serve a representational function and that affect human cognitive performance" (Norman, 1991, p. 17). A CDs analysis considers the fit between the representation and the cognitive requirements of the task; that is, the ability of the representation to support relevant

TABLE 16.1 Summary of CDs Framework

Cognitive Dimension	Definition
Abstraction	Can elements be encapsulated? If so, to what extent?
Closeness of mapping	How directly can the entities in the domain be expressed in the notation? Does the notation include entities that match the key concepts or components of the domain?
Consistency	When some of the language has been learned, how much of the remainder can be inferred? Are similar features of structure and syntax used in the same way throughout?
Diffuseness	How many symbols or graphic entities are required to express a meaning?
Error-proneness	Does the design of the notation induce "careless mistakes"?
Hard mental operations	Does the notation use mechanisms such as nesting and indirection that require mental unpacking or "decoding"? For example, are there places where the user needs to resort to gestures or additional annotation to keep track of what is happening?
Hidden dependencies	Is every dependency overtly indicated in both directions? Is the indication perceptual or only symbolic?
Premature commitment	Do developers have to make decisions before they have the information they need?
Progressive evaluation	Can a partially complete representation be executed or evaluated to obtain feedback on "how am I doing?"
Provisionality	Can indecision or options be expressed?
Role expressiveness	Can the reader see how each component relates to the whole, and what the relationships between the notational elements are?
Secondary notation	Can developers use layout, color, and other cues to convey extra meaning, above and beyond the "official" semantics of the language?
Viscosity	How much effort is required to perform a single change? How much effort is required to perform multiple changes of the same type? Does making one change then have the "knock-on" effect of requiring other changes?
Visibility	Is every part of the notation simultaneously visible—or is it at least possible to juxtapose any two parts side-by-side at will? If the notation is dispersed, is it at least possible to know in what order to read it?

Note: Drawing from Green, T.R.G. in Diaper, D., Hammond, N.V. (eds), *HCI'91: Usability Now*, Cambridge University Press, pp. 297–315, 1991; Green, T.R.G., Petre, M., *J. Vis. Lang. Comput.*, 7, 131, 1996.

reasoning (and to avoid obstructing it), and in doing so it enables issues and trade-offs to be identified. Hence, a CDs analysis allows us to critique the notation, critique the process, and reveal aspects of the relationship between the process and the notation.

Studying what goes on at the whiteboard has implications for *tool development*—especially in terms of providing continuity between different representations of design and different stages of design. It also has implications for the *quality* of software design. On the one hand, effective tools may facilitate the understanding and exploration of the important issues and requirements of the problem, encouraging effective, creative solutions. On the other hand, if assumptions, discontinuities, or misconceptions are embedded at the early design stage, then they will be problematic later.

The next sections set this study in the context of some related work and offer a general characterization of how the three pairs of designers used the whiteboard, as background to the analysis. The CDs framework is then introduced, and the analysis is presented.

16.2 BACKGROUND: DESIGN STUDIES OF EXTERNAL REPRESENTATIONS

Many researchers have portrayed the importance of external representations in design, where they are used both to support design reasoning and as a medium of communication among designers, both in design generally (e.g., Fish and Scrivener, 1990; Schön, 1988) and in software design in particular (e.g., Flor and Hutchins, 1991; Cherubini et al., 2007). Such research tends to focus on the sketches used to generate, explore, and communicate design possibilities, whether individual or shared, on paper or on whiteboards.

16.2.1 Sketching in Design in General

Sketching—externalizing design ideas in informal representations—plays an important role in the design process, regardless of the discipline. Sketching provides a quick, fluid, and flexible means to extend memory by off-loading ideas from mind to paper (Newell and Simon, 1972), but it also adds value by making ideas available to external inspection and perception, thereby supporting thought processes and idea development (Tversky, 2002; Schütze et al., 2003). Goldschmidt (1991) describes a "dialectic" between the designer and a sketched representation that contributes to the processes of design, evaluation, and reflection. Schön (1983) described this cycle of drawing, understanding, and reinterpretation as having a "reflective conversation" with the material. Such conversations can lead to "unexpected discoveries" (Suwa et al., 2000), because the ambiguity in sketches can be a source of creativity, encouraging the discovery of unintended features, relationships, and consequences. "One reads off the sketch more information than was invested in its making" (Goldschmidt, 1994, p. 164). The process of reinterpreting a sketch can draw attention to details that were not necessarily intended, or not attended to, when they were drawn—which can, in turn, prompt a reexamination and reinterpretation of the problem. Sketching plays a role in design communication with others, as well as with oneself, by making ideas visible externally. Feedback and dialog inform the design process. Ferguson (1992) identifies "talking sketches" as well as "thinking sketches," calling attention to the utility of sketches in supporting the discussion of an idea, as well as in supporting the design-thinking process. The literature characterizes sketching as central to designing, emphasizing that the process of sketching is itself important, and that its role is therefore more than just producing sketches.

16.2.2 Sketching in Software Design

Although the software engineering literature has tended to concentrate on formalisms such as specifications and code, there are some studies of sketching in software development and in user interface design. Damm et al. (2000) derived design guidance for their Knight electronic whiteboard tool from two field studies of software developers using CASE tools and whiteboards; their analysis focused on cooperation, action, and use. Dekel and Herbsleb (2007) studied the participants of 3 years of OOPSLA DesignFest events, observing how software design teams use and manipulate sketches. Cherubini et al. (2007) interviewed eight software developers, identifying nine recurring scenarios in which drawings were produced by the developers. They then surveyed over 400 more developers, using those scenarios as the contexts for their questions. Walny et al. (2011) interviewed eight graduate students and academics who use sketches as part of their software development and analyzed the life cycles and transitions of example sketches as part of the software development process. The studies of sketching in software design are strongly resonant with the more general literature, which they typically cite, articulating sketching as part of a process of design exploration. Cherubini et al. (2007) summarize a framework of goals for sketching from the literature: *to share* thoughts by making them visible to others; *to ground* communication, clarifying interpretation though conversation; *to manipulate* mental models, while reducing the burden on memory; and *to brainstorm*. Certain features of sketching tend to recur in the different studies of sketching in software design, such as mixing notations, the ad hoc nature of some representation, and the advantages of low fidelity, which are discussed in the following sections.

16.2.3 Juxtaposition

A mixture of notations is commonplace in design (Goel, 1995). Software designers make different types of sketches when they explore a design problem, often juxtaposing the different types (Cherubini et al., 2007). Petre (2009b) articulates this as "expressive juxtaposition": mixing disciplines, representations, and levels of abstraction within one sketch and hence maintaining more than one "view" or alternative at the same time. Sometimes, the change of type of sketch signals a change of context or perspective, as designers consider alternatives, explore different facets or different levels of abstraction, and consider different issues (Myers et al., 2008; Petre, 2009b). Goel (1995) identified two types of transition that take place when designers work with sketches: vertical transitions (shifting between levels of abstraction) and lateral transitions (switching from one idea to another, or attempting different design strategies). Petre (2009a) described how the ease of juxtaposition and transition makes it easy to direct attention, to tease out discrepant understandings, and to provide alternative descriptions or explanations—all important for a team understanding of a design proposition (Figure 16.1).

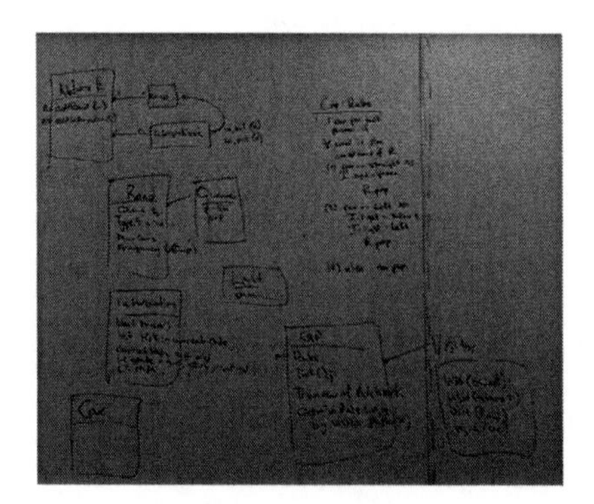

FIGURE 16.1 An example of juxtaposition, showing two levels of abstraction of the "Cop" object (Adobe).

16.2.4 Fluidity and Improvisation

Some sketches refer to formalisms (e.g., "use of freeform notations that borrows and utilizes several well-recognized UML constructs as idioms," Dekel and Herbsleb, 2007, p. 278); others may be informal or improvisational (Petre, 2009b). What is sketched is economical; sketches may omit content and connections outside the current focus, which may be added later as the focus shifts (Damm et al., 2000). Designers improvise their own notations and evolve their sketches across many canvases (Dekel and Herbsleb, 2007). Software designers have been observed to create models that are specialized to support their understanding in a specific context (Petre, 2009a). The notations for such models may emerge as part of a process of discovery and adaptation, in which designers alternate opportunistically between symbols until some consistency of use emerges. Sketches that begin as one form of representation may be altered into another form, as the problem-solving process progresses and interpretations shift. Early sketches are typically low in detail and visually imprecise; sketches are revisited and refined over time, reflecting a clearer organization and structure, and a firmer commitment to design decisions (Dekel and Herbsleb, 2007) (Figure 16.2).

16.2.5 Flexibility, Provisionality, and Low Fidelity

In sketching, speed and flexibility matter. Freedom from prescription, from syntactic and other rules of formalism help designers to work fluidly with ideas. In contrast, Goel (1995) argued that forcing people into one notation is harmful to creativity. Designers tend to "incorporate relevant information and omit the irrelevant" in their sketches (Tversky, 2002), using only as much detail as is necessary to advance their thinking. Low-detail sketches allow ideas to be expressed quickly and modified easily (Cherubini et al., 2007), and low fidelity may offer advantages in early design, leaving room for alternative interpretations (Goel, 1995). "Provisionality," the freedom to ignore some things, to indicate some with placeholders, and to express lack of a firm decision, is a useful feature of software design sketches (Petre, 2009b). In contrast, too much fidelity too soon may form a barrier

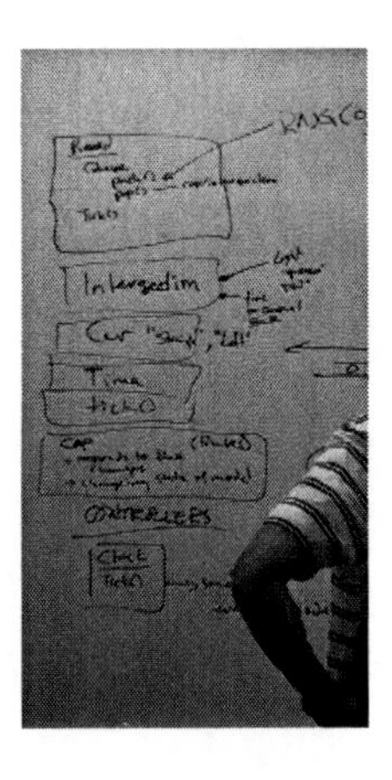

FIGURE 16.2 An example of improvisational representation, combining list and box-and-line representations (Adobe).

to change, causing a designer to be less likely to reconsider the ideas underlying it, resulting in a less exploratory and less broad search for candidate solutions (Wong, 1992; Fish and Scrivener, 1990).

16.3 REPRESENTATIONS DESIGNERS DREW ON THE WHITEBOARD

This section provides a general portrait of what each pair of designers rendered on the whiteboard, enumerating the representations that each pair produced, as a foundation for the subsequent CDs analysis.

The AmberPoint designers produced (Figures 16.3 and 16.4):

- A map—a simulation visualization or street grid, with graphical elements
- A box-and-line diagram (entity relationship [ER] diagram)
- Lists
- Tables
- A formula
- A sketch of a car
- A timing diagram

The Adobe designers produced (Figures 16.5 and 16.6):

- Drawings of an intersection, with indications of cars, etc. (scenario sketch)
- Sketches of possible interfaces, showing actions by drawing them
- Boxes and lines/arrows
- Basic unified modeling language (UML) (boxes and lines)
- Lists
- A list of boxes (system architecture)

FIGURE 16.3 Overview of the AmberPoint whiteboard.

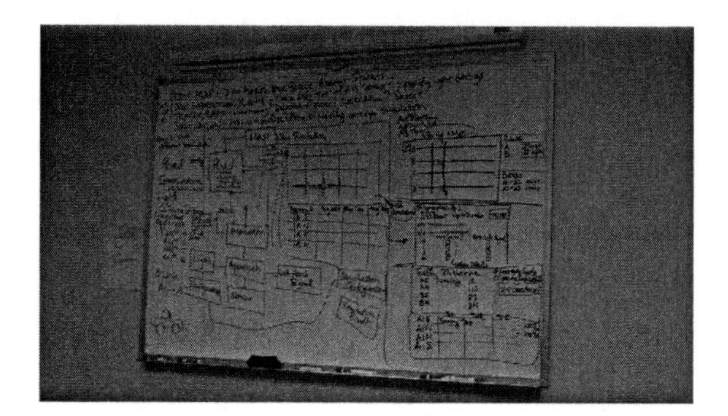

FIGURE 16.4 Overview of the AmberPoint whiteboard.

- A bullet list of user stories
- Pictures of interface elements: clockface and slider
- Slotted queue
- A list of natural language user stories, for example, "As a user, I want to ... create an intersection"
- Pseudocode
- A graph

The Intuit designers produced (Figures 16.7 and 16.8):

- Lists
- Networks (words, sometimes circled, and arrows = entities with properties)
- Intersection sketch
- Pseudocode

FIGURE 16.5 Overview of the Adobe whiteboard.

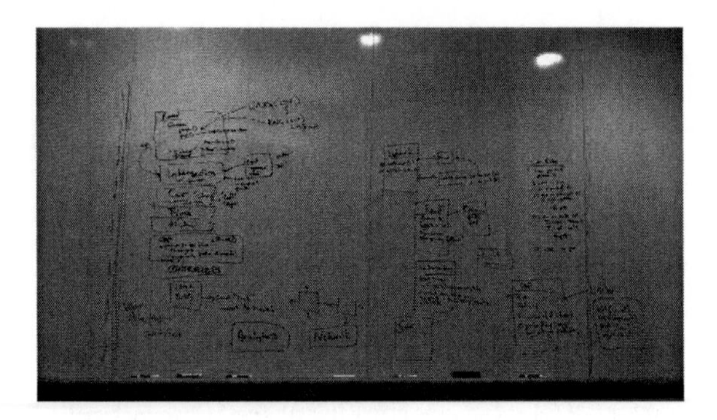

FIGURE 16.6 Overview of the Adobe whiteboard.

FIGURE 16.7 Overview of the Intuit whiteboard.

There was some consistency among the pairs. All three observed pairs produced a representation of the system architecture and a representation corresponding to the simulation. All three drew intersections in some form (i.e., sketched part of the problem domain), either within the representation of the simulation, or separately alongside it. All three made lists, sometimes preparatory to producing other representations.

It seems that these designers used a limited (albeit varied) repertoire of representations, a number of them related to existing formalisms (such as ER or UML diagrams, or pseudocode). This is consistent with a previous empirical study that analyzed a substantial corpus

FIGURE 16.8 Overview of the Intuit whiteboard.

of "idea capture" representations—informal notes and sketches which software designers made for their personal use and for discussion at meetings (Petre, 2009b). That study observed that, "when allowed to choose freely (i.e., to grab a pencil and the back of an envelope, or to step over to a whiteboard), designers use a limited set of representations, some of which refer to the formal representations from their discipline, and others of which are either generic or extremely versatile."

Considering the whiteboards in the context of the design discussion and the gestural interaction highlighted phenomena that were not evident from the whiteboard representations alone. These phenomena, concerning how the use of representations changes during discussion, and the role of gesture in using and understanding the representations, are discussed in turn in the following text.

16.3.1 Role Change and Morphing of Representations

Something that came across from considering the representations in the context of the dialogs (that was not evident in the previous corpus study) was that a representation that might appear to fall into one category might take different roles during the dialog. For example, the street grid produced by the AmberPoint pair can be interpreted as a scenario sketch (a way to think about the domain) or a design (a representation of an intended display), or even (during parts of their dialog) as an abstract model of the system. In other cases, a representation of one type might evolve or morph into a representation of another type. For example, a list might be transformed with boxes and arrows into a hierarchical structure (Adobe). This is consistent with Dekel and Herbsleb's (2007) observation, concerning teams using UML in collaborative object-oriented design, that "In some cases, they began constructing a diagram of one type, only to have it morph into another…" (p. 266).

16.3.2 Use of Gesture to Amplify Representation

The whiteboard dialogs featured frequent and specific gesturing. Whiteboards are vertical, affording a good overall view for a pair of designers, with close similarity of orientation. The large, fixed surface facilitates gesturing with two hands and oversight during the production of scribbles. All three pairs used annotation freely. In addition, all used drawing actions such as circling or pointing to add emphasis or clarify meaning during discussions—effectively, a form of gestural annotation. Certain gestures (e.g., pointing from one thing to another to indicate the relationship between two elements on the board) were used repeatedly and consistently. The explicit use of gesture to amplify what was drawn was significant, allowing designers to reinforce meaning without cluttering the board. In effect, the gestures were part of the notational system in use during the design discussions.

These phenomena are taken into account in the CDs analysis that follows (Figure 16.9).

16.4 CDs ANALYSIS

The CDs framework (Green, 1989, 1991; Green and Petre, 1996; see Table 16.1) was devised to consider information artifacts, representations, and notational systems in general. The framework was based on a deep analysis of notational examples and empirical results

FIGURE 16.9 An example of gesturing used to show the relationship between elements of two representations during discussion (Adobe).

(Petre, 2006). It has been applied most familiarly to visual programming languages (e.g., Green and Petre, 1996), graphical user interfaces (GUIs) (e.g., Blackwell and Green, 2003), and tangible user interfaces (Edge and Blackwell, 2006), among others. In such examples, CDs have been used to improve the design of notations, interfaces, or environments. The framework focuses on the artifact–user relationship, in order to consider the implications and trade-offs of representational features in the context of user activities.

The framework provides a vocabulary for discussing how artifacts are understood and used (the vocabulary and definitions are given in Table 16.1). CDs provide a broad-brush approach to considering how representations are used for a purpose. Various refinements have been proposed over the years (for examples see the CDs resources website maintained by Alan Blackwell [2012]), but typically the vocabulary is used as an analytic lens, with each CD considered in turn, and relationships/trade-offs among the CDs considered subsequently. That is how it has been applied here; the analysis proceeded as follows:

1. A detailed repeated review of the videos and transcripts provided for each of the three pairs, adding annotations to the transcripts concerning the use of representations and, where relevant, the use of gesture. Note was made of key moments in the discussion, such as disagreements, differing interpretations of representations, questions about meaning, and so on.

2. Cataloging which representations were used by each of the three pairs (as presented in Section 16.3), with attention to the order of their generation, and to the evolution and refinement of the representations.

3. Consideration of each CD in turn, in light of the examples and evidence in the videos.

4. Consideration of the relationships/trade-offs among the CDs in this context.

The results are presented in the following text. Each "cognitive dimension" is considered in turn. The discussion of each dimension is "layered": first, the general contributions of the whiteboard medium are considered; then, observations from each videotaped session are discussed (where relevant). Where the consideration of a dimension was not particularly informative for a given pair of designers, no observations are offered. The CDs vocabulary is indicated in italics.

16.4.1 Abstraction

The amount and granularity of *abstraction* are determined by the usage, rather than by the medium. The whiteboard has the potential to support a wide range of abstraction, as well as to express encapsulation.

16.4.1.1 AmberPoint

The material on the whiteboard was, fundamentally, a collection of abstractions of user requirements, interface design elements, and system components. Information was encapsulated at a variety of levels of abstraction; the available range of abstraction was flexible, from high-abstraction views of the whole system, through to requirement or component specifications of relatively low abstraction, with a granularity that related to the perspective (e.g., architecture, interface, or component views) that the designers were considering at the time. Elements were explicitly grouped and labeled as groups (e.g., three tables were grouped with a partial box and labeled "build simulation" [1:27]).

16.4.1.2 Adobe

The pair relied on familiar notations and abstractions, for example, objects (and associated properties) and later UML. They made explicit use of boxes to encapsulate and demarcate, and the encapsulated elements were typically labeled (sometimes afterward) and referred to by their label (e.g., drawing a box around the model list [0:14]). They also used graphical abstractions, for example, a box (sometimes three-sided) to indicate a queue. The pair reflected on their abstraction structures, periodically considering alternative structures or alternative abstractions (e.g., "Is this all Cop?" at [0:30], "up to this point we've said intersections contain the roads, but beginning when you just create a road there's no intersection, so roads exist outside of the intersection" [1:38]).

16.4.1.3 Intuit

The pair thought explicitly about the abstraction structure: "Let's think about containment a bit here" [0:30]. They drew more than one structure. They drew a first structure, starting from two lists labeled "Data" and "Abstract[tion]," to which they added arrows and extensions, until it turned into a network (first half of the session). Then they drew again at the top of the board, starting from Map, when they were reconsidering the structure [0:31].

16.4.2 Closeness of Mapping

Since the users choose what to represent, and they are free to use their own language, the whiteboard notionally has the potential to *map closely* to the problem domain.

16.4.2.1 AmberPoint

The basic starting point is close to the application: the designers drew a direct representation of a (simplified) traffic system—a map. Throughout their discussions, they considered which domain concepts to represent, and to what level of detail, for example, how cars were to be represented. Although the notional street grid was mapped closely, the representation of cars/traffic on that map was more or less closely mapped, depending on whether the designers represented traffic as numbers, densities of lines, or dots (with a diagrammatic aside resulting in a cartoon of a car in the lower left corner). The choice of representational detail, such as the use of color, often mapped directly onto the domain (e.g., color-coding status to correspond to traffic lights), or onto conventions from other contexts (e.g., thickness of line representing density), but some were ad hoc and had to be acquired.

The designers also discussed the closeness of mapping from different perspectives, considering how easily students might interpret the proposed representation in terms of their domain models, and also considering adopting an alternative representation that might map more closely for engineers: "I kind of want to build an ER diagram so we could talk to the engineers too" [1:36].

Some things had more than one representation, and the designers made direct gestural references to the correspondences between different representations on the board.

16.4.2.2 Adobe

The designers were concerned about the closeness of mapping in the simulation for the users—they talked about a representation of cars that "jumps" cars from intersection to intersection as unrealistic: "When it went through, it would pop to the front of the queue in the next intersection. Which is not necessarily a bad thing but it's not realistic" [1:26].

16.4.2.3 Intuit

This pair, too, started from domain concepts, quickly drawing an intersection (which they called a "real-world model") as a means of considering their mapping between domain concepts and program objects ("Do cars equal traffic?"). They alternated between their intersection representation and their developing list of constructs, and they discussed how they intended to interpret elements of the domain within their model, calling attention to differences between the model and the domain (e.g., between a continuous road system with traffic, and a discontinuous view of transitions between intersections).

16.4.3 Consistency

There is no inherent *consistency* in whiteboard use. Hence, any observed consistency arises from the designers' practices.

16.4.3.1 AmberPoint

The language used was simple and reasonably consistent. Tables were consistently set out, with headers and a clear layout. Meanings and usage were reinforced, and labels were altered to make them consistent in different representations or with the design dialog (e.g., labels added and "parameters" revised to "patterns" at [1:28]).

16.4.3.2 Adobe

This team had similar usage, with simple, consistent representations. Refinements and adjustments were made to improve consistency (e.g., [58:21] M2 directed M1 to correct what he was writing to "offramp," consistent with their usage). They discussed usage explicitly (e.g., "So I'll use a solid dot to mean ownership and then an open dot to mean referencing" [1:42]).

16.4.3.3 Intuit

As the representation of the conceptual structure or system architecture emerged on the left, the usage varied. It is not clear that the elements were used consistently (e.g., sometimes lines connecting to properties of the entities had arrows, sometimes not; color was used apparently arbitrarily, depending on which pen was to hand; entities were circled, but so were instances). Moreover, the designers redrew elements already represented on the board, without reference to the earlier version.

16.4.4 Diffuseness

Diffuseness associates with the representation rather than the medium per se. In discussions in front of a whiteboard, it might be considered that terseness and selective, indicative representation are characteristic, with an underlying principle that detail and rationale are best communicated through discussion and feedback, not by complex notation. The advantage of the whiteboard for many is its immediacy and flexibility, whereas the generation of complex, constrained notation might slow down the flow of the dialog. So, at one level, diffuseness is low, because a great deal of meaning may be expressed in a few marks. At another level, diffuseness may be high, because the capacity of a sparse representation to capture complex meanings and structures might be limited, and so multiple representations are brought into play.

16.4.4.1 AmberPoint

There were high levels of diffuseness. The designers had difficulty expressing the necessary complexity for some of the elements in the street grid, for example, intersections. Because they were unable to devise a concise, localized system for representing all the elements contributing to the status of an intersection (the state of each of the lights, the value of sensors, the state of the traffic, and flow through the intersection), and because their discussion tended to conceptualize the intersection differently at different times (e.g., as one intersection, or as four approaches), and because the whiteboard represented different perspectives on the problem and system (e.g., interface, simulation visualization, system architecture, and actions), some elements such as intersections were represented in a number of places.

16.4.4.2 Intuit

The use of the whiteboard was less orderly. Additions were made where there was space, with an explicit link from the entity being elaborated to the addition (e.g., properties of the entity). The effect was that information pertaining to the entities was distributed around, albeit with explicit connecting lines or arrows. At one point, the designers redrew the architecture on another section of the board.

16.4.5 Error-Proneness

The whiteboard has no safeguards against error, and is therefore error-prone. The informality and the potential for invention and improvisation potentially contribute to error-proneness, although this can be mitigated if the designer uses the whiteboard in a disciplined way that introduces safeguards. The organizing principle underpinning the notations or representations used may be adhered to or compromised. It is the social discipline—the secondary notation governing how the annotations are used, how the layout is adhered to, and so on—which safeguards against errors being made, not the medium.

16.4.6 Hard Mental Operations

Hard mental operations are usually associated with a combination of indirection (such as a reference elsewhere that must be decoded) and nesting. Hence, the level of hard mental operations in whiteboard scribbles depends on the degree of reliance on referencing and nesting in the representations. The diffuseness of whiteboard representations and the use of multiple related representations might entail a significant amount of relating and dereferencing. Complex or numerous relationships between multiple representations might be hard to express directly because of the limitations of space and configuration on a whiteboard, and so keeping track of things might rely on additional annotation or gesturing.

16.4.6.1 AmberPoint

The notational simplicity of these examples of whiteboard use meant that the representation per se did not require hard mental operations. However, the interpretation of the system from the whiteboard was not supported, and trying to do so might well be a hard mental operation. The hidden—or unexpressed—dependencies between representations, the implicit references between perspectives, and the abstraction of the representations, all contributed to pushing the burden of keeping track of deep structures onto the user, and that process of interpretation and decoding may entail hard mental operations.

16.4.6.2 Adobe

Again, the use of multiple related representations entailed references and relationships that must be understood. The team used gesture consistently and persistently to emphasize the relationships during their discussion. They walked back and forth across the board (between the emerging architecture and the domain model) during discussions in which they tried to unpick the consequences of their interpretation of the domain features—referring to things that were not represented, reasoning along a chain of implications, and trying to keep track of the relationships between implications.

16.4.6.3 Intuit

As with the other pairs, the use of multiple related representations entailed references and relationships that must be understood. The two gestured to emphasize relationships, and they added annotations during their discussion to amplify and reiterate relationships and meanings.

16.4.7 Hidden Dependencies

Notionally, the whole whiteboard is *visible*, and therefore anything represented on it is visible (i.e., not hidden). Hence, *dependencies are "hidden"* when they are not expressed (e.g., when one element in a dependency relationship is not represented yet, or when a relationship is not made explicit using a device such as a connecting arrow or labeling) or when one element is viewed in isolation. The use of multiple related representations tends to result in hidden dependencies between the views, unless the relationships between the representations are expressed overtly.

16.4.7.1 AmberPoint

Given the diffuseness—and the multiple representations and multiple perspectives—it is no surprise that there were dependencies that were discussed but not represented directly. Some dependencies were indicated and reinforced gesturally. At various times, the designers gestured over the board, indicating relationships between different representations, or more specifically showing that one element was represented in different places, or that a change in an element in one representation should be reflected in a change in another representation. They also indicated a desire to make explicit links in the proposed interface between the different representations, which would reduce the number of hidden dependencies. At points in the discussion (e.g., [1:33]) they drew arrows to link representations.

16.4.7.2 Adobe

Again, the pair used multiple representations and multiple perspectives. Although they rarely drew explicit links between representations, they did use gesture very consistently to indicate and reinforce relationships throughout their dialog. They also used gesture to indicate elements that were not drawn on the board, to indicate relationships beyond what was represented.

16.4.7.3 Intuit

As in Sections 16.4.7.1 and 16.4.7.2, the use of related representations entailed references and relationships that must be understood. The pair used gesture and annotations to make some dependencies explicit. However, the representations that this pair produced were relatively sparse, and so there were elements not represented that nevertheless featured in discussion and reasoning, indicating hidden dependencies.

16.4.8 Premature Commitment

The development of the design depends on discussion and exploration of meaning, hence *premature commitment* in application terms is not an issue—nothing is committed during the session. Moreover, the lack of premature commitment interacts with provisionality and relatively low viscosity (see Section 16.4.13); the three reinforce each other and are reinforced by the social context.

16.4.8.1 Intuit

This pair made a conscious effort to avoid premature commitment as they explored the problem space, repeatedly pulling themselves back from discussing or determining too much detail too early, and asking repeatedly "What is important?"

16.4.9 Progressive Evaluation

The whiteboard does not support *progressive evaluation*. Determining the "rightness" of functionality or design is achieved through the design dialog, the social process. All three teams undertook a loose form of progressive evaluation during their discussion, considering "what would happen" in different situations if they implemented the model they were constructing. Hence, the evaluation pertained to their logic (it was as though they were executing the proposed model) rather than to the externalized representation (they did not have code to run). During these evaluation discussions, they typically annotated or gestured over their diagram of the simulation. Sometimes, they added additional representations around the diagram to amplify an issue.

16.4.9.1 Adobe

The Adobe team undertook some progressive evaluation, discussing how the state of the simulation at a given intersection would change at a sequence of "ticks." They depicted two states of the intersection, T0 and T1, and spoke and gestured about what T2 would look like as well without representing it.

16.4.9.2 Intuit

The pair spent long periods simply standing in front of the board, considering. They evaluated their design constructs progressively, considering the consequences of their interpretation of those constructs (e.g., focusing on intersections rather than roads), and assessing coverage. Periodically, they would identify a construct and review its operation (e.g., "How do signals work?").

16.4.10 Provisionality

Because of the inherent flexibility of the whiteboard, the qualities of handwriting, and the ability to wipe marks away with a finger, the whiteboard conveys *provisionality*. Provisionality may be expressed directly, for example, by the way something is written, but it is also embodied socially in the design dialog which treats everything as provisional and flexible during the session.

16.4.10.1 AmberPoint

The male designer conveyed a reduction in provisionality around [1:06–1:09], when he rubbed out the tables on the right and redrew them much more precisely.

16.4.10.2 Adobe

The pair explicitly considered different interface options, with each designer drawing an example of how the simulation might be created, and the two rubbing out and reexpressing elements in those examples. They revised freely (e.g., [0:21], when M2 rubbed out "Cop," replacing it with "Sim," and then rubbing out "Sim" and replacing it with "Cop" again; or [0:23], when M2 redrew the "Road" box and rewrote its contents, or [1:30] when they took "observe" out of the user story and then reinstated it after discussion).

16.4.11 Role Expressiveness

The whiteboard can accommodate *role expressiveness* if it is afforded by the representation—at least as long as there is space on the board to accommodate the representation clearly. In other words, *role expressiveness* is potentially limited if the representation itself omits components or does not make the relationships between parts explicit. Role expressiveness may also be limited by the constraints of the whiteboard, for example, if the limitations of what can be sketched on a single whiteboard mean that elements that would ideally be represented are omitted or erased in order to manage the space, or if the board becomes too cluttered to be legible. On the other hand, the flexibility in juxtaposing different representations, in annotating with text or graphics, and in *secondary notation* can increase the *role expressiveness* of the whole of what is sketched, augmenting any given notation or representation.

16.4.11.1 AmberPoint

The use of multiple representations and multiple perspectives, and the attendant diffuseness and hidden dependencies, tended to reduce role expressiveness in this instance. On the other hand, the association of different representations with different perspectives contributed to role expressiveness, as did the design dialog, the reinforcing use of gesture to indicate relationships, and those explicit links that were drawn.

16.4.11.2 Adobe

Role expressiveness relied greatly on the dialog and gesturing. For example, consistent and repeated gesturing between the objects and the intersection drawings [0:19] emphasized the overall relationships. Some use of representation enhanced expressiveness. For example, explicit linking between the entities and properties in the structural representation was expressive of the overall relationships. Behavioral relationships were captured largely in the dialog and accompanying gesturing. In another example, analytics were dispersed through the structure, associated with objects, but they were distinguished by color, showing their location throughout the structure. However, other usage was less expressive. For example, the role of the Cop was not fully expressed on the whiteboard.

16.4.11.3 Intuit

The representations were relatively sparse, there were many discussions that made little reference to the representations, and design elements were discussed that were not represented on the board. Limited visibility contributed to limited role expressiveness in this case. Late in the discussion, one designer pointed to a construct on the board and asked: "Which 'time' is this?"

16.4.12 Secondary Notation

Whiteboard scribbles tend to use incomplete notations and ad hoc combinations. The whiteboard provides enormous flexibility in how information is conveyed, but its use typically relies on natural language for capturing meaning, and on social process—that is, the design dialog—to codify the semantics and fill in information that is not represented. Designers may use the whiteboard to display an existing notation (such as ER or UML diagrams) with its associated semantics and conventions, or they may invent a new notation for the moment. Use of the whiteboard means that the representation system is extensible: designers may add different types of representation, elements such as color, and conventions at will. Even for existing notations, designers may use variations (whether within the notation or beyond its conventional use) in layout, labels, and other cues to reinforce meaning, as a *secondary notation.*

The potential use of multiple representations adds to this versatility; designers are free to juxtapose, contrast, or combine different representations or notations. They are free to borrow conventions from one representation in order to apply them to another, perhaps to smooth their simultaneous use, to extend the receiving representation, or to amplify meaning, for example, by emphasizing the relationships between the representations by extended use of the borrowed convention.

In the absence of a well-formed primary notation, or in the novel combination of several notations, there is a need to negotiate semantics and establish conventions of use as part of the design dialog. Hence, *secondary notation*—the conventions and practices that amplify expressiveness and improve clarity—takes on considerable importance in the context of whiteboard use.

16.4.12.1 AmberPoint

Annotation is straightforward and is used repeatedly. It is clear that both designers value secondary notation. One comments that: "It's messier than my usual whiteboards" [1:47]. The other designer amplifies that the first is respected in the organization because of his command of expressiveness and clarity at the whiteboard: "When Jim is present, only Jim does the board, as you see his hand-writing, his ability to use markers is remarkable, so no one approaches the board, he's very good" [1:47].

16.4.12.2 Adobe

The pair adopts a number of conventions that assist understanding, including the use of color as part of the notation, for example, noting analytic information in red in the object boxes (which are otherwise black); the use of naming, with monitoring for consistency during the session; and the use of gesture to fill in elements and relationships not represented.

16.4.12.3 Intuit

Although M1 used different colors at different points in the discussion, it is unclear if they are associated with meaning. Additional lines (e.g., the vertical squiggle to the left of "Length") were added for emphasis during the discussion, but were left in place subsequently.

16.4.13 Viscosity

The whiteboard is highly flexible, and therefore exhibits low *viscosity*, at least when it is sparsely populated. Marks are accessible, easily changed, and flexible, and the designers routinely made swift small changes by wiping off a mark with a finger and then revising the representation. Annotations and small changes can be added at low cost, at least until the clarity is impaired and the whiteboard must be rewritten. Making a systemic change, on the other hand (e.g., changing the color coding, or adding a new category of information), could require altering each instance in turn, a form of repetition viscosity. Making changes that require more board real estate may require moving existing material to make space, a form of knock-on viscosity. Representations, even ad hoc ones, may be used in a disciplined manner. Such discipline is self-imposed and self-enforced, or socially imposed and socially enforced. Such discipline may increase the viscosity of the changes, in order to preserve the consistency and conventions.

16.4.13.1 AmberPoint

Annotations were added freely. Given the relative lack of explicit expression of the dependencies, amending, substituting, or removing any given element was straightforward. There were many examples of low-viscosity change, including the following:

- Labels were added to the street grid [13:36]
- A colored bracket was added to the street grid, and an associated annotation "A1 → A2" was made below it [0:16]
- A formula was written and partially rubbed out [0:25]
- A "north arrow" was added [25:26]
- The box representing the input form was expanded to include more information at [33:45] and again at [36:01]

Viscosity even of such small changes can increase when the whiteboard fills up. For example, when the designer rubbed out and redrew the car entity and added a traffic configuration entity, there was a squeeze, and he worked to make the arrows clear [1:44]. There were also examples of *high-viscosity* change that caused anxiety, such as rubbing out the street grid [1:29] in order to make space for other material.

16.4.13.2 Adobe

Again, local changes (e.g., rubbing out, revision, additions, annotation) were made freely, and there were many examples of low-viscosity change. The use of conventions such as color coding (e.g., the use of red for analytics) might have led to repetition viscosity, should a systemic change have been required, but this did not occur. There were also examples of high-viscosity change, when large sections of the board were rubbed out to make way for new material, but in this case without apparent anxiety: "Yeah, it's a little messy. Maybe we can use this space" [1:39].

16.4.13.3 Intuit

Viscosity did not seem to be an issue for this pair, who produced sparse representations that fit easily within the available space. They annotated freely and erased little.

16.4.14 Visibility

Everything on the whiteboard is notionally *visible* at the same time. However, the reconfiguration of parts, such as the juxtaposition of elements at will, is not available except by redrawing. Hence, visibility is compromised. Visibility declines as soon as the board is filled, because adding new material requires squeezing in material in suboptimal configurations, or rubbing out some of the existing material.

16.4.14.1 AmberPoint

One designer repeatedly expressed anxiety about visibility—and about the implied potential loss of information when the board changed—and apparently paused to take photographs before any substantial material was erased from the board. The designers tended to compensate for the lack of ability to juxtapose by gesturing to related parts, making the relationship salient in that moment. There is no clear order of reading of the board as a whole, although the designers within the session had some sense of the order of generation. The placement of elements on the board once the board was partially filled was clearly ad hoc and opportunistic, a matter of grabbing the available space.

16.4.14.2 Adobe

The designers made reference to things that were not drawn and hence not visible. Again, they used gesture to compensate for the lack of juxtaposition and for unexpressed dependencies and relationships.

16.4.14.3 Intuit

Everything the designers drew was visible, and their sketches were sparse enough that they fit easily within the available space. However, as discussed in Section 16.4.11 on role expressiveness, the designers discussed things that were not drawn and hence not visible.

16.5 REFLECTION ON CDs ANALYSIS

This is not a definitive analysis; the CDs analysis takes us only so far in understanding the issues of representation on the whiteboard. A summary of the analysis is compiled in Table 16.2. What it reveals most clearly is how individual practices can influence dimensions such as *diffuseness, hidden dependencies, role expressiveness*, and *consistency.* It also reveals constraints that associate with the medium: *error-proneness*, lack of *progressive evaluation*, and some limitations of *visibility*—and hence indicates where the cognitive burden lies in using these artifacts. On the other hand, it also identifies characteristics such as *provisionality*, variable *viscosity*, and some affordance of *juxtaposability* that contribute to fluid design discussions. These characteristics were also highlighted in earlier work on the nature of sketches that designers produce (Petre, 2009b). Thus, the CDs analysis allows us to consider, in a principled way, why—and to what extent—these representations work

TABLE 16.2 What CDs Analysis Highlights about Whiteboards

Cognitive Dimension	Whiteboard Assessment Based on SPSD Designers' Use
Abstraction	Amount and granularity of abstraction are determined by usage. Evidence of abstractions at different levels, use of boxes to encapsulate, and use of graphical abstractions indicate a flexible range of abstraction and an ability to express encapsulation. However, abstraction is shaped and disciplined by the dialog that agrees and maintains those abstractions—the flexibility in using abstractions comes at the cost of maintaining that understanding.
Closeness of mapping	Designers produced representations close to the problem domain; representational detail related to meaning; gestural references maintained the mapping between different representations of the same thing. There was some explicit discussion about the closeness of mapping (especially Adobe).
Consistency	Consistency is a matter of usage. Two pairs (AmberPoint and Adobe) managed of consistency (e.g., of layout, labels, graphics, and visual elements), taking care to refine representations during the session to improve consistency.
Diffuseness	High diffuseness: multiple representations, multiple perspectives, different conceptualizations of intersections, and information distributed over the board. Some annotation is added (e.g., connecting lines) to reduce diffuseness.
Error-proneness	The whiteboard has no safeguards against error; all safeguards rely on social discipline associated with use.
Hard mental operations	Multiple representations and diffuseness entail references and relationships that must be decoded. Users must keep track of deep structures. Two pairs (AmberPoint, Adobe) reinforced structures and relationships persistently with gesture.
Hidden dependencies	Multiple representations and perspectives meant that many dependencies were not expressed explicitly. Some were indicated gesturally. Some explicit links were added between representations during discussion.
Premature commitment	Not particularly relevant in this context; nothing is "committed" during such preliminary design sessions. Lack of premature commitment interacts with provisionality and relatively low viscosity within a fluid, exploratory social context.
Progressive evaluation	The whiteboard does not support progressive evaluation directly; any evaluation is achieved through the social process of the design dialog.
Provisionality	Provisionality is conveyed effectively (e.g., through qualities of rendering and the ability to erase with a swipe of a finger) and is reinforced by the exploratory nature of the design dialog. A reduction in provisionality was signaled explicitly by one pair (AmberPoint), by precise redrawing.
Role-expressiveness	Role expressiveness is not supported well, especially in the context of multiple representations and multiple perspectives, but relies instead on usage, dialog, and gesturing. Conventions such as the use of color and explicit linking between related elements improved role expressiveness.
Secondary notation	In the absence of a well-formed primary notation, or in the novel combination of several notations, there is a need to negotiate semantics and establish conventions of use as part of the design dialog. Hence, secondary notation takes on considerable importance in the context of whiteboard use. Two pairs (AmberPoint and Adobe) made systematic use of secondary notation.
Viscosity	Low viscosity when sparsely populated, but systematic changes and changes that require more board real estate than is available can entail high viscosity. High-viscosity changes (e.g., AmberPoint rubbing out the street grid to make space) can cause anxiety.
Visibility	Visibility is high when the board is sparsely populated; everything is visible at the same time. However, juxtaposition is often only available by redrawing; adding to a crowded board can impair visibility, and the need to "clear space" can mean a loss of information/loss of visibility (addressed by AmberPoint by photographing the board before a major change). Designers referred to things that were not drawn and hence not visible.

in context, and where to look for obstacles. Looking at the representations in the context of the design dialog allows us to identify some of the ways that the designers enhance the whiteboard, through gesture, through disciplines of use, and by adding extra facilities, such as photographing the board before a major change.

One challenge for the CDs analysis has always been the scope of the analysis: notations have properties not just in their own right, but also as supported within notational environments, and as interpreted by social practices and conventions. Indeed, the development of CDs was based on the observation of "notations in action," of notations in use. The CDs analysis clarifies the interrelationships between the medium and representation and highlights the role of the users' representational practices. Applying the CDs framework while also taking the dynamic of the session into account draws attention to the fluidity of the transition between different forms of representation on a given whiteboard, and to how gesture is used to make links between representations explicit and to reinforce them. Importantly, the analysis emphasizes the role of the design dialog in completing, clarifying, refining, and reinforcing the representation use during the sessions. For example, the considerable flexibility in abstraction is shaped and disciplined by the dialog that agrees and maintains those abstractions. Representational practice maintains the consistency of the representations. But the CDs analysis is inadequate to analyze that social context fully.

What are the implications for software design support and tool design? The analysis highlights some of the qualities of "sketching on a whiteboard" that are beneficial in early design and contribute to fluid design discussions—qualities that would be desirable in an early design tool. These include some qualities of the *sketches* that resonate with earlier work (Petre, 2009b): fluidity, expressive juxtaposition, and provisionality. But this latest analysis also adds to the list, by extending our attention to the *dynamic of the design session*, to the dialog with and around the sketched representations, and identifying phenomena that emerge as important to the process: gesture (as a means to clarify and extend meaning), morphing (from one type of representation to another, as understanding develops and focus shifts), and history (leaving marks that refer to previous iterations and to the process of exploration).

16.5.1 Gesture

It is no surprise that diagrams convey much richer meanings in the context of the dialog that generates, refines, elaborates, and interrogates them. Whiteboard notation is richer when gesture extends it, with an impact on dimensions such as *hidden dependencies* and *juxtaposition*. Might an electronic whiteboard or other design-sketching tool manage to preserve some of that "gestural notation"? One example is the "highlighter" function in the Calico tool (Mangano and van der Hoek, 2012) that allows the user to add a sketch element—such as an arrow between two parts of the board, comparable to one of the common gestures in the video—that is visible for a period, but then fades away. Another example might be some form of annotation layer, which can be hidden once the discussion has ended, or perhaps recalled on demand. An open question is how much of such gestural language is useful over time, or whether capturing it would clutter the sketch, potentially impairing its economy and utility?

16.5.2 Morphing

The transition of a given sketch from one form of representation to another over the course of a design session was a striking feature of the sessions, and it has been observed elsewhere as well. Dekel and Herbsleb (2007) described "morphing" as part of the opportunistic selection and use of diagram types in their study: "In some cases, they began constructing a diagram of one type, only to have it morph into another, while on other occasions the diagrams turned into a collection of fragments related only by their relevance to a particular design issue" (p. 266). In their description, morphing complements expressive juxtaposition as part of the fluid development of a sketch over the course of a design session. Walny et al. (2011) also draw attention to the transitions of and between diagrams during design. The fluid transitions between representations, and between freedom and degrees of formalism, are an important quality of sketching that the literature identifies as crucial to design exploration and creativity, and the ability to borrow and move freely between "idioms" of formalisms is of particular interest in software design. Tools such as Knight (Dekel and Herbsleb, 2007) aim to address the transition from informal to formal, but most are oriented to a particular formalism, such as UML. However, the evidence "morphing" suggests that support tools should also accommodate transitions between types of representation—between references and familiar formalisms—as well.

16.5.3 History

The notion of history, of leaving evidence of the process by which representations evolved and emerged, of the order of generation, poses similar challenges. Software designers at the whiteboard often cross through, rather than erase, hence leaving an indication of "history." The designers in this study took additional steps, such as photographing the board before a major change, to preserve previous views. The whiteboard is a constrained space, limiting the current design representation to what can fit in that space, requiring management of that space. Similarly, an electronic tool might encompass a bigger notional space, but it will still be constrained by what can be viewed at a given time. One strategy for management is the photograph; that is, preserving material that is erased to make space. Another is making the space bigger, while preserving the ability to achieve a single overview, for example, via zooming facilities or some form of encapsulation (cf. Guimbretière et al., 2001). Each of these preserves history and aids memory, but each also comes with some cognitive cost, in terms of organizing and managing more or larger views, and in terms of navigating through the different representations and understanding them as part of a history. Although preserving sketch history is something designers tend to consider desirable, it is not clear to what extent it is actually beneficial or would actually be used if available. How often is it truly important to remember the order of generation—and how often is it beneficial (e.g., less constraining) to forget it? Are some transitions more important to capture, for example, transitions associated with design insights or key decisions? Is there a balance to be achieved between remembering and forgetting?

16.5.4 Persistence

This study focused on sketching at the whiteboard during early design, on the process of sketching, and on how it shapes and reinforces the representations generated. Yet, whiteboards also persist; often the product of discussions remains on the board afterward, and is available for further scrutiny and discussion until the board is reused. The attention this study draws to some notion of sketch history is a reminder about the further use of whiteboards in the software development environment, which are familiar features, both as standard discussion tools in meeting spaces and as familiar elements of the workspace. Previous empirical studies of professional software development (Petre 2004, 2009a, 2009b) observed that traditional teams employ a rich repertoire of informal external representations, offering differing perspectives on the software; those on shared whiteboards tend to be concerned with requirements, functionality, conceptual structure, and software architecture, with a certain amount of planning information (usually in the form of lists or annotations) juxtaposed. Even when developers are not engaged in discussion "at the whiteboard," they are observed standing around, glancing at it, and gesturing toward it. Hence, whiteboards generated during design discussions also act as, what Cockburn (2002) calls, "information radiators": displays posted where people can see them as they work or walk by, and which present relevant information, improving group communication with fewer interruptions.

16.6 CONCLUSION

In summary, the CDs analysis has provided some leverage in understanding the issues of representation on the whiteboard, providing some insight into the impact of individual practices, constraints that associate with the whiteboard medium, and hence into elements that contribute to cognitive burden—or contribute to fluid design exploration. But the analysis also emphasizes the importance of seeing representations *in use*—particularly in understanding early design, and in understanding the impact of both the activity of sketching and its outputs. Clearly, further work is needed, both in terms of considering the relationship of the CDs framework to representational practice, and in terms of studying the role of sketching in early software design.

ACKNOWLEDGMENTS

The author gratefully acknowledges the support of the National Science Foundation (NSF grant CCF-0845840). This work was also supported by a Royal Society Wolfson Research Merit Award. The workshop would not have been possible without the willingness of the professional designers from Adobe, AmberPoint, and Intuit to work on the design task and to allow the recording of that work to be examined in public. Thanks also to Nicolas Mangano, Andre van der Hoek, Simon Holland, and Sheep Dalton.

REFERENCES

Blackwell, A. [accessed 27 March 2012] Cognitive dimensions of notations resource site. http://www.cl.cam.ac.uk/~afb21/CognitiveDimensions/.

Blackwell, A.F. and Green, T.R.G. (2003) Notational systems—The cognitive dimensions of notations framework. In J.M. Carroll (ed.), *HCI Models, Theories and Frameworks: Toward a Multidisciplinary Science*. Morgan Kaufmann, San Francisco, CA, pp. 103–134.

Cherubini, M., Venolia, G., DeLine, R., and Ko, A.J. (2007) Let's go to the whiteboard: How and why software developers use drawings. In M.B. Rosson and D. Gilmore (eds), *CHI*. ACM Press, New York, pp. 557–566.

Cockburn, A. (2002) *Agile Software Development*. Addison-Wesley, Upper Saddle River, NJ.

Damm, C.H., Hansen, K.M., and Thomsen, M. (2000) Tool support for cooperative object-oriented design: Gesture based modeling on an electronic whiteboard. In G. Szwillus and T. Turner (eds), *SIGCHI Conference on Human Factors in Computing Systems*. ACM Press, New York, pp. 518–525.

Dekel, U. and Herbsleb, J.D. (2007) Notation and representation in collaborative object-oriented design: An observational study. In R.P. Gabriel, D.F. Bacon, C. V. Lopes and G.L. Steele Jr. (eds), *OOPSLA '07*. ACM Press, New York, pp. 261–280.

Edge, D. and Blackwell, A. (2006) Correlates of the cognitive dimensions for tangible user interface. *Journal of Visual Languages and Computing*, **17**, 366–394.

Ferguson, E. (1992) *Engineering and the Mind's Eye*. MIT Press, Cambridge, MA.

Fish, J. and Scrivener, S. (1990) Amplifying the mind's eye: Sketching and visual cognition. *Leonardo*, **23**(1), 118–126.

Flor, N.V. and Hutchins, E.L. (1991) Analysing distributed cognition in software teams: A case study of team programming during perfective software maintenance. In J. Koenemann, T.G. Moher, and S.P. Robertson (eds), *Empirical Studies of Programmers: Fourth Workshop*. Ablex, Norwood, NJ, pp. 36–64.

Goel, V. (1995) *Sketches of Thought*. MIT Press, Cambridge, MA.

Goldschmidt, G. (1994) On visual design thinking: The vis kids of architecture. *Design Studies*, **15**(2), 158–174.

Goldschmidt, G. (1991) The dialectics of sketching. *Creativity Research Journal*, **4**(2), 123–143.

Green, T.R.G. (1989) Cognitive dimensions of notations. In A. Sutcliffe and L. Macaulay (eds), *People and Computers V*. Cambridge University Press, pp. 443–460.

Green, T.R.G. (1991) Describing information artifacts with cognitive dimensions and structure maps. In D. Diaper and N.V. Hammond (eds), *HCI'91: Usability Now*. Cambridge University Press, pp. 297–315.

Green, T.R.G. and Petre, M. (1996) Usability analysis of visual programming environments: A 'cognitive dimensions' framework. *Journal of Visual Languages and Computing*, **7**(2), 131–174.

Guimbretière, F., Stone, M., and Winograd, T. (2001) Fluid interaction with high-resolution wall-size displays. In *Proceedings of the 14th Annual ACM Symposium on User Interface Software and Technology (UIST)*. ACM Press, New York, pp. 21–30.

Larkin, J. and Simon, H. (1987) Why a diagram is (sometimes) worth ten thousand words. *Cognitive Science*, **11**(1), 65–100.

Mangano, N. and van der Hoek, A. (2012) A tool for distributed software design collaboration. In *Proceedings of the ACM Conference on Computer Supported Cooperative Work Companion*. ACM Press, New York, pp. 45–46.

Myers, B., Park, S., Nakano, Y., Mueller, G., and Ko, A. (2008) How designers design and program interactive behaviors. In *IEEE Symposium on Visual Languages and Human-Centric Computing (VL/HCC)*. IEEE Press, Piscataway, NJ, pp. 177–184.

Newell, A. and Simon, H.A. (1972) *Human Problem Solving*. IEEE Computer Society Press, Washington, DC.

Norman, D.A. (1991) Cognitive artifacts. In J.M. Carroll (ed.), *Designing Interaction*. Cambridge University Press, pp. 17–38.

Petre, M. (2004) Team coordination through externalised mental imagery. *International Journal of Human-Computer Studies*, **61**(2), 205–218.

Petre, M. (2009a) Insights from expert software design practice. *ESEC/FSE'09*. ACM Press, New York, pp. 233–242.

Petre, M. (2009b) Representations for idea capture in early software and hardware development. Technical Report 2009/12, Centre for Research in Computing, The Open University.

Petre, M., van der Hoek, A., and Baker, A. (2010) Studying professional software designers 2010: Introduction to the special issue. *Design Studies*. DOI: 10.1016/jdestud.2010.09.001

Schön, D. (1983) *The Reflective Practitioner: How Professionals Think in Action*. Maurice Temple Smith, London.

Schön, D. (1988) Design rules, types and worlds. *Design Studies*, **9**(3), 181–190.

Schütze, M., Sachse, P., and Römer, A. (2003) Support value of sketching in the design process. *Research in Engineering Design*, **14**(2), 89–97.

Suwa, M., Gero, J., and Purcell, T. (2000) Unexpected discoveries and S-invention of design requirements: Important vehicles for a design process. *Design Studies*, **21**(6), 539–567.

Tversky, B. (2002) What do sketches say about thinking? AAAI Spring Symposium, Sketch Understanding Workshop, Stanford University. AAAI Technical Report SS-02-08.

Walny, J., Haber, J., Dörk, M., Silllito, J., and Carpendale, S. (2011) Follow that sketch: Lifecycles of diagrams and sketches in software development. Sixth IEEE International Workshop on Visualising Software for Understanding and Analysis (VISSOFT).

Wong, Y.Y. (1992) Rough and ready prototypes: Lessons from graphic design. In *Posters and Short Talks of the 1992 Conference on Human Factors in Computing Systems* (SIGCHI). ACM Press, New York, pp. 83–84.

CHAPTER 17

Concern Development in Software Design Discussions

Implications for Flexible Modeling

Harold Ossher, Rachel K. E. Bellamy,
Bonnie E. John, and Michael Desmond
IBM T.J. Watson Research Center

CONTENTS

17.1 INTRODUCTION

In software engineering, concerns capture the essential concepts of a design [1]. They are the primary motivation for organizing and decomposing software into manageable and comprehensible parts. Many kinds of concerns may be relevant to different developers in different roles, or at different stages of the software life cycle. For example, the prevalent kind of concern in object-oriented programming is the class, which encapsulates abstract data types. The appropriate separation of concerns has been hypothesized to reduce software complexity, improve comprehensibility, promote traceability, facilitate reuse, noninvasive adaptation, customization, and evolution, and simplify component integration.

Though all software engineers are familiar with the separation of concerns [2–4], definitions of the term *concern* are hard to come by and are often inadequate. Following Sutton and Rouvellou [5], we use the IEEE definition of concerns as "... those interests which pertain to the system's development, its operation or any other aspects that are critical or otherwise important to one or more stakeholders" [1].

In this chapter, we report on an exploratory study of the life cycle of concerns during three design sessions, each session involving a pair of designers working at a whiteboard. Each pair worked on designing a traffic simulation system for educational purposes, based on the requirements laid out in a design prompt. In studying the concerns in the design sessions, we therefore studied various "interests," including key concepts in the domain (e.g., road), user-interface (UI) elements (e.g., visual map), model elements (e.g., road class), quality of service issues (e.g., number of intersections supported), and so on.

Identifying concerns is a process of conceptual development. We are interested to see how concerns develop during design discussions. Our investigation consists of an informal analysis of the three software design sessions, focusing on how representations drawn on a whiteboard are used during the identification and development of concerns. This focus limits the scope of the analysis such that obvious issues such as running out of space on the whiteboard are not discussed.

Not surprisingly, we observed a variety of representations being drawn on the whiteboard. Some of them, such as class hierarchies and UI sketches, are supported by modeling tools (e.g., unified modeling language [UML] tools), but the manner in which they were drawn and evolved was not consistent with the support offered by such modeling tools. There is a growing body of research investigating a more flexible approach to modeling information during the software life cycle, such as domain models, requirements, and designs [6–11]. *Flexible modeling tools* aim to support the fluid process that occurs during concept identification and development, by blending the advantages of free-form approaches, such as presentation and document editors, tablets, and whiteboards, with those of modeling tools. These are described in more detail in Section 17.4. We are also interested to see how the manner in which concern representations were developed in the design sessions aligns with the goals and characteristics of flexible modeling tools.

The structure of this chapter is as follows. First, we describe the analysis methodology used, including how we coded for concerns. Next, we discuss what we observed about the design process, and we conclude by discussing the implications of our findings for flexible modeling tools.

17.2 METHODOLOGY: CONCERN MODEL AND ENCODING

Many of the concerns that must be addressed by developers arise during a requirements analysis; others arise as the development proceeds. In this study, the design teams were given a written statement of the design problem (the design prompt), which served as a requirements document. We treated it as the initial source of concerns. The design sessions dealt with the concerns that were raised in the design prompt and introduced new ones.

In this section, we describe the concern model and how it was produced initially from the design prompt, and the encoding of the design sessions and how the concern model evolved during this process.

17.2.1 Producing an Initial Concern Model

We began by producing an initial concern model from the design prompt [5,12]. One author analyzed the prompt phrase by phrase, using both the language of the prompt and knowledge of the domain to identify concerns to produce a draft. As the subsequent discussion will reveal, there turned out to be considerable subtlety involved in this process, despite the simplicity of the design prompt.

Initially, a simple list of concerns was produced, mapping closely to the phrases in the design prompt. For example, consider the first phrase describing the requirements: "Students must be able to create a visual map of an area." Taken as a whole, this phrase expresses a concern that the UI provides users with the ability to create a visual map. The subphrase "visual map" expresses a concern that there is a visual map. Consideration of the domain indicates that the visual map depicts a network of roads and intersections. We thus ended up with five concerns in our first version of the concern model derived from this single phrase:

1. Road
2. Intersection
3. Network of roads and intersections
4. Visual map
5. Creation of visual map

It quickly became clear that concerns pertain to different domains. This is illustrated by the preceding list. The first three concerns pertain to the physical domain: roads and intersections in the real world, knowledge of which allowed everyone to understand the problem. The last two concerns are in the UI domain, since they involve the visual map presented to the user. The fourth involves the presentation to the user whereas the fifth involves giving control to the user.

After an initial discussion and refinement of the draft, two of the authors examined short portions of two of the session transcripts (AmberPoint and Intuit), associating concerns with utterances as described in the following text. This was to perform calibration encoding, aimed at ensuring consistency across coders. It soon became clear that the list of concerns was becoming too clumsy. The designers often dealt with the same concept across different domains. The last three concerns shown earlier are of this form: they all address the concept of a map (or the underlying network of roads and intersections), but in three domains: physical, UI presentation, and UI control. The design prompt was short enough that the list based on it was just about manageable (45 concerns), but once we began examining the transcripts, the list started becoming the unwieldy cross-product of the set of concepts and the set of domains (we expected we would need over a hundred concerns).

We therefore decided to restructure the concern model to reflect this two-dimensional nature directly. Together, we went through the concerns we had already identified, and separated them into (concept, domain) pairs. We then discussed the concepts to determine which were subsumed by others, and pared the list accordingly. In the end, we reduced the initial concern model based on the design prompt to 15 concepts and 5 domains, shown in Table 17.1 as originating from the design prompt.

A concern henceforth means a (concept, domain) pair. In some contexts in the rest of this chapter we emphasize concepts, in other contexts we emphasize domains, with the understanding that we are really discussing (different aspects of) concerns.

17.2.2 Encoding the Design Sessions

We encoded the design sessions for three kinds of information: concerns being dealt with, activities being performed by the designers, and the types of representations drawn by them on the whiteboard. The transcripts were used to encode the concerns, with the videos themselves consulted when necessary to resolve uncertainty. The videos were used directly to encode the activities and representations. After describing these in the next three sections, we discuss issues of consistency and reliability.

17.2.2.1 Concern Encoding

With the initial concern model in hand, three of the authors each encoded a transcript of a design session (Adobe, AmberPoint, or Intuit). The encoding involved associating one or more

TABLE 17.1 Original Concept and Domain Concern Codes

Concepts	Domains
• Avoid crash producing combinations	• Physical domain
• Intersection	• Model
• Lane	• UI control
• Lane sensors	• UI presentation
• Map	• Quality of service
• Protected left-hand turn	
• Road	
• Road length	
• Traffic density	
• Traffic flow	
• Traffic flow simulation	
• Traffic light	
• Traffic light behavior	
• UI layout	
• UI modes	

concerns with each utterance. Two different granularity issues arose immediately. The transcripts provided are broken up into fairly large chunks, divided only by speaker transitions, and time boundaries were provided only at this level of granularity. A single such chunk often deals with multiple concerns. However, for this exploratory study, we simply mapped each chunk to all the applicable concerns. This resulted in some potentially interesting detail being lost, such as the sequence in which the concerns were mentioned, and which speaker activities accompanied which of the concerns. It is worth noting that even in the case of fine-grained utterances, it is sometimes necessary to associate multiple concerns with a single utterance, as in the case of the design prompt phrases discussed earlier.

The second granularity issue is the granularity of concepts, and hence concerns. For example, consider the following statement about traffic lights in the Intuit session: "So each one's got a state either red, yellow, green." "Traffic light state" makes sense as a concept, introduced by this statement. However, the statement is also covered by the more general concept "traffic light behavior." In this and similar situations, we chose to use the existing, coarser-grained concept, and only to introduce new concepts in cases that did not seem to be covered at all by the concepts identified previously. This made the encoding task more manageable, but at the cost of losing information. A fine-grained approach, in which the "traffic light state" concept would be introduced and modeled as contributing to "traffic light behavior" would lead to a more precise concern model [5,13], and would be of benefit in a study of the concern life cycle: it would be interesting to examine what fine-grained concerns are raised in the course of discussing a coarse-grained concern, and whether and how they become important or ignored. This is left for future work.

The other author encoded the summary presentations given by the groups at the end of each session, to determine what concerns surfaced during these presentations.

The encodings of the concerns were performed using spreadsheets, with one utterance per row. The times, speakers, and utterance text were imported from the transcripts. The end time of an utterance was defined as the start time of the next utterance, an approximation that allows us to study durations, but loses detail about silences. Additional pairs of columns were used as needed to record concerns as (concept, domain) pairs. This was done with reference to lists of concepts and domains—the concern model—that were maintained in a shared GoogleDocs document. In two cases (AmberPoint and Intuit), the document was updated with new concepts and, very occasionally, domains, so that only concepts and domains in the concern model were used as codes; in the third case (Adobe), a separate list of new concepts was maintained.

17.2.2.2 Activity Coding

We wished to capture the activities performed by the designers as they were working at the whiteboard, to enable us to correlate these with the concerns being discussed. Table 17.2 shows the activity codes we used. These were developed incrementally during the encoding process, and shared using the GoogleDocs document previously mentioned.

To perform this encoding, the design session videos were imported into ELAN (http://www.lat-mpi.eu/tools/tools/elan), a video annotation tool. The videos were viewed in ELAN, and the activities were coded. Unlike the coding of the concerns in the transcripts,

TABLE 17.2 Final Activity, Activity Domain, and Representation Codes

Activity	Activity Domain	Representation
• Reading design prompt	• Physical domain	• List
• Pondering	• Model	• Diagram
• Marking	• UI	• Paragraph
• Gesturing	• Use cases	• Sketch
• Erasing	• Metrics	• Table
	• Presentation	• UI controls

the coding of the activities was not restricted to the timings of utterances, but could use begin and end times as needed.

The activity domains were encoded based on what was represented on the whiteboard, whereas the concern domains were encoded based on what was said, though actions and representations were consulted to resolve uncertainty. Also, a single UI domain was used for activities, rather than trying to distinguish presentation from control, since this could generally not be done based on what appeared on the whiteboard.

17.2.2.3 Limitations

We attempted to ensure that codes were being used consistently. We had several face-to-face discussions early on, discussing and refining the initial concern model. We then performed calibration encoding on small portions of the session transcripts, with face-to-face discussions of the results, leading to adjustments. As the encoding proceeded, we had occasional discussions about what we were finding. We used the GoogleDocs document to share the evolving sets of codes as they grew, though in the case of the Adobe session, a separate list of new concepts was made. We used spreadsheet facilities (data validation or autocomplete) and post-processing as an aid to ensuring that only codes from the lists were used.

As noted previously, however, concern encoding proved subtle, with most cases open to interpretation. The primary difficulties were with concepts, especially deciding whether an utterance really introduced a new concept or if it should be associated with an existing, coarser-grained concept. There were also cases, however, where the domain encoding remained unclear. In particular, it was sometimes difficult to tell whether the designers were talking about the physical domain or the model. We consulted the videos to try to resolve such issues, rather than relying just on the transcripts; but even so, some cases were unclear. Similar issues of interpretation were also discovered for activity encoding.

The encoding that we did would really be just the first step of a full study. We would need considerable further discussion to produce a coding manual, with definitive lists of codes and explanations of how to use them, how to handle granularity issues and simultaneous activities, and so on. This coding manual would then be used to encode a larger body of data, with further discussion and enhancement if we found too many cases that were not properly covered by the manual. We would then have multiple coders analyze the same data and iterate on the process until a satisfactory inter-rater reliability (e.g., Cohen's Kappa > 0.8) was achieved. Given the data set and the time frame of this exploratory study, such a full approach was not possible. We did have an agreed-upon initial concern model from the design prompt, but we did not have time to iterate sufficiently during the encoding to have well-defined additional concerns. The result is that we do not have inter-rater reliability, and, since our understanding evolved considerably during the encoding process, there are probably some inconsistencies within the encoding done by each of us. Any patterns, similarities, or contrasts observed in the visualizations should, therefore, be treated as suggestive only, pointing to areas that might be worthy of more rigorous study.

17.3 VISUALIZING THE CONCERN ANALYSIS

The approach we took to exploring the coding we had done of the design transcripts was to create visualizations to help us see patterns in the data. We used several tools to create these visualizations. Our initial visualizations were done using ManyEyes (http://manyeyes.alphaworks.ibm.com/manyeyes/), which provides a variety of interactive visualizations. In some cases, we found that, at that time, ManyEyes did not provide the visualizations we needed. The primary example was time lines, showing the sequences of concerns and the related activities and representations. To visualize these, we used MacShapa (http://www.itee.uq.edu.au/~macshapa/, now available as OpenSHAPA at http://openshapa.org/).

With these visualizations in hand, we looked to see how concerns were used in the design process and how they related to the specific design activities and representations used. Knowing that this study has many limitations, we looked for broad effects that we could not explain due to limitations in the method or differences between the coding done by the coders. In the following section, we describe our exploration of the visualizations and what we noticed about concerns in the design discussions.

17.4 CHARACTERIZING SOFTWARE DESIGN DISCUSSIONS

17.4.1 Concerns Evolve, They Do Not Only Originate from Requirements Documents

As noted earlier, each concern is represented as a (concept, domain) pair; for example, the concept "lane" is paired with the "physical" domain in one concern, and in another concern—the "(lane, model)" concern—the concept "lane" is paired with the "model" domain. We created a visual time line of the concepts that appeared in each transcript, which is shown in Figure 17.1.

Figure 17.1 is divided into three sections, one for each of the design teams (Adobe, AmberPoint, and Intuit). Within each team, each thin horizontal line represents a single concept (ordered the same vertically for each team for easy comparison). The thick bars around the thin lines indicate when concepts are being discussed. The concepts in the top section of the three design team time lines were inferred from the design prompt. The concepts in the second section were added by all three teams during their design sessions; the concepts in the third section were added by two of the three teams during their design sessions; and those in the bottom section were added only by the team in whose time line they appear (the Intuit team had no such concepts). Although some variation is undoubtedly due to encoder variability, the concepts added by all three teams suggest that not all valid concepts can be inferred from design requirements (in this case, the design prompt).

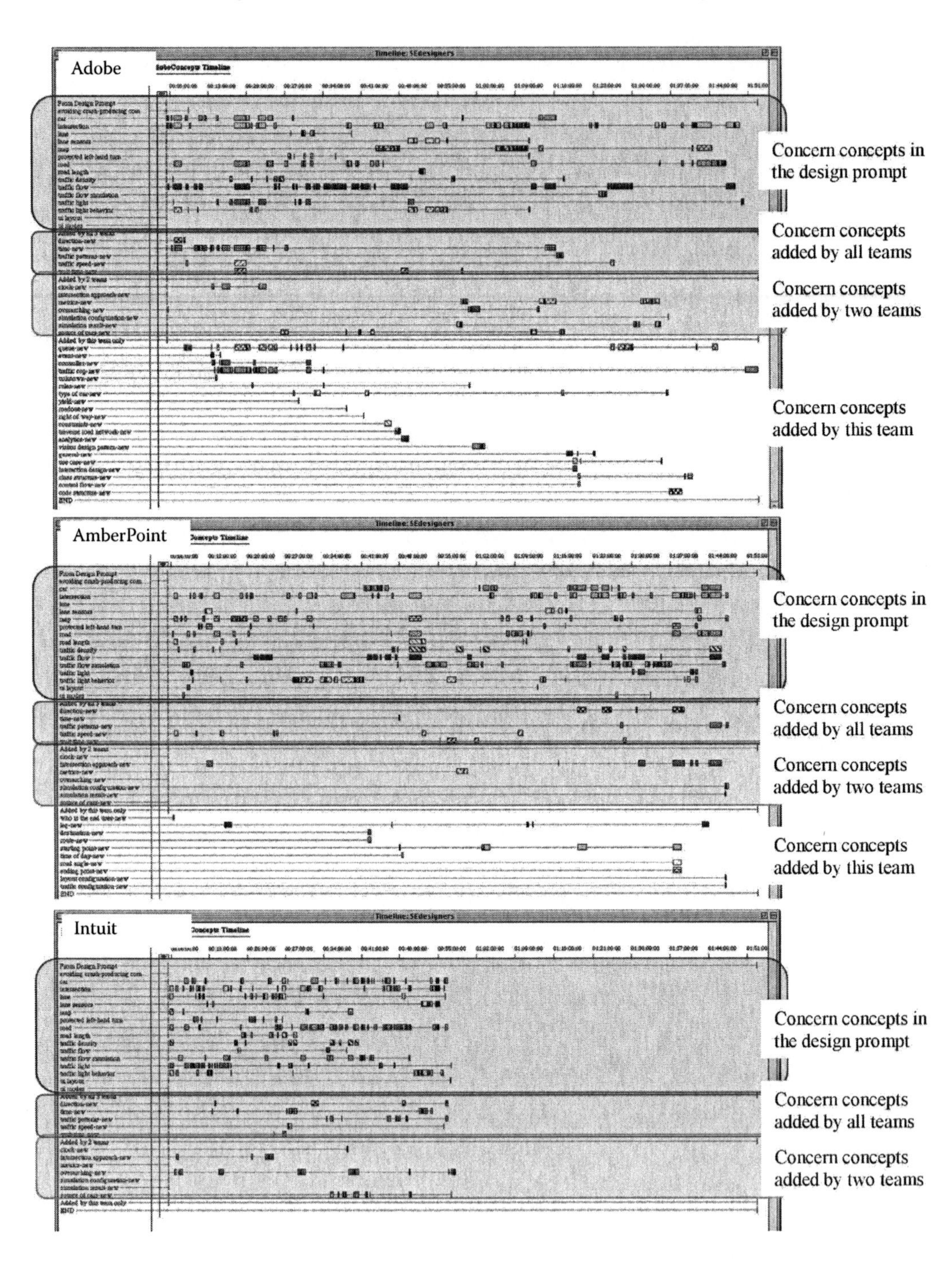

FIGURE 17.1 Time lines for concern concept codes, showing their origins. Each thin horizontal line represents a concept, and the bars on a line indicate when that concept was discussed.

However, when the frequency with which the concepts were actually discussed is examined (Figure 17.2), surprisingly the majority of the concepts discussed arise from the design prompt. The time lines for the concern domains (Figure 17.3) show that they do not evolve as dramatically as the concepts do over the course of the design sessions. In fact, only "use cases" arose as a new domain in all three sessions and only the Adobe team discussed anything else: code complexity and the presentation they would have to give at the end of the design session. These are, in fact, domains that are common to software engineering in general.

17.4.2 Concern Concepts Are Revisited throughout the Design Process

Concern concepts were repeatedly revisited throughout the discussions of all the teams, as can be seen in Figure 17.1. An examination of the concept time lines of all three sessions shows constant and frequent jumping around among concepts. This is especially true of the major concepts introduced in the design prompt. There are a few cases where the same concept is discussed for a few minutes at a stretch, but not for very long, and even then usually in part concurrently with other concepts.

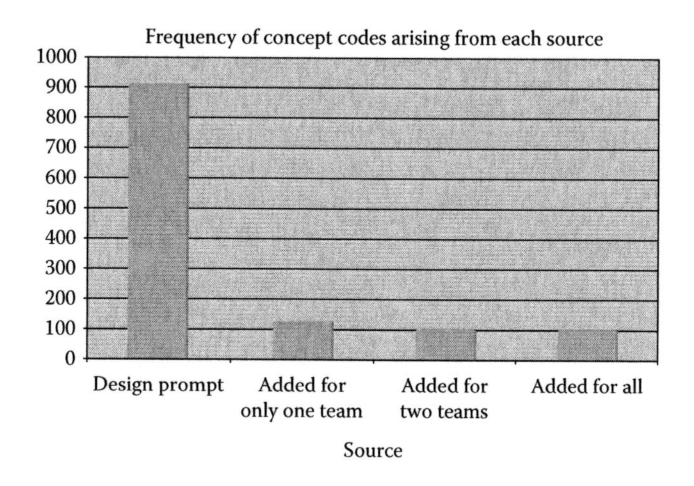

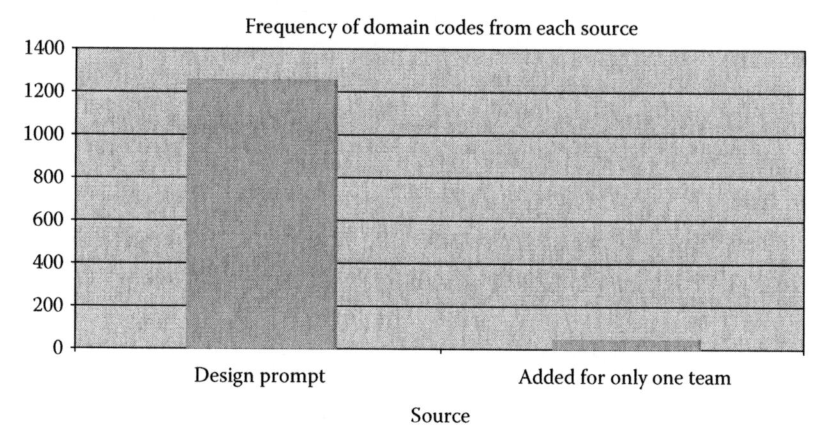

FIGURE 17.2 Frequencies for concern codes.

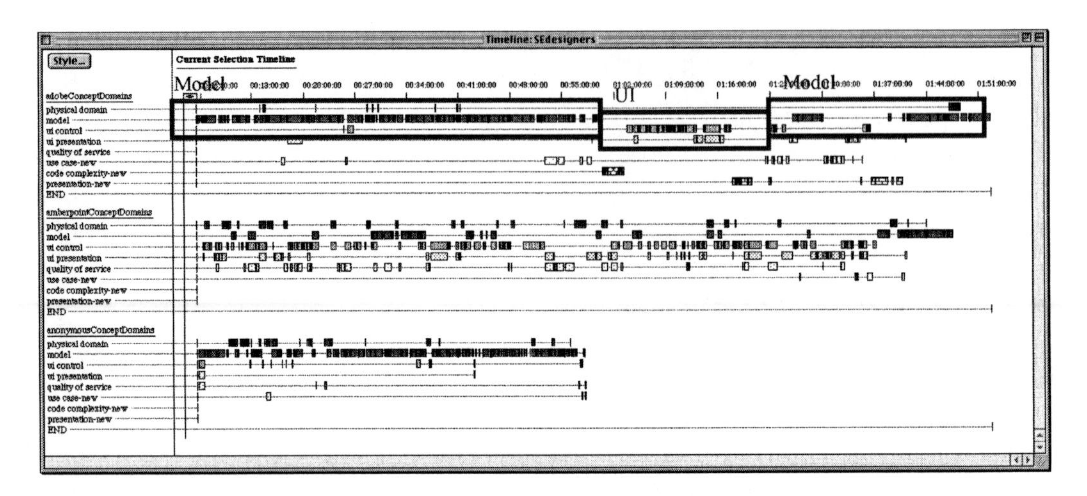

FIGURE 17.3 Time lines of the domains that were discussed by all three teams.

Some concepts, however, were seldom discussed, in some cases only once. This is especially true of new concepts introduced during the design sessions. The white space to the right (jagged right margin) indicates that the discussion of a number of concepts is dropped fairly early during the session, never to be resumed. While some of this may be attributed to encoding differences, some concerns may be resolved and therefore not talked about again. The time lines help focus our attention on when these resolutions might have come about, but further work is necessary to confirm this hypothesis.

Revisiting concepts, going back to concepts that have previously been discussed is characteristic of discussions in general. They are not linear and well structured, but messy and ill-structured. During these design conversations, the designers were struggling to understand the domain, and create a well-structured representation of the key concerns. This going back and forth is possibly related to the highly interrelated nature of concerns. Although the designers went back and forth between concepts during the discussion, surprisingly they did not explore multiple possible organizations by drawing them on the board. All of the teams used a different, but a single main organization for their design, and they developed this early on.

17.4.3 Concern Domains Are Different for Teams with Different Backgrounds

The time lines for the concern domains (Figure 17.3, where each horizontal line represents a domain) reveal that the domains discussed are different across the three groups. The Intuit team barely touched on any aspect of the UI, mentioning it only at the very end. The Adobe team had clear phases of discussing mostly the model (Figure 17.3, 00:06:00:00–00:55:00:00), then mostly the UI (Figure 17.3, 01:03:00:00–01:15:00:00, both control and presentation), then back to mostly the model (01:25:00:00). However, there is a much larger difference between the AmberPoint team and the other two: AmberPoint discusses all of the domains inferred from the design prompt all the way through, with heavy emphasis on the UI control and presentation and substantial time on the physical domain, model, and quality of service. After finding this pattern, we inquired as to the backgrounds of the different team members and indeed, although all the participants had programming in their backgrounds, both members of the AmberPoint team currently held positions as "interaction designers" while the Adobe and Intuit teams were software engineers, software architects, or software project managers.

17.4.4 Multiple Representations Are Used

It is clear from the analysis of the designers' activities that they made considerable use of the whiteboard. This is shown in the time line in Figure 17.4. Marking and gesturing

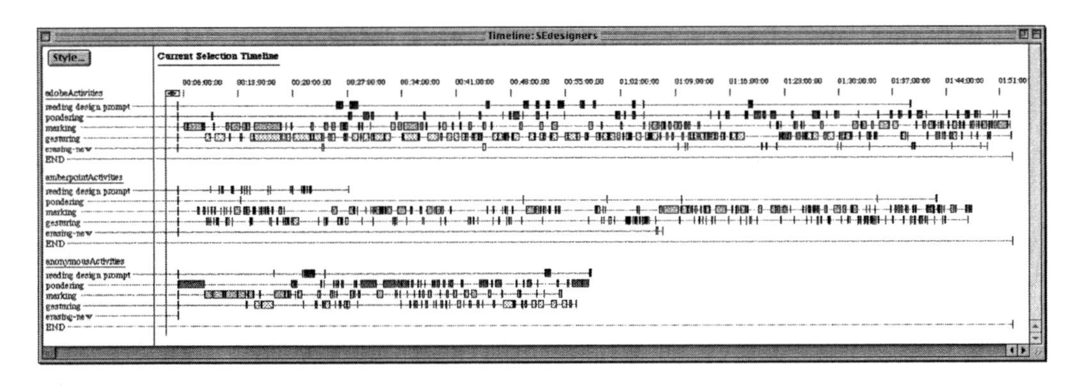

Activity domain

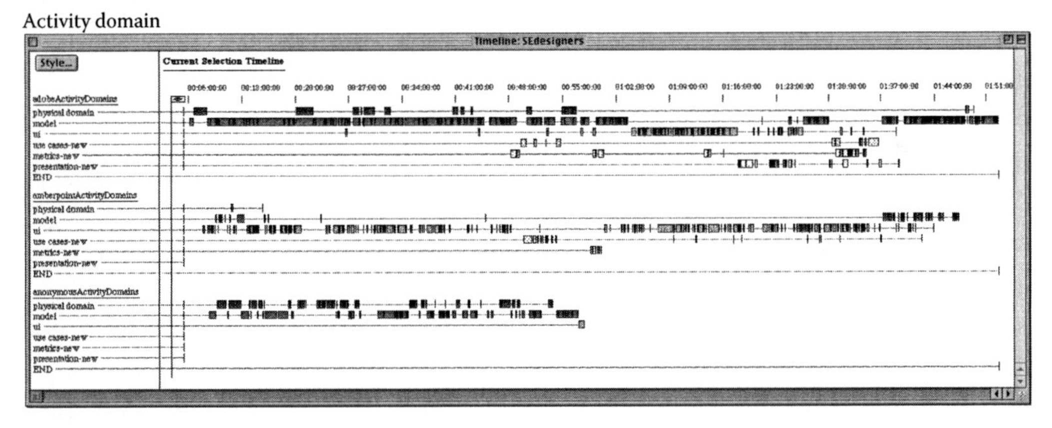

Representation

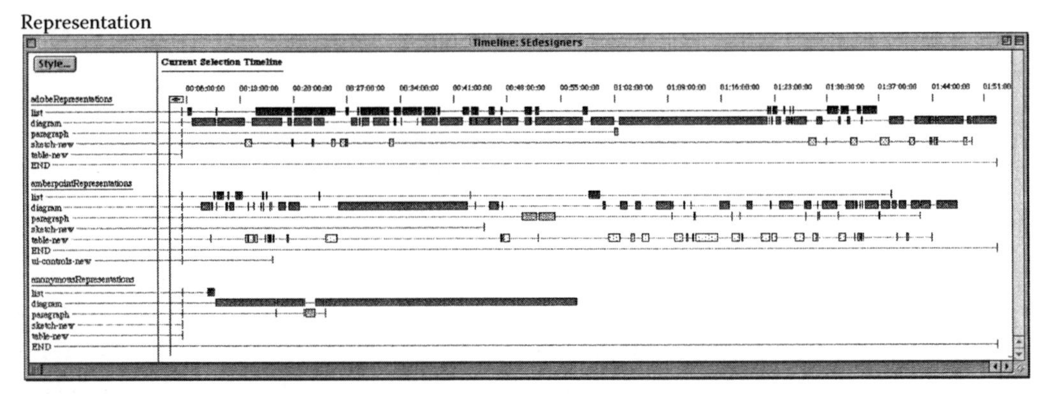

FIGURE 17.4 Time lines for each activity, activity domain, and representation codes for each of the teams.

occupied a considerable proportion of their time in all cases. In addition, much of the time was spent pondering, with the designers looking at what was drawn on the whiteboard, though in some cases we could not tell what, if anything, they were looking at. A variety of representations, from informal sketches and lists to somewhat more structured diagrams, were used. They covered most of the domains, with domain emphasis, unsurprisingly, seeming to match that in the transcripts.

We also noticed the use of cross-representation annotations such as those used by the AmberPoint team and shown in Figure 17.5. In the center of the AmberPoint board shown in the figure, there is a labeled arrow annotation pointing between the UI presentation diagram and the UI control table.

17.4.5 Representations Evolved Becoming Increasingly Formal and Detailed

The designers did not make a picture and then talk about it; rather they continually updated the picture as they talked. This interleaving of conversation and marking is typical and is shown in Figure 17.6. This figure shows, for each team, that activities centered on lists and diagrams occurred while the design teams talked about a single concept. In the figure for each team, we show two concepts: first the "intersection" concept and second the "road" concept.

We also observed that the representations used by the designers evolved over time. We observed that lists commonly became trees and diagrams. Adobe's class list evolved into a class diagram. Intuit's physical diagram became a UI diagram. AmberPoint's entity list became an entity tree.

Often, details were added to the representation as the conversation led to deeper understanding. Figure 17.6 shows the interplay between two important concepts: "intersections" and "roads." We focus on the beginning of the time line where we see six instances of the designers moving between representations (indicated by the large arrows). Some of these are examples of the use of multiple representations and some are instances where the designers were changing the form of a representation. The inset shows how the Intuit team changed their list into a diagram by drawing arrows between items in the list. As they created the list, they came to realize that relationships existed between the list items, so they then drew the arrows to record these relationships in the representation.

Representations also became more formal and detailed over time. For the AmberPoint and Adobe teams, formality appeared as the designers thought about what they would need to express in order to successfully and clearly communicate their design to others. Returning to Figure 17.5, the numbers indicate the phase of activity in which each representation appeared. Notice that the entity–relationship diagram was the final diagram created by the AmberPoint team and the Adobe team. The Adobe team also created a list of use cases to aid communication when they gave their presentation. It was clear that these two teams thought of the creation of the presentation as a separate step, and this was not integrated with the conceptual exploration that they were doing on the whiteboard. When making diagrams and lists to present, they needed to refer to what they had previously done, but they had to do the work of rewriting.

Adobe

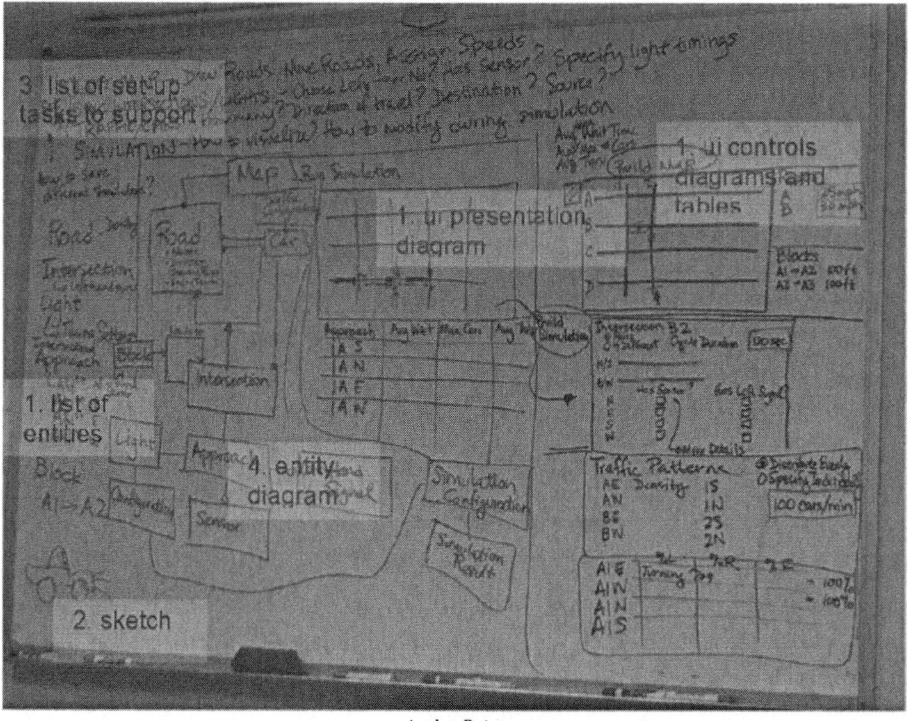

AmberPoint

FIGURE 17.5 Contents of whiteboards showing representations that were created. The numbers indicate the order in which representations were predominantly used.

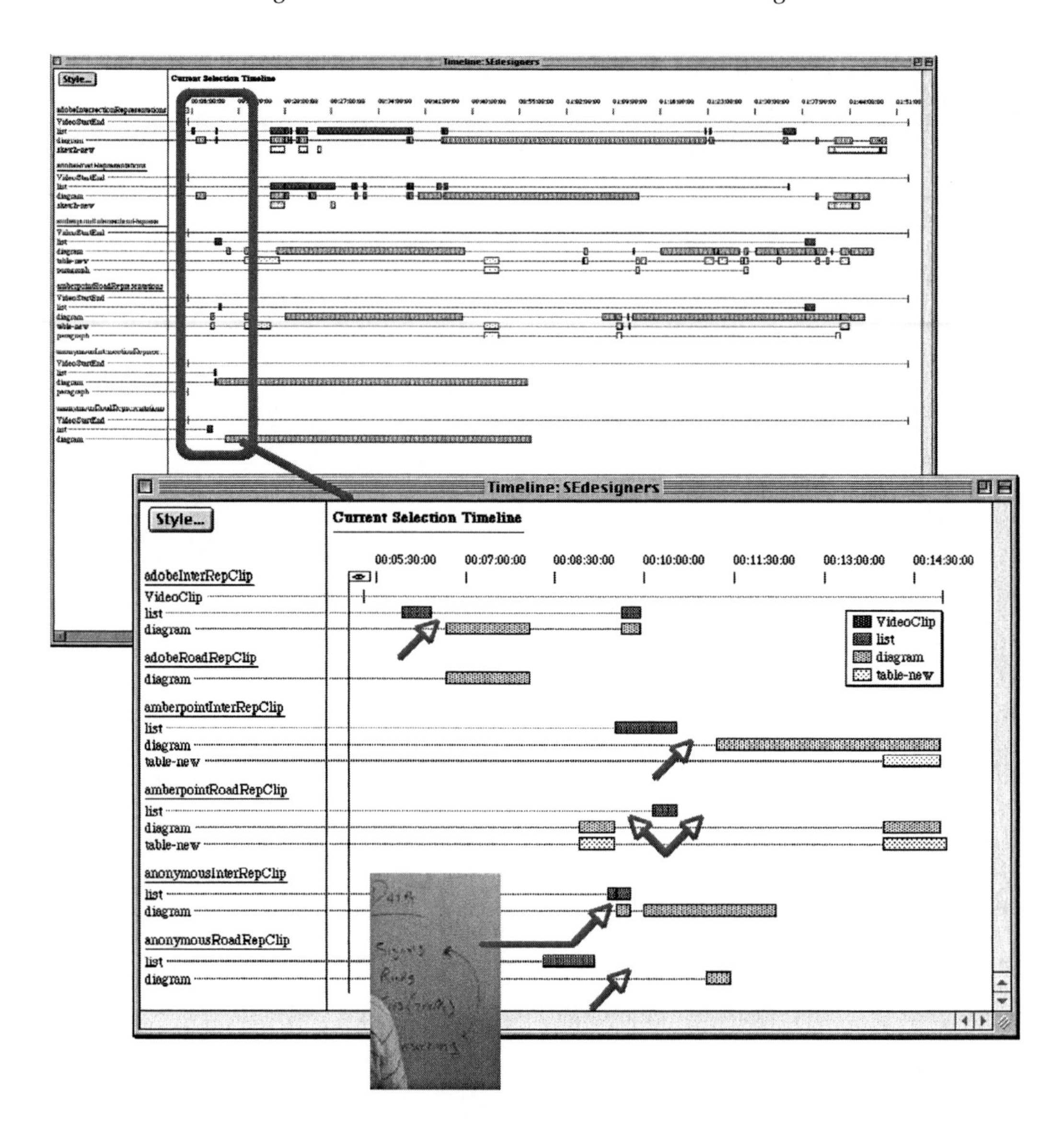

FIGURE 17.6 Time lines showing the interleaving of conversation and marking on the whiteboard.

17.5 DISCUSSION

For the three examples of software engineering design discussions explored in this chapter, we observed that the process of concern development and the use of representations was nonlinear. Concepts in concerns gradually evolved over time, whereas domains remained largely stable. The designers made use of multiple different representations to capture their ideas during the design process, and they used these representations informally and imprecisely. These findings align with those of Dekel and Herbsleb [14], who also found that ad hoc representations were used during the early phases of design by collaborative teams. We also noted that the designers moved toward more formal and precise representations as they thought about how to present and communicate their designs to others who had not participated in the design discussion.

We should note, however, that the exploratory study described in this chapter is limited in several ways. First, the three design discussions we observed were not real-world instances of design discussions, but were artificial tasks staged specifically for the purpose of study. Time was limited and unrealistically focused, the problem was not real world, and the designers were on camera and knew that they were being recorded for others to watch. There are also limitations to our analysis of the videos, as discussed earlier. Our results are preliminary; there is no inter-rater reliability in the coding. All comparisons are only "eyeball" comparisons.

Despite such limitations, we believe that our findings suggest useful implications for tools to support software engineering discussions at the whiteboard. In particular, such tools need to allow for the creation and evolution of concerns, the informal use of representations, the use of multiple representations of concerns, and the evolution of representations from informal representations during exploratory design to more formal representations that can be used to communicate the design to others. These are characteristics found in flexible modeling tools.

In the next section, we turn our attention to flexible modeling and discuss the relationship between the characteristics of flexible modeling tools and the concern life cycle revealed by our exploratory study and described in the previous text.

17.6 FLEXIBLE MODELING

The traditional modeling tool (e.g., a UML tool) supports the creation and manipulation of models of a particular kind (e.g., UML models), that is, models that conform to a particular metamodel (e.g., the UML metamodel, which describes the legal structure of UML models). It is usually impossible to use such a tool to create anything not explicitly permitted by the metamodel. This facilitates checking, analysis, and consistency management, which are valuable in formal modeling situations, but such tools are too restrictive for convenient use in exploratory contexts. There is a growing body of research investigating a more flexible approach to modeling [6–11]. *Flexible modeling tools* aim to support the fluid process that occurs during concept identification and development. Since concerns capture the essential concepts of design in software engineering, concern identification and evolution is a process of conceptual development that could be aided by flexible modeling. In this section, we discuss the implications of our earlier analysis for flexible modeling tools.

Flexible modeling tools combine the advantages of free-form approaches, such as office tools (informal tools such as presentation and document editors), tablets, and whiteboards with those of modeling tools. A flexible modeling tool allows users to work freely and easily with visual elements, yet be able to associate visual characteristics with elements of an underlying model when necessary [16,17]. For example, rectangles might be interpreted as classes in a class diagram, and colors might be mapped to visibility modifiers. These associations can be created incrementally, as and when desired. To the extent that they exist, they enable the automatic construction and maintenance of an underlying model and the management of consistency as the user draws. Given suitable domain-specific definitions, the tool can provide guidance and checking, but without requiring strict conformance to a rigid metamodel. Users can thus move smoothly between informal exploration and modeling with varying degrees of formality and precision.

Our interest in flexible modeling developed from our observations that business users working on a prerequirements analysis prefer to use office tools, such as Microsoft Word, Powerpoint, and Excel, rather than modeling tools [16]. Office tools are usually used for many good reasons. However, these tools are semantics free, are limited in organizing information, and do not have an underlying domain model. Consistency can only be maintained manually, and even small changes can take a lot of effort to propagate. Practitioners reported that entire days are often spent on the mechanics of maintaining consistency, which is not only time consuming and error-prone, but it also disrupts flow. Migrating results from these tools to downstream modeling tools, which are founded on underlying domain models, is also a manual process, and traceability is usually lost [16].

Flexible modeling tools allow the user to move smoothly between informal exploration and modeling. In our architecture [16], this is done by providing the following specific features:

- A visual layer providing multiple views, in which the user can work with much of the freedom of office tools.
- An underlying model consisting of related visual and semantic submodels, enabling visual cues to be associated with entities or relationships in the semantic submodel (loosely speaking, allowing visual cues to be given semantics).
- A forgiving approach to domain-specific guidance, with structure definitions specifying the structural constraints used to check for structural violations and to provide assistance. When provided as a package, they effectively define a metamodel. Models that violate them can, however, be created, manipulated, and saved.
- Refactoring support to allow convenient reorganization, such as converting a list to a tree or a tree to a class diagram.
- A presentation layer supporting the synthesis of presentations from working views.

This architecture was partially reified in the Business Insight Toolkit (BITKit) [16–18], a flexible modeling tool for the standard desktop. Flexible modeling can also be supported in a whiteboard medium, as demonstrated by Calico [19].

Flexible modeling appears to support the concern life cycle we observed in the design discussions. Tools that embody flexible modeling allow the visual and underlying ("semantic") representation details to evolve in concert; guidance reminds users of possible structural violation, but it does not enforce precision; and the presentation layer enables the synthesis of a presentation from the conceptual explorations that have already happened during the discussion.

This study has also suggested more specific possibilities for any particular flexible modeling tool. For example, given that the majority of the concern domains arose in the design prompt, and that many concern concepts also originated in the prompt, a flexible modeling tool initially populated with concerns derived from a concern analysis of the requirements documents might be helpful as a starting point for design discussions. There is an interesting trade-off between the helpfulness of such an initial population and the disadvantage that it might bias the designers and focus their thinking prematurely. This remains an issue for further study. Flexible modeling tools do make it easy to change things (even the types of entities, which are hard or impossible to change in traditional modeling tools [16]), so decisions that turn out to have been made prematurely are not commitments, and can be rectified.

Our analysis also revealed the importance of cross-representation annotation; flexible modeling tools will need to enable this style of annotating representations. In the architecture on which BITKit is based, such annotation is immediately possible in the visual layer, giving the user the freedom to create, view, and manipulate such annotations. Understanding the implications for the semantic submodel requires further research.

Of particular interest is the evolution of the representations found in the study. While existing flexible modeling tools do support diagramming as a way of creating a new view (e.g., building it up from a palette of visual elements), a more typical style of representation evolution observed in this study is for an existing, well-known representation, such as a list, to evolve into another well-known representation, such as a tree. Flexible modeling tools need to provide the underlying framework that makes such an evolution as trivial for a user of the electronic medium as it is when drawing on a whiteboard. The refactoring layer proposed in our architecture (but not actually implemented in BITKit) and the *softmodes* capability in Calico that allows scraps of drawings to be converted to structured entities such as lists and class definitions [15] are promising steps in this direction, but significant research is still required to approach the necessary level of function and convenience.

17.7 CONCLUSIONS

We investigated the concern life cycle during design conversations. The design sessions studied were initiated by a design problem, and were supported by a whiteboard on which designers drew representations as they jointly designed. We observed that concerns evolved and were revisited repeatedly throughout the design process, that multiple representations of material pertaining to the concerns were used, and that these representations evolved, becoming increasingly detailed and formal. We then discussed how flexible modeling tools might support concern evolution as it occurs in such design conversations. We concluded that they show promise in this area, particularly due to their ability to allow free-form diagramming blended with more formal modeling. We hope that, even though this work is preliminary and our study is exploratory in nature, it provides a basis for future investigations of how electronic media can better enable conceptual development during design.

ACKNOWLEDGMENTS

We would like to thank Sandy Esch from CMU for advice on using MacShapa, and the anonymous reviewers for many helpful comments. This research was sponsored in part by the National Science Foundation under grant CCF-0845840.

REFERENCES

1. IEEE, IEEE Recommended Practice for Architectural Description of Software-Intensive Systems. IEEE Standard 1471-2000, Approved September 21, 2000.
2. E.W. Dijkstra, On the role of scientific thought. In Dijkstra, E.W., *Selected Writings on Computing: A Personal Perspective*, Springer-Verlag, New York, 1982, pp. 60–66.
3. D. L. Parnas, On the criteria to be used in decomposing systems into modules. *Communications of the ACM*, 15(12), 1972, 1053–1058.
4. P. Tarr, H. Ossher, W. Harrison, and S. M. Sutton Jr, N degrees of separation: Multi-dimensional separation of concerns. In *Proceedings of the 21st International Conference on Software Engineering*, IEEE, Los Angeles, CA, May 1999, pp. 107–119.
5. S. M. Sutton Jr and I. Rouvellou, Modeling of software concerns in Cosmos. In *Proceedings of the 1st International Conference on Aspect-Oriented Software Development*, ACM, New York, April 2002, pp. 127–133.
6. Flexitools 2010 workshop position papers. http://www.ics.uci.edu/~tproenca/icse2010/flexitools/position.html.
7. Flexitools 2011 workshop position papers. http://www.ics.uci.edu/~nlopezgi/flexitoolsICSE2011/position.html.
8. Flexitools@SPLASH 2010 workshop position papers. http://www.ics.uci.edu/~nlopezgi/flexitools/position.html.
9. D. Kimelman, H. Ossher, A. van der Hoek, and M.-A. Storey, SPLASH 2010 workshop on flexible modeling tools. In *Proceedings of the ACM International Conference Companion on Object-Oriented Programming Systems Languages and Applications Companion*, ACM, New York, October 2010, pp. 283–284.
10. H. Ossher, A. van der Hoek, M.-A. Storey, J. Grundy, and R. Bellamy, Flexible modeling tools (FlexiTools2010). In *Proceedings of the 32nd ACM/IEEE International Conference on Software Engineering (ICSE'11)*, Vol. 2, ACM, New York, May 2010, pp. 441–442.
11. H. Ossher, A. van der Hoek, M.-A. Storey, J. Grundy, R. Bellamy, and M. Petre, Workshop on flexible modeling tools (FlexiTools 2011). In *Proceedings of the 33rd International Conference on Software Engineering (ICSE'11)*, ACM, New York, 2011, pp. 1192–1193.
12. M. P. Robillard and G. C. Murphy, Concern graphs: Finding and describing concerns using structural program dependencies. In *Proceedings of the 24th International Conference on Software Engineering*, ACM, New York, May 2002, pp. 406–416.
13. M. P. Robillard and G. C. Murphy, FEAT: A tool for locating, describing, and analyzing concerns in source code. In *Proceedings of the 25th International Conference on Software Engineering*, ACM, New York, 2003, pp. 822–823.
14. U. Dekel and J. D. Herbsleb, Notation and representation in collaborative object-oriented design: An observational study, *SIG-PLAN Notices*, 42(10), 2007, 261–280.
15. N. Mangano and A. van der Hoek, Softmodes: A transition tool between the informal and formal, CASCON Workshop on Flexible Modeling Tools, November 2009.
16. H. Ossher, R. Bellamy, I. Simmonds, D. Amid, A. Anaby-Tavor, M. Callery, M. Desmond, J. Vries de, A. Fisher, and S. Krasikov, Flexible modeling tools for pre-requirements analysis: Conceptual architecture and research challenges. In *Proceedings of the ACM International Conference on Object Oriented Programming Systems Languages and Applications*, ACM, New York, October 2010, pp. 848–864.

17. H. Ossher, R. Bellamy, D. Amid, A. Anaby-Tavor, M. Callery, M. Desmond, J. de Vries, et al., Business insight toolkit: Flexible pre-requirements modeling. In *Proceedings of 31st International Conference on Software Engineering*, Companion Volume, 16–24 May, 2009. ICSE, pp. 423–424.
18. H. Ossher, D. Amid, A. Anaby-Tavor, R. Bellamy, M. Callery, M. Desmond, J. De Vries, A. Fisher, S. Krasikov, I. Simmonds, and C. Swart, Using tagging to identify and organize concerns during pre-requirements analysis. In *Proceedings of the 2009 ICSE Workshop on Aspect-Oriented Requirements Engineering and Architecture Design (EA '09)*. IEEE Computer Society, Washington, DC, 2009, pp. 25–30.
19. N. Mangano, A. Baker, M. Dempsey, E. Navarro, and A. van der Hoek, Software design sketching with Calico. In *Twenty-fifth IEEE/ACM International Conference on Automated Software Engineering*, ACM, New York, September 2010, pp. 23–32.

IV

Human Interaction in Design

CHAPTER 18

The Craft of Design Conversation

Alan F. Blackwell

University of Cambridge

CONTENTS

18.1 INTRODUCTION

Software design is a representational craft. By "representational," I mean that the purpose of software designs is communicative—software designs are information structures encoded in representational form, to be read by authors, collaborators, users, and machines. This much is uncontroversial. The word "craft," however, is less often used in respect of software design. On the contrary, it has frequently been suggested that crafts constitute primitive modes of professional practice, to be superseded by the planned, managed, and regulated engineering discipline that is more appropriate to the industrial activity of software development.

This group of chapters illuminates the ways in which the social processes of the software design profession are enmeshed with the nature of software design as a representational craft. The themes that stand out for me are the nature of the design as a linguistic artifact, the use of representational conventions to structure that artifact, and the accommodation of various social constraints, including not only the interaction between the designers, but also their engagement with the anticipated users and with the experimental setting in which the research corpus has been collected.

18.2 CRAFT PERSPECTIVE

At the time of writing this chapter, I am nearing the end of a 10-month software development project, undertaken during a sabbatical research leave. My reflections on the chapters in this section are unavoidably influenced by my own current experiences of software development. In one sense, my personal project does not reflect the main concerns of the research reported in this book. Rather than being socially situated, my work has been both personal and isolated—extremely so, as I have deliberately taken my family to live in the forest, in a remote location with no broadband or cell communications, and a 45-minute drive across a mountain range to the nearest store. In this setting, my design work is communicative only to the extent that I must communicate with myself, and engagement with potential users is minimal—certainly less frequent than the once a week that we travel to town to buy supplies.

However, this rather old-fashioned software project does illuminate the craft aspect of software design. The system that I have been developing is intended for use by those in creative professions, to accommodate their ways of working (Woolford et al. 2010). This has provided an opportunity for experimental reflective practice, in which the software has been developed using working methods derived from those of its intended users. In particular, the functionality of the system, rather than being specified in advance of its implementation, has been allowed to evolve from the reflective practice of its development. Such practice is conventionally regarded as "undisciplined," in complete contrast to the good practices of software engineering. Furthermore, craft-based design is often a private endeavor—large elements of the software architecture have emerged at the keyboard, or when leaning back with my eyes closed. The user interface (UI) has been prototyped on-screen in response to the emerging technical capacities, and in 10 months, the number of times when it has been necessary to resort to pencil and paper (usually to resolve complex geometry) can be counted on the fingers of one hand.

18.3 CONTEXT OF COLLABORATION

In contrast, the research in this volume draws attention to the nature of collaboration, and of design as a collaborative enterprise. Professional design work carried out by an individual is still communicative, and indeed collaborative to the extent that conversation is always collaborative—shared representations, or grounding of discourse, must be established with clients, suppliers, and not least users. However, the structural and elaborative craft processes of software design, when performed by an individual, are not accessible to observation and analysis. The experimental settings studied in this volume are constructed precisely because they require that design processes be elaborated—in order to be shared by study participants, but more importantly in order to be recorded, transcribed, and analyzed. This results in the kind of collaboration that Détienne describes as "symmetrical," in which the participants alternate roles, alternating the stances of creation and reflection that would otherwise be undertaken by one participant. Fully symmetrical collaborative design is an ideal object of study, because no participant is jealously preserving the rights to particular kinds of knowledge and skill. In commercial design practice, such situations are, of course, less common—design meetings more typically involve engineers with artists, senior technicians with trainees, analysts with programmers, and so on. The participants in these studies, having been extracted from the constraints of their usual professional context, are able to act more as ideal professional peers—mutually polite and expert partners.

A further respect in which this experimental context is not fully representative of typical professional software design is that, as with many controlled studies conducted in other design professions, this corpus investigates the early "ideational" stages of design work, rather than the commissioning, contract negotiation, detailing, regulatory and construction management, and the other essential elements of professional practice. Nevertheless, the ideational stages are often regarded as being the kind of activity that is most characteristic of design (perhaps because designers themselves find it one of the most enjoyable). However, the ideational nature of the research exercise means, paradoxically, that there is less iteration in the process. Because there is no real client, and no real implementation constraint, the research context is more characteristic of a stage in a waterfall process—the participants do not expect to negotiate requirements with the researchers, and they try to produce a design outcome that is a final deliverable—however imperfect, they know that it will never be revised after the end of the experimental session. This type of practice, although unavoidable in a controlled experimental context, is also one that is often criticized in academic prescriptions of engineering design processes (for its lack of practicality), or in poorly managed engineering companies where design work is "thrown over the wall" for testing and evaluation rather than allowing true engagement of the designers with the customer and user needs.

18.4 COMMUNICATING WITH USERS

But it is users, even more than researchers and the designers themselves, who are the primary focus of representation in these exercises. Nakakoji and Yamamoto (Chapter 22, this volume) observe that the session under analysis is a typical exercise in interaction design, and, indeed, the nature of the design brief has ensured that the major part of the necessary work is interaction design, rather than (say) network protocols, embedded machine controls, or database architectures. This has resulted in a design process where representations of the UI, discussions of the user goals, and elements of the software architecture are "intertwined," as they say. This is, in fact, a relatively recent phenomenon in the software industry (at least, it is recent within the last 20 years). For many years, UI issues were discussed separately from technical concerns, especially under the reign of the analyst/programmer paradigm in software project management. Even today, many user researchers and "wireframe" designers make their contributions to design without ever writing or discussing code. The design approach observed in these transcripts was pioneered in the object-oriented programming era in the form of scenario-based design methods, and in Jacobson's more formalized use cases (Jacobson et al. 1992), where the representational core of the software is derived from consideration of the problem domain from a (more or less) user perspective.

As a result of this evolution in software design practice, the codependence between the user and technical descriptions of the problem has become increasingly highlighted. Given

McDonnell's (Chapter 20, this volume) skepticism about understanding other minds, we must ask how a designer can ever succeed in anticipating the goals of a user. Can engineers really think themselves into the heads of nontechnical users? When they engage in "linguistic technicalizing," are they refining, or perhaps distorting, the uses of the system? Is this the means by which so many software artifacts are relegated to Bucciarelli's "object world" (1994) and are alienated from their social context? To the extent that UIs are themselves semiotic constructs (de Souza 2005), technical interaction design is precisely the development of a representational language to be shared by users and engineers. As the two are rarely in the same space, this poses a challenge for conversational analytic approaches to the problem—these representations of UIs are constructed in the absence of those with whom the conversation is to be conducted.

18.5 NAMES AS THE MATERIAL OF SOFTWARE

In the absence of users, these imagined conversations center around names, expressed not only through labels and lists of menu commands in UI prototypes, but also in the internal elements of the software model. Expert software designers often emphasize the importance (and the difficulty) of creating appropriate identifier names, whether for functions, objects, or variables (McConnell 1993). The current analyses show many points in which concepts are refined by renaming them (Nakakoji and Yamamoto [Chapter 22, this volume] call this the refinement of terminology), or alternative names are proposed and debated. In my own recent experiment with craft-style development, I have been impressed by the centrality of simple renaming as the most powerful element of the "refactoring" functionality available in contemporary development tools (in my case, IntelliJ IDEA). The majority of classes and local identifiers in my project have changed their names at least once since their initial creation, as implementation and testing clarify their precise significance within an overall design. This is a classic example of craft "conversation with materials," where the design concept is refined while engaged in craft manipulations, providing visual or tactile sensory associations that are complementary to verbal descriptions of the requirements and system structure. In my own experience, where system functionality is being derived from technical capability, names initially emphasize their technical origin, perhaps in textbook descriptions or library functions. As the names evolve, they better reflect the intended usage of the system—rather the reverse of typical scenario-based design processes in which the terminology is captured from the user domain before being technicalized. In my view, naming is underappreciated in the research discourse on software development tools. Because identifiers are interchangeable, with no consequence for the compiler or the CASE tool design, they have seldom drawn the attention of tool developers—even language key words are dismissed by theoreticians as "semantic sugar," trivializing the centrality of language in human discourse.

It is the emphasis on names that makes Joel-Edgar and Rodgers' analysis particularly interesting (Chapter 21, this volume). In provocatively taking social network analysis tools, and applying them to technical components, this chapter rediscovers the actor network theory as an engineering infrastructure. In the analytic process of developing ad hoc thesauri and ontologies from the design discourse, these authors blur the distinction between studying design and doing design. This could indeed be a starting point for a new generation of language-aware software development tools. However, the key question is whether the structure that emerges from such an analysis clarifies, or perhaps obscures, the developing conceptual structures of the designers themselves. As names are refined or replaced in accordance with evolving design concepts, can an automated analysis pluck these from the designers' minds? Even more ambitious is the promise that user discourse or domain discourse could be automatically processed and refined as a direct source of software design entities—a promise that has been extended from the earliest days of object orientation, but has often been frustrated by the reality that most objects in most systems relate to aspects of the internal mechanism that are of little interest to users.

All of these writers, from whatever perspective, have identified the significance of names. A consideration that I find particularly stimulating is the way in which naming must be deferred, or held in tension, to maintain creative spaces. Nakakoji and Yamamoto (Chapter 22, this volume) observe the way in which collaborators "defer detail," each maintaining their own private terminology while proceeding with other aspects of the design. Here is a context in which the challenges of conversation provide a creative impetus, a genuine "conversation with the material" of language. As noted by McDonnell (Chapter 20, this volume), software designers can resemble poets in their need to remain uncertain about the meanings of words—Keats' classic "negative capability" in the creative process. The much-studied and celebrated role of ambiguity in the design sketch (Fish and Scrivener 1990; Goldschmidt 1999) supports a reinterpretation in the visual medium, but few software representations have been able to emulate this sketch function. Instead, it is the creative use of language that forms our tools.

18.6 DEALING WITH DIFFERENCE

Wherever ambiguity and creative tensions exist, the capacity for disagreement also exists, as observed by several of these authors in the transcribed conversations. Although some of the disagreements appear transient, ultimately to be resolved if the design were to be completed, it also seems plausible that, as suggested by Nickerson and Yu (Chapter 19, this volume), tension between the user perspective and the technical perspective will often be a permanent feature of design discourse. However, the conclusion that Nickerson and Yu draw from this is rather different from my own. I agree with their observation that differing design perspectives (especially on multidisciplinary teams) can arise from differences in the ultimate objectives. But I would argue that these differences are a design resource, to be expected and celebrated in the context of interdisciplinary collaboration. Nickerson and Yu's suggestion that such fundamental differences should be resolved with assistance from tools led me to imagine an interaction along the following lines: "Now I understand—the real cause of our differences is that you believe there is actually a God, while I don't. Let's just settle which of us is right on that question, before we continue with any other work." In fact, my own research into the social processes of interdisciplinary collaboration suggests that those trained as designers are often unusually skilled at continuing work in the face of such differences (Blackwell et al. 2010). I like to think of this social aspect of design work as another kind of "conversation with materials," where the social material is the fabric of various belief systems and personal commitments. When I return from my isolated development project, it is these kinds of conversation that I most look forward to.

18.7 CONCLUSION

These concerns, and this kind of analysis, should be central to the work of a reflective practitioner. Schön's prescription of a reflective practice involves not only reflection in the moment (Détienne's meta-conversational roles, or Joel-Edgar and Rodgers' meta-analysis of an emerging domain model [Chapter 21, this volume]), but reflection on the tools and experiences of the designer, extending to the designer's own professional stance and career (Schön 1983). Research of this kind is perhaps most valuable to the extent that it allows us to reconsider the nature of our representational craft, not subjecting it to a priori theoretical and methodological discourse (whether technical or humanistic), but rather allowing it to develop as a living professional design practice.

REFERENCES

Blackwell, A., Wilson, L., Boulton, C., and Knell, J. (2010). Creating value across boundaries: Maximising the return from interdisciplinary innovation. Research Report CVAB/48. London: NESTA.

Bucciarelli, L.L. (1994). *Designing Engineers.* Cambridge, MA: MIT Press.

de Souza, C.S. (2005). *The Semiotic Engineering of Human–Computer Interaction.* Cambridge MA: MIT Press.

Fish, J. and Scrivener, S. (1990). Amplifying the mind's eye: Sketching and visual cognition. *Leonardo 23*(1): 117–126.

Goldschmidt, G. (1999). The backtalk of self-generated sketches, *in* J.S. Gero and B. Tversky (eds), *Visual and Spatial Reasoning in Design*, pp. 163–184. Cambridge, MA/Sydney, Australia: Key Centre of Design Computing and Cognition, University of Sydney.

Jacobson, I., Christerson, M., Jonsson, P., and Overgaard, G. (1992). *Object-Oriented Software Engineering: A Use Case Driven Approach.* Reading, MA: ACM Press/Addison-Wesley.

McConnell, S. (1993). *Code Complete: A Practical Handbook of Software Construction.* Redmond, WA: Microsoft Press.

Schön, D.A. (1983). *The Reflective Practitioner: How Professionals Think in Action.* New York: Basic Books.

Woolford, K., Blackwell, A.F., Norman, S.J., and Chevalier, C. (2010). Crafting a critical technical practice. *Leonardo 43*(2): 202–203.

CHAPTER 19

Going Meta

Design Space and Evaluation Space in Software Design

Jeffrey V. Nickerson and Lixiu Yu

Stevens Institute of Technology

CONTENTS

19.1 INTRODUCTION

Design is an exploratory activity, and there are many ways to get lost or stuck. There is no preexisting map—the map is built through the design process. Just as in any form of exploring, it is difficult to take stock, to *go meta* and consider the situation from a different level of abstraction—that is, to engage in metacognition (Metcalfe and Shimamura 1994). But we will argue that this is an essential part of any design process.

We will use the opportunity provided by access to transcripts of professional software design activity to illustrate a normative model of design conversation, comparing the model to what happens in a particular design conversation. Our goal is to provide a normative process theory for design that might be used to improve the training of software designers.

In analyzing the transcripts, we will pay particular attention to the form that representation—verbal, diagrammatic, and gestural—plays in shaping design conversation. First, however, we will develop a theory of the design process.

19.2 DESIGN PROCESS

19.2.1 Background

We will integrate two different perspectives on design, one formal and the other natural. The formal view sees design as a search through the design space for solutions that satisfy certain criteria. This formal process has many adherents, including Simon (1996), whose description of the design process as a search was, in part, motivated by the goal of implementing design computationally. Since Simon, much progress has been made toward this goal, and examples and critiques of computational design abound (e.g., Deb et al. 2002; Gero and Kazakov 1996; Nadin 1996).

In an alternative view of design, things are rarely clear and distinct—search is not the right metaphor, but instead sense-making better communicates the design process (Schön 1983; Visser 2006; Weick 1995). There are many works that integrate elements of both views; for example, Winograd (1987) and Purcell and Gero (1998) model conversational stages formally, describing what actually happens and should happen in analytic terms, but with particular attention paid to the role of context and interaction in design. Brooks (2010) critiques the Simon-like formal approaches to design while at the same time offering alternative formal ways of describing the design process.

A notable body of work on the design process has taken place under the term *design rationale*; the root of this idea lies in Rittel's (1980) work, which set out to document design conversations to prevent endless arguments caused by previously resolved issues coming up for debate again and again. A large contribution of the work on design rationale has been an in-depth examination of the concept of design space (Lee and Lai 1991). Yet, even after years of study, the concept admits many variations and misunderstandings; so next, we describe in detail what we mean by such a space and the role that this, and related spaces, play in design conversation.

19.2.2 Spaces of Design

We will start first with a simple model of design activity, one that will be embellished upon later. In this process, there are two stages: a design stage and an evaluation stage (Figure 19.1).

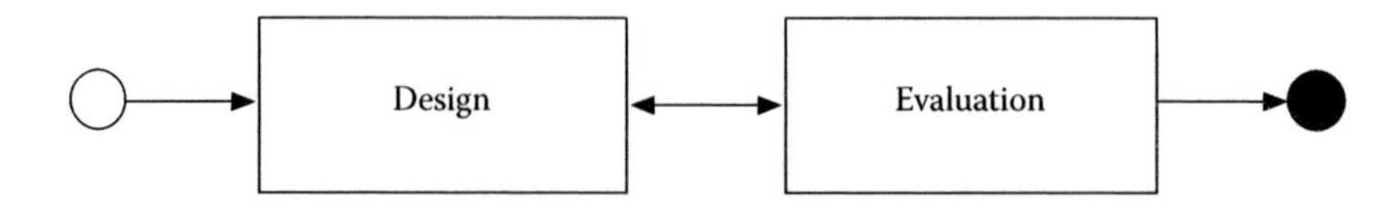

FIGURE 19.1 A simple model of a design process. The open circle represents a start state, and the closed circle represents an end stage.

Several designs are created and then evaluated. After evaluation, the design process continues, seeking to improve on the best of what has been done before, or seeking to explore the *design space* more fully. The designs are compared in an *evaluation space*. This continues until one or several adequate solutions have been found. This model is commonly discussed in practice, and, is, indeed, a process that can be automated: variations are generated by computer, sometimes using a search with random components, and one or several designs are chosen in the end (Deb et al. 2002). Moreover, it is also a simple description of the brainstorming process (Osborn 1963), in which ideas are generated and later evaluated, leading to a second stage of idea generation.

What do we mean by *design space* and *evaluation space*? We will begin with a simplified example from the domain of package design, a domain in which our physical intuitions will provide a grounding for the abstractions we want to discuss. Specifically, we imagine ourselves as consultants who are given the task of designing a cereal box for a food company. Suppose we conclude from the conversations that we have only two dimensions of freedom. Since the volume of the box is fixed, we can define solutions by changing the width or the height of the box (the depth can then be derived). Height and width are quite literally dimensions. That is, all of our possible designs can be expressed by the different values of the height and the width dimensions (Figure 19.2).

The important thing to understand about a design space is that every point in the space represents a possible design. The point defines a position along a set of dimensions, the axes of the space. As we will see later, in most cases there are many more than just two dimensions—and the dimensions are not necessarily independent of each other. But here, it is simple to understand how the concept of a design space emerged, and how, in some situations, it has a clear geometric analogy.

Even in this simple design space, there are an infinite number of boxes possible: any point on the plane is an alternative design. How are we to judge these boxes? We can think of each design, each point, as having a location in a different space, called the evaluation space. The axes in this space are not design dimensions such as height and width, but instead are evaluation dimensions, criteria.

Imagine that the cereal company has told us that there are two objectives that we should keep in mind. On the one hand, they would like to use as little cardboard as possible in the box. This is a "use of material" criterion, with less use of material represented at the top of the axis, and more use represented at the bottom. On the other hand, we would like

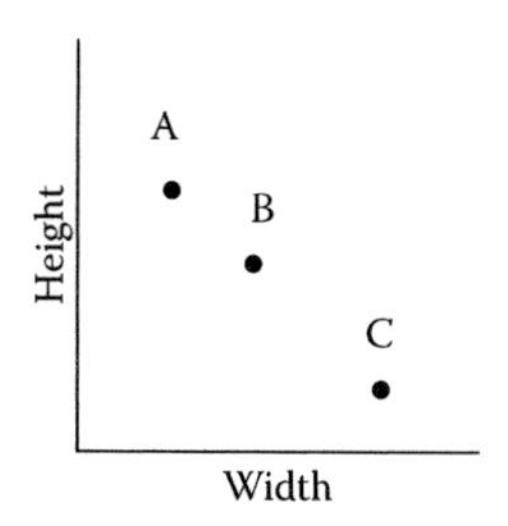

FIGURE 19.2 A design space for a box design; each point, labeled A, B, and C, represents a design.

to help supermarkets maximize their shelf space. This is our "use of shelf space" criterion, with better utilization represented on the far right in Figure 19.3, and worse utilization represented on the far left. We can take each design and place it in this space by calculating each measure. From basic geometry, we know we will minimize the use of cardboard with a cube-shaped box. But, we find through discussions with the supermarkets that we will maximize shelf space with a tall box, which they prefer to wide boxes. From this, we can calculate the scores for each design along the two dimensions of the evaluation space and plot them.

The evaluation space, we note here, has two dimensions. How do we know which designs are better? There are two techniques available. One technique weights the value of each criterion, and computes an overall utility score by adding the weighted sums. The advantage of such an approach is simplicity, but quite often it is hard to come up with accurate weights—and stakeholders sometimes manipulate the weights to achieve a desired objective (Moore et al. 2003). So, often, a second approach is used, in which we do not try to reduce the designs to a single utility measure, but instead compare designs in the space, and eliminate designs that are clearly worse than other designs. For example, Box C shows up lower and to the left of each of the other boxes in Figure 19.3, which indicates that it is less economical and less space efficient than Boxes A and B: we say that the design is dominated by Box A and Box B. Judging between Box A and Box B will depend on the relative importance of shelf space versus material—we say that Box A and Box B are members of the Pareto optimal set (Keeney and Raiffa 1993).

As designers, we may want to generate several alternatives—A, B, and a few others—and produce prototypes to figure out the actual production costs and shelf use metrics. Thus, we might use the results of the first evaluation to go back to the design space and generate several more designs, in the upper left quadrant of the design space that we have now identified as producing more valuable designs. That is, we use the first evaluation to guide our search through the design space.

We use this example because it gives us a good intuition about the design space. The dimensions we chose were physical dimensions in space. But dimensions may be categorical. For example, perhaps we have a choice between packaging cereal in a box or in a plastic bag. This choice can be thought of as a dimension, but not a continuous physical one. Instead, the dimension only has two values. We can think of such design spaces as vectors made up of dimensions, each of which has a type, and certain types have numerical attributes and

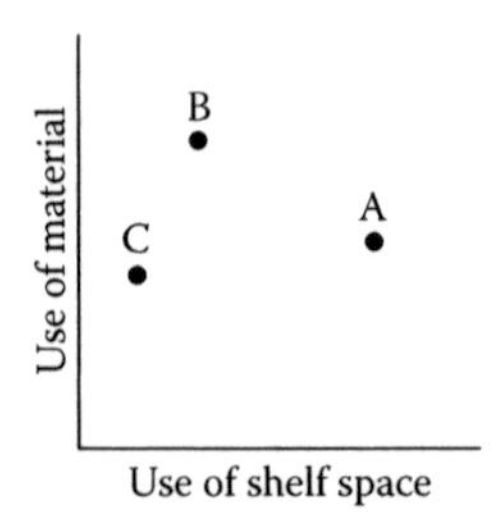

FIGURE 19.3 The evaluation space based on two criteria.

other types have categorical attributes. More concretely, we can represent such spaces as a spreadsheet, in which each column is a dimension and each column has a type—sometimes numbers, sometimes categories—which determine the attributes that can be used. A design, in such a case, is constituted by setting one value for each column.

However, this representation does not capture the dependencies between dimensions. For example, if the product is a box, the material can be a type of cardboard, but if it is bag, it can be a type of resin. That is, the design space begins to resemble less a multidimensional box and more a tree, because certain decisions change the options of future decisions. Moreover, the tree has two types of nodes: those we have been describing, in which one of a set of mutually exclusive options must be chosen, and a different type, in which all paths are to be followed independently. For example, once we have decided on the type of box, we need to figure out both the dimensions and the material: we need both. This way of specifying design is explained in Brooks (2010). Figure 19.4 shows our example, now expanded.

A design can be represented by a full traversal through the tree. For example, a design can be completely described as a box with height 8, width 6, and depth 3, and made of cardboard A. Alternatively, it can be described as a bag with height 7, diameter 5, and made of resin A. These designs can also be evaluated in the same way as before, which might lead us to focus on one set of subbranches in the design space.

As we examine our options, we may find that our evaluations favor bags, but we learn from the company that past sales records indicate that customers favor boxes. We might decide that we are missing criteria in the evaluation space.

There are criteria that are called functional—usually related to the problem at hand—and others that are called nonfunctional—that transcend a particular problem. These are sometimes called the "ilities" and "ities." For example, usability, flexibility, maintainability, and security are often used when evaluating software (Chung and do Prado Leite 2009). There are two other nonfunctional criteria that are also popular, and will come up in the illustrative case. The first is simplicity, which has been found to be a near-universal criterion in all forms of design (Case and Pineiro 2006; Chopra and Dexter

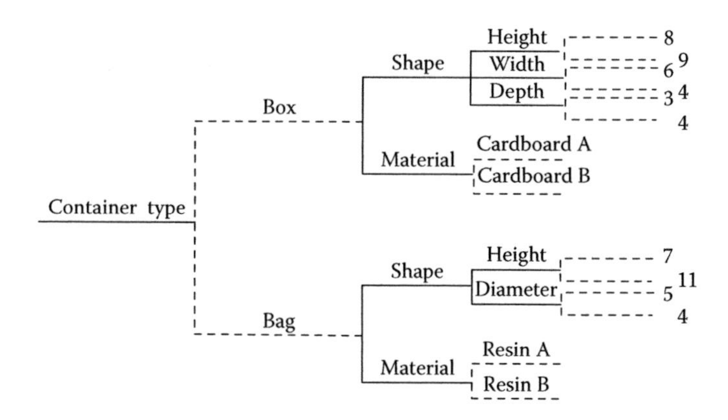

FIGURE 19.4 The design space represented as a tree. The dotted lines indicate mutually exclusive options (follow one) and the solid lines indicate independent options (follow all); this notation is based on Brooks (2010).

2007; Schummer et al. 2009; Maeda 2006). The second is modularity (Baldwin and Clark 2000; van der Hoek and Lopez 2011), which is related to an economic objective, lowering the cost of development. These criteria are important because they can be used as proxies for other criteria. For example, it may be hard to assess the usability of an interface before building and testing it, but it may be relatively easy to assess the simplicity of an interface, and so a designer may, over time, bias toward simple interfaces as a result of experiencing usability problems when complex designs were implemented.

Returning to our example, it may be that after having conversations with customers, we learn that they find boxes more usable: they are easier to pour from and position on shelves. This discovery of a new criterion would lead us to rescore the existing designs and explore the box part of the design space more thoroughly.

To sum up, we use two levels of abstraction in thinking about design and evaluation spaces. At one level, we can focus on generating or positioning points in the spaces. At another level, we can focus on generating the axes of the spaces. These levels are at a higher level of abstraction—if we shift to these levels, we are going meta. Then, our two kinds of space and our two levels of abstraction suggest that we might model the design conversation around four topics. We expand on this idea next.

19.3 TYPES OF DISCUSSIONS

Design dialogs are casual, but they often exhibit identifiable states. Such goal-directed dialogs can be modeled using speech acts (Searle 1969) and related techniques such as the language/action perspective (Winograd 1987). Winograd's technique sees conversation as an activity that can be expressed as a state transition diagram, in which speech acts move the participants from state to state, toward an action. The technique has generally been applied to processes that involve promises and commitments. The early stages of software design, however, do not require commitments. Instead, agreement on the structure of a system is sought. With this in mind, we create a provisional state transition diagram for software design conversations in the exploratory stage.

In our previous discussion, we delineated two spaces at two levels of abstraction, identifying four potential topics of conversation. One is about example designs. Another is about the dimensions of the design space. A third is about the evaluation of individual designs. And a fourth is about the criteria to be used to perform the evaluation. Then, design in the early stages might involve a rapid shuttling across these four conversations.

The conversations may inform each other. The discussion about the criteria for evaluation may guide the evaluation conversations, and may also influence the articulation of the design space, which will lead to the discovery of particular designs. Indeed, this is the definition of a top-down design process. For example, an evaluation criterion for software that calls for the persistence of the data might lead to a design dimension called *datastore*, which would ensure that all generated designs have a module in charge of creating files or interacting with databases. In a top-down design, the conversation starts with the evaluation criteria, including the functional requirements, which, in turn, leads to conversations about which components (the design dimensions), are necessary to implement the functions, which, in turn, leads to a set of designs (Figure 19.5).

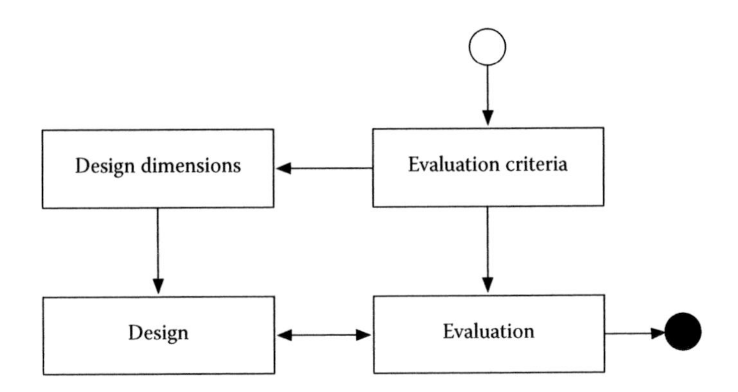

FIGURE 19.5 A top-down design state machine.

Conversely, we can consider a bottom-up design process that starts with a design (Figure 19.6). The discovery of a particularly compelling design instance may lead to an expansion of the design space, which will lead to augmented evaluation criteria. This is a bottom-up design.

While the idea that criteria shift during design may strike some as antithetical to a rational design process, finding new design objectives is often part of early design (Gero and Kannengiesser 2006). Eliminating design objectives is also important because the problem statement or specification for large systems may contain conflicting requirements (Rittel and Webber 1973), so the designer may need to loosen or disregard some constraints. In open source design, this kind of bottom-up process where examples drive the evaluation criteria is respected: top-down approaches are seen as marketing driven, whereas bottom-up approaches are seen as engineering driven and are more likely to produce useful results (Nickerson and zur Muehlen 2006).

However, in design practice, the situation is much messier than either the top-down or bottom-up process implies. Most often, design processes involve some top-down thinking and also some bottom-up thinking intermixed (Brooks 2010). Moreover, designers often disagree with each other when the evaluation process is reached. That is, they may have instinctively different reactions about whether or not a proposed design is good.

In such a case, designers could just accommodate each other by tabling the dispute and returning to the design stage. But long term, it may make more sense to reconsider the evaluation criteria as soon as a dispute arises: it could be that the different designers are

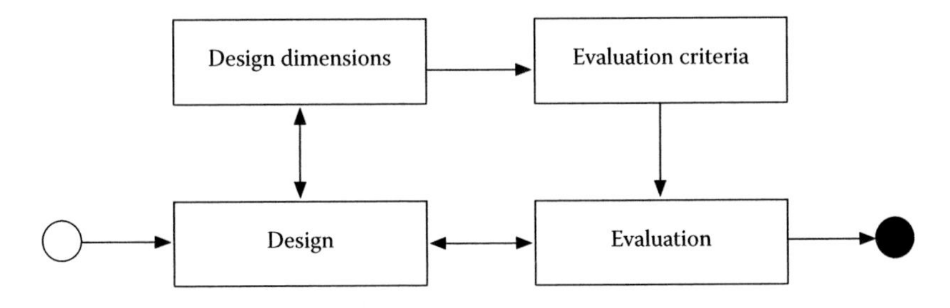

FIGURE 19.6 A bottom-up design state machine.

using different criteria—that is, they have different objectives in mind. How can this happen? Specifications can be interpreted differently. Or the nonfunctional criteria differ, with designers preferring certain "ities": for example, one designer may prefer simplicity in a design, and another may prefer realism, and hence they may have different utility functions. Especially when there are multiple criteria, designers may be collapsing their evaluation into one reaction using a utility function that they are not even conscious of. If designers are schooled in multiobjective techniques, they may evaluate along several dimensions and consider the Pareto optimal set—but, as Brooks (2010) points out, many designers appear to operate on a simple summative utility model. By forcing the conversation back to the evaluation criteria, it increases the chances that such perspectives are uncovered, and makes it less likely that the dispute will reoccur in the next design evaluation.

Thus, a normative model would incorporate both the top-down and bottom-up pathways. And the model would constrain designers to always revisit criteria in the event of an evaluation dispute. But when there is agreement, then most likely there is also agreement on criteria and so it makes little sense to discuss criteria: instead, more designs may be generated by returning to the conversational state. This model in shown in Figure 19.7.

Combining top-down and bottom-up approaches, a two-way link is shown between design dimensions and design. Evaluation criteria, in the form of functional requirements, can drive design dimensions, and discovered design dimensions might lead to enhancing the evaluation criteria. And the design process can start bottom up or top down—by starting with design, or by first having a conversation about the evaluation criteria. The central aspect of this model is the dispute—the model forces designers in a dispute state to reconsider the evaluation criteria, to go meta. What can be done in such a metaconversation? If designers can see that different criteria are held by each, then the evaluation criteria might be expanded, and designs can at least be differentiated as dominated or dominating. This may lead to designs that score well on all criteria; at a minimum, the dominating designs will be the set that pleases at least some designers, and may be a good set to take to the system's stakeholders for consideration. What such an identification of differences allows is for a set of candidate designs to be developed fully: without such an identification, it is possible that the fight over evaluation will prevent any one design from being fully developed.

In looking at the design transcript, we will be particularly interested in disputes that may happen in the evaluation process. Design can be an argumentative process (Rittel 1980), and we hope to uncover here how such arguments takes place and what they yield.

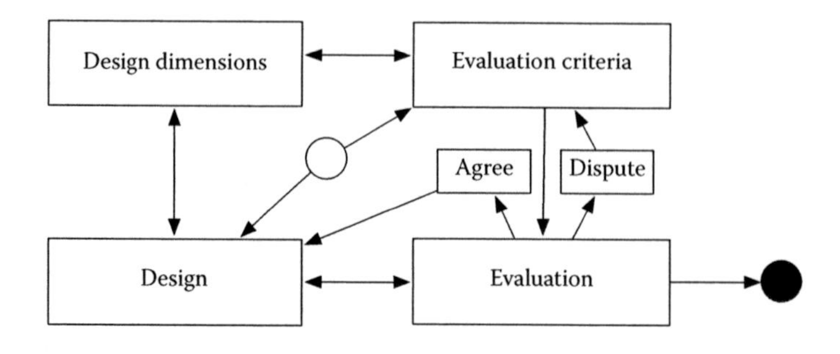

FIGURE 19.7 A normative model for design conversation.

We are also interested in the ways that arguments relate to criteria and the way that criteria relate to perspectives. In the world of spatial design and spatial cognition, one can quite literally shift perspectives. For example, one might view a map from above, or one might describe the way one traverses a route (Emmorey et al. 2000). These shifts may also correspond to the use of diagrams to show plans, and the use of gestures to show traversal (Tversky et al. 1999, 2009). In the world of software design, perspective may correspond to the point of view of actors and objects. That is, designers may consciously try to see the world from the perspective of these objects. In doing so, they may understand better how to model the system. On the other hand, they may fixate on one particular viewpoint. Sometimes, viewpoints are mutually exclusive, as in a Necker cube: we can only see things from one perspective at a time. In such cases, design dialogs may help designers to take on roles and explore the space of different partitions and relationships. And perspective changes may be signs of disputes, and, ultimately, of evaluation criteria.

With this as a background, we now analyze the AmberPoint design conversation. Our goal is to compare this instance of a design conversation to the normative model we have built—to better understand what really happens in design processes versus what we think should happen. This particular conversation, shared as part of the Studying Professional Software Design (SPSD) Workshop in 2010, has been analyzed by other researchers, and in particular depth by McDonnell (2012), who documented the dispute at the heart of the conversation. She reaches the conclusion that the design collaboration was effective, whereas we, by contrast, see the collaboration as a failure, albeit one from which much can be learned.

19.4 AMBERPOINT DESIGN CONVERSATION

19.4.1 Background

The conversation is between Ania and Jim, who work for the company AmberPoint. Both have over 20 years of software design experience. Ania has a background working with software developers, and Jim has a background working with users. They are presented with a written problem definition, the design of a traffic simulator, and they proceed to sketch out a design in the course of a little less than 2 hours. The interaction was videotaped and transcribed, and supplied to participants in the SPSD Workshop by the organizers.

Approximately 12,000 words were spoken over the course of 108 minutes. Each person spoke about 150 times, so the average turn was about 40 words. No one turn used over 300 words. Thus, the conversation involved a series of short speeches, consistent with the concept of turn-taking (Sacks et al. 1974). Most of the drawing was done by Jim, and Jim was in the frame more often than Ania, but Ania gestured on and enhanced the diagrams at various points in the talk.

19.4.2 Opening Themes

The first conversation is about the end users of the simulation, identified as students and a professor by Jim: "Students need to be able to build these roads." This is a functional requirement, and so the conversation starts off top down, delineating one functional criterion for evaluation. Ania says: "So it's like a drawing tool that can allow them to lay down these intersections." In other words, Jim identifies a particular object, the road, and Ania elaborates this by using the word "intersection." She is, in a sense, finding a design exemplar already—she mentions *a drawing tool*. She next says: "I suppose there's some traffic speeds at which these cars travel." She has introduced another object, a car, and an associated characteristic, speed. She is defining design dimensions, albeit by adding to the image of a particular design. He responds: "I don't know if they can set the speeds. They can set the density." He is referring to the problem definition that mentions the ability to set the density of traffic but not the speed. He is evaluating her introduction of the concept of speed, and questioning it on the basis of parsimony: he is observing that the specification of the problem does not call for setting speeds. So, already the two designers are in a dispute over a proposed design.

This initial interaction will be expanded upon over the next hour or so, as they debate in different forms whether or not it is necessary to model the cars themselves. Density is an attribute of a road, not a car, but speed is a characteristic of a car, and so this opening conversation delineated two alternate perspectives with two different centers. Jim drew the diagram on the whiteboard shown in Figure 19.8, and the diagram was used to visualize intersections in the real world as well as the user interface. It was almost as if the whiteboard became the screen. This diagram remains throughout the design process, assisting the designers to think about and argue their ideas, but perhaps also fixating them at the design and evaluation levels of the potential conversation, rather than the metalevels.

While at the very beginning the designers cycle through all four states, what does not happen after this initial conversation is a debate about the design criteria. Instead, the designers table the dispute and return to the design state, sketching out an example model and interface. After Jim says, "They can set the density," he immediately follows with, "Want to draw something?" According to our model, this is not normative—it introduces

FIGURE 19.8 The diagram used to represent the intersections and the user interface.

a transition from dispute back to design. By contrast, another view, consistent with that of McDonnell (2012), would see the comment as a positive move, an accommodating strategy, a way of proceeding while minimizing conflict. We will see in the end the result of this accommodation.

But first, we need to understand the debate. In the intersection-centric view, cars are controlled by intersections, and in the car-centric view, cars have start points, destinations, routes, positions, and speeds. Both perspectives have something to recommend them, but it is not apparent to the two designers the strengths of the other's beliefs and arguments until later in the conversation.

In keeping with Brooks' notation, the designers have implicitly uncovered a central choice in the design process, shown in Figure 19.9. That is, a central choice revolves around the representation of traffic. The two views are mutually exclusive: once one is chosen, the model and the interface choices afterwards will be very different. The designers do not, however, explicitly draw this diagram: their work on the whiteboard creates an instance of a particular design, rather than exploring the set of decisions that must be made in the design process. What might have happened if they had created the aforementioned diagram? It might have been possible to follow through on two alternate designs, and then, while comparing them, to work out the hidden evaluation criteria that might be behind their disagreement.

The designers of the other two videos in the corpus, who work at Adobe and Intuit, do not explore the issue in as much depth as Ania and Jim. They decide that the cars can more or less be controlled by the intersections, consistent with Jim's viewpoint in this AmberPoint video. But there is much to be said for the car-centric approach. Many simulators do model the car in the way that Ania recommends. Indeed, the first object-oriented programming language, Simula, included an example of a car wash in which the cars themselves were modeled (Birtwhistle et al. 1979), and discrete event simulations usually model the autonomous objects, and, in particular, cars when modeling traffic (e.g., Benekohal and Treiterer 1988; Resnick 1997). So, while we characterize the AmberPoint collaboration as a failure, it is, in our eyes, more successful than the other two design collaborations, in that the AmberPoint team came closest to developing the kind of design that is prevalent in the literature on discrete event simulation.

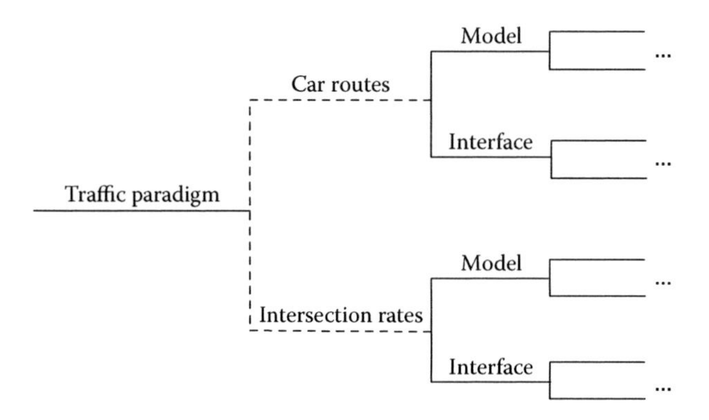

FIGURE 19.9 The implicit model of Jim and Ania's conversation.

Returning to the conversation, we see how Ania argues her case:

ANIA: Well, so one is you want to change the layout of the map, two is you want to change the parameters you gave us in terms of speed and timing, right? And three, you want to run it, meaning little dots are moving, showing you how the traffic is flowing.

Here, she is sticking to the idea of speed, and she is imagining cars as dots moving on the simulation. Later, we will find out that Jim is thinking that the road itself could just change color to indicate congestion—that the cars themselves would not need to be drawn.

19.4.3 Argument

For the next 35 minutes, the designers proceed to work through how signals will work at intersections. Then, Ania reintroduces the car idea [0:41:23.8]:

ANIA: Now, with regard to the actual cars, okay. So, you start simply. You want to pump a single car through the system and you want to be able to specify where it starts and where it's going/

JIM: I think it sounded like from here you pump in traffic from the edges of the streets.

ANIA: I understand but what does that mean? In other words, the traffic, you know what does that mean to pump in traffic?

JIM: You'd set a rate, like this many cars come in per minute going this direction.

ANIA: But you want to build up on the complexity of it, so I want to start off and say I have a car starting at A that wants to end up on the diagonal of, you know, D and whatever, if there was a fourth street or something, right? So it seems to me that in order to do that, each car has to have a destination, a starting point, and a destination—the route it needs to take. Because if I just start pumping, I say hundred cars and they just go.

JIM: Yeah, I don't know how you'd figure out like what percentage of cars are turning left.

ANIA: Exactly, so how many cars are going this way, are they making right, so this traffic increases because this is the suburbia and they're all making a right over here.

JIM: I don't know so much that it'd be done at a car-by-car basis, but I think there's additional information on each intersection about what percentage of cars do you expect to be going straight, making a left turn, and making a right turn.

This is a conflict about the design: we can see the conflict through the language, and, in particular, the use of the word *but*. Jim says that you can stay with the intersection, with the distribution of cars at the intersection, and Ania says that, realistically, cars have destinations: the destinations are set ahead of time, not at an intersection.

He is right that it is possible to set the distributions at intersections, feed in cars, and let the intersections manage the traffic distribution. But she is right that this is not a realistic model of what happens in the world. Perhaps because she understands that his way might be simpler to implement, she makes the following design argument [0:43:08.7]:

ANIA: That why I thought that maybe, you know, just like we have pallets for these things, maybe we have—in other words so you can customize your car but there's some cars that you know, again, Volkswagon vs. Mercedes that have certain characteristics. You know, I always go straight, or I always make a left turn whenever possible. I always make a right turn when possible, you know I make left and the right. So this way, you can simulate something that easy again. Because otherwise, how do you tell these cars where to go? You see. Okay, I say I want fifty cars on the grid, but I don't want them all going the same way.

JIM: I just think at each intersection, there's like a random number generator. That's the distribution of a given car that comes up to this intersection from the west to the east, you know 80% go forward 10% go left and 10% go to the right. Something like that.

ANIA: But then you have no control over that.

JIM: Yeah, it didn't sound like you need control at the individual car level, I guess that's all.

ANIA: Well, by controlling it at an individual car level you end up controlling the flow. You need some control with the flow, so what kind of maneuvers, you know how do you program a little car? How do you say, whether each car or—

JIM: Are you saying there's actually a car out there?

She is certainly saying that, and he is beginning to realize that she is making an argument from a different perspective. The dialog continues [0:44:47.0]:

ANIA: I'm entertaining this idea, whether there should be a car out, and there are different types of cars you know.

JIM: Yeah, put it up there.

ANIA: I don't know, talk through this with me. Because on one hand, as you said, you can generate these, but you still need to be given some [inaudible]. You need to characterize the cars. ...

JIM: Yeah, I think you would have to do that with a combination of the density at certain hours and for a given intersection, a percentage, what you expect to go left right or straight for that hour.

ANIA: So you say, I want 50 cars like this.

JIM: No, no, then I think you would say between these hours, 50 cars start at B and then given the percentages, they're going to turn at 9 am.

ANIA: So that's what I was getting at, so you need direction, south. Where they're going, start, end, where they want to end, right? Because if it's starting where is it going? Or it doesn't have an end, it just follows.

JIM: I think it doesn't have an end, you can just give it a density and it just follows these rules. Because I think, in viewing, like the simulation would be so hard for the students to set up if they had to design a separate car for each.

ANIA: So it has to have a starting point, and then you set something about time. How this can adjust itself depending on time.

JIM: Potentially, I don't know if it needs to be that complicated but I could see, I mean realistically—

Jim thinks that the introduction of the car makes the model overly complex for what it is required to provide, because the problem statement does not say that the designers need to model cars, it just says that they need to model density. He thinks that the model can be built at a higher level, based on specifying a distribution at the intersection, without modeling the actual routes of the cars. Thus, his objection to her scheme is based on the criterion of practicality. She says the cars can be modeled en masse, but the cars should not be controlled by the intersections. They should have directions and routes.

They come back to the issue again [1:02:34.5]:

ANIA: But Jim did we finish with the cars kind of? Did we have a [inaudible] for the cars that you would just be saying, I want fifty cars?

JIM: I think I felt like each of these has a density, like number of cars per— ….

ANIA: … you would say left, south going, you know, whatever, that you just put in these percentages and say, I will have three cars starting at D West, and they're going to go this way.

JIM: I think this, though, is part of the intersection, the percentage that turn at the intersection.

ANIA: Really? You see I thought it was, because what I thought would be that the car would be its own thing. Cars are independent of intersections I feel, so you want to leave the intersection. You want to change the behaviors of the cars without changing the behaviors of the intersection, so it's the cars [see Figure 19.10] that has the knowledge of where it's going and how it's behaving. So my idea was that, you know, a car would start here and it will note that at every opportunity it will turn left and then right and then left and then right, that's the characteristic of this car. This car would just go straight. You know?

In Figure 19.10, Ania is turning away from the board toward Jim. She gestures, perhaps as if she is becoming the car. This is an example of taking on a perspective through gesture. She portrays a route, not a survey, view of the traffic problem.

This example suggests that gestures as well as language may be important in delineating changes in perspective, as well as indicating disputes. By contrast, the problem with the diagram as they drew it is its specificity: it keeps them in the design conversation. The gestures, the turning from the board, can potentially be used to transition to different states.

Ania is making a point related to modularity: cars are independent of intersections. That is, once an intersection has its traffic light timing set up, the user might want to change the nature of the overall traffic flow and see its effect on all the intersections. The user does not necessarily want to reset the distribution of left turns at every intersection. Thus, she now has a strong argument that her model will aid the user.

Later, Jim brings the issue up again [1:18:24.8]:

JIM: Okay, then cars. This is where we still have a little doubt. This is kind of the editing-setting mode, and then for each road, you can specify the density: how many cars per minute, hour, are traveling through.

FIGURE 19.10 Gesture movement vectors expressing the car's perspective.

ANIA: Yeah, but I would take more, so I would take the map drawing as one thing, but the intersection specification and the car specification, I would take it more as a part of the simulation. So you build your roads and you build all of that, and you've got your distances and your speeds and all of that. Now you're getting ready to simulate.

So, he wants to just specify density, and she wants to have a separate car specification. Now she makes a new argument [1:20:06.1]:

ANIA: So I would start simple, Jim. So why don't we start out by saying how many cars do I want in the system, and for now we will assume that they are evenly distributed starting at all possible entries, you know, lets make it simple, hmm? So in other words if I say I want one hundred cars, I'll get, you know, however there are prongs outside divided by that and that how many will start. Start simple. And again, just to start off, I would say, you know, just to even see whether the traffic is moving at all, they're all going straight. The simplest possible scenario. You have X cars starting even going straight. Is the traffic moving? Because there is no point in, you know, I would think that's how you would approach you know, is it even moving at that point? And then if it's moving well, however you measure that, you can start building complexities into the cars, into the traffic.

JIM: Anyway, so then you could say how many permanent. Or you could just say 100 and distribute evenly.

ANIA: That's what I was saying, just to make it simple.

JIM: Because if you want to specify it, you know, that's like a big-ass dialog. We would have to think about it some more, because there's a lot of, right?

Ania is using the criterion of simplicity, which seems a fitting tactic in response to Jim's objections to the complexity of specifying car movement. But this highlights a difficulty of nonfunctional requirements: one person's idea of simplicity may be different from another's, and without pinning down what the criterion actually is and how to measure it, it can be for designers to talk past each other.

Indeed, Jim is still not in agreement [1:23:54.0]:

JIM: This is something where I'd go back to the customer and try and figure out, how did they collect this data, you know like Professor E must have statistics about, you know San Francisco traffic, like does she have it in terms of at each intersection, this is the percentage of cars going each direction, or does she have, between nine and five, this many thousands of cars are trying to get from this point to this point, you know. So look at what kind of data she has available to her that her students are used to working with, because otherwise we're [inaudible] how this data is really working.

His point is the following: it would be nice to know where all cars go, but it is likely that we do not know that. We may have sensors at intersections, in which case we do know which percentage turns left or right. Jim, at least, is recognizing that this is really an issue of the interpretation of requirements—it is a criteria discussion. However, here, the artificial nature of the design task may interfere. Jim suggests going back to the (fictional) user. In a real design situation, they might be able to do exactly that: call the professor, and find out whether or not there is a need for the simulation to model cars; whether, in fact, the parameters exist to create such a model, and if not, if car routes can be inferred from intersection data. Indeed, developers often resolve difficulties interpreting specifications by going right back to the stakeholders and asking.

The topic is raised again [1:26:18.0]:

ANIA: This is cars?

JIM: Yeah, percentage of cars we're doing it off, rather than do it at the intersection. This is why we need to check with the professor again. This is part of the simulation.

It's either we build up a whole set of cars or we say at each intersection, this is the percentage of

ANIA: Yeah, yeah I don't know the level of control we need on the traffic— how it's going, its not clear to me.

She is making the argument of modularity again: from the user's perspective, it seems better to model cars separately. She makes it clearer [1:27:28.6]:

ANIA: Yeah, and in some ways, I would say, within the simulation you may want to keep, again, separate the intersection from the cars. Because you want to have this map, with this kind of intersections, and pump different kind of cars through it. So you can mix and match.

JIM: So you want to leave one of these constant and change the other.

ANIA: Exactly, exactly.

JIM: This lighting seemed to work for, you know, this [inaudible] of traffic.

ANIA: So this has to be like the main object. Your map is a named object, your intersection configuration is a named object, and your traffic patterns are named object. So you can say, I want to use this map, with this intersections, with these traffic patterns.

JIM: So then this was the percent? What was this? This was the density, and then this was the turning percentages, direction percentages, okay. Okay. So then what I want to do, you build the simulation, you run it in which you use this map and then we have a graphic way of showing where all the traffic is backing up, and we also have some kind of table that shows you the average maximum wait-time, average maximum number of cars and average through each intersection.

ANIA: So what do we do? We build a flex application that has little dots moving? I mean how do you visualize this simulation?

JIM: Yeah, yeah or we could just color the road bolder if there's traffic waiting there.

Even in the visualization, he is thinking that they can avoid having cars. She comes back to the issue yet again [1:42:36.2]:

ANIA: Yeah, but then there are also the cars, which kind of [inaudible] them into sections need to be aware of, because that's where we know what's happening, and we want the cars themselves also to know where they are at. So there is a relationship between the road, very dynamic relationship, between the car, the road, and the intersection. Because car is always on the road, on some road. And some percentage of the time the cars are in intersection. So cars always on the road, and sometimes in intersection. And the intersection tells the car what it can or cannot do at any given time. And the car needs to tell the intersection what it wants to do. So the car knows what it wants to do, where it's going and the intersection tells the car, yeah.

JIM: It might ask the approach what the traffic configuration is.

ANIA: So the car is going on the road, it approaches the intersection.
JIM: Yeah, that's kind of where this stuff lives. We haven't really, this is where we find some conflict.

Jim is fighting for practicality, parsimony, the letter of the problem statement. Ania is fighting for realism and modularity. Simplicity is invoked by both: it is simpler to avoid specifying the traffic patterns of cars, but it is also simpler to reuse intersection timings and just vary the traffic patterns, rather than reconfigure all the distributions at each intersection.

These different proposed models are in a Pareto optimal set: neither dominates the other. Which is more desirable depends on how strong the different criteria are in the minds of the designers. The reason the design session fails is the designers never explicitly discuss their nonfunctional criteria. If they had, they might have spent less time in dispute and instead worked through two alternate designs, which, in the fictional world of the problem, stakeholders could run their own evaluation on.

McDonnell (2012) is correct that the designers are very accommodating of each other. She also points out that the situation is artificial, and that with a real design problem they might have found ways of resolving the conflict after the session. All of these are good points. But while certain kinds of conflicts are counterproductive, so are certain kinds of accommodation. By not having a conversation about the evaluation criteria, and in particular the nonfunctional criteria, the designers relegate themselves to a dissatisfying dispute that might have been either better understood or even resolved by going meta.

19.5 DISCUSSION

We formed a normative model of design conversation, which posited that designers should move through four different design states, two of them metalevel states that involve discussion about design dimensions and evaluation criteria. We also posited that conflict may occur in the evaluation stage as a result of designers emphasizing different evaluation criteria.

We analyzed one instance of a design conversation. In the conversation, there was indeed a conflict over the way that automobile traffic should be represented. The conflict surfaced early in the conversation, and was returned to many times over the 108 minutes of the dialog. Thus, at least in this instance, design conversation appeared to function as a state machine, with a small set of states visited and revisited.

The reasons for the conflict can be inferred from the dialog. Both designers invoked different criteria to bolster their argument. Jim used an argument based on practicality: the specification did not include a call to manipulate individual cars, only traffic densities. These could be most easily controlled at the intersections, objects that were necessary anyway to control the signals. In addition, it was likely that statistics about the traffic flow were probably collected at intersections, and these statistics can drive the simulation. He also invoked the criterion of simplicity: the model would be easier to build than the alternative. Ania thought that traffic should be modeled as sets of individual cars, with starting points and destinations, or routes, built in. She argued that this is a more realistic approach, as, in fact, cars are not directed to destinations as a result of their encounters with intersections.

Later, she argued her point based on the criterion of modularity: the patterns of lights at an intersection might be independent of the mixture of traffic that flows through the intersection. She also invoked simplicity: it might be easier to configure a set of cars than to fiddle with the percentage of left-hand turns at a large number of intersections.

Over the course of the conversation, each designer gained a better understanding of the alternative models. However, the conversation did not lead them to uncover the latent reasons for their respective preferences. Instead, they were mustering better arguments to bolster their positions.

We were also interested in the fine-grained interactions of the designers. The design conversation involved very short segments of speech punctuated with sketches and gestures. The gestures sometimes involved pointing to the whiteboard, and sometimes the gestures came off the board. These off-board gestures sometimes corresponded to a change in perspective. Diagrams served as fixation points. They kept the conversation pegged in a state of design. Relegating part of the board to drawing the evaluation space, or showing the decision points in the design process, might have encouraged the designers to move more often into the metastates of the conversational model. Negating words such as "but" often indicated that they were in disagreement over an evaluation. From these observations, we can form a conjecture about what might improve software design conversation.

19.6 TRAINING SOFTWARE DESIGNERS

This chapter developed a normative process model of design conversation. But we saw that the designers we studied did not follow it. We cannot know if following the model would have helped, but instead we can define an experiment to discover if, with other designers, training based on the model would make a difference in the quality of the designs produced. We present a set of training actions that, based on our proposed process model and the observations made from the transcript, might improve the output of an exploratory software design conversation. The action set consists of:

1. Training designers to visualize design spaces using the method described in Brooks (2010). That is, designers would be shown how to draw decision trees with exclusive branches as well as independent branches and how the traversals through the trees represent designs.
2. Training designers to visualize evaluation spaces based on the idea of multiobjective optimization in which designs are plotted along multiple dimensions and evaluated by inspecting for Pareto optimal sets (Keeney and Raiffa 1993).
3. Training in a new form of design session visualization, in which portions of whiteboards would be allocated to design space and evaluation space diagrams.
4. Training in facilitation, in which conflict cues would be recognized as signals to move into metastates. In particular, trainers would act out linguistic and gestural cues that indicate disagreement over design evaluation, so that designers could practice recognizing them and signaling other designers to transition to a discussion of evaluation criteria.

How might such training be validated? Teams of designers, matched for experience levels, might be randomly assigned to a training condition and a control condition. The control condition might consist of training in a commonly taught design methodology such as object-oriented analysis. After the training, the teams might engage in design exercises, and the resulting designs would be evaluated. If the training proved effective, then subsequent tests might be arranged to understand which subset of the four different actions in the training was responsible for changes in design outcomes. Moreover, the differences between the processes shown in each condition might be used to refine the design conversation model.

19.7 CONCLUSION

When conflict occurs in a design process, the underlying root of the conflict may be a difference in evaluation criteria. Making such criteria more explicit, and mapping potential solutions in a space defined by these criteria, might speed up the design process by allowing designers to find a set of possible designs rather than fighting to convince others that one design is the best.

Design can begin bottom up as well as top down, but at some point, a metaconversation needs to happen between the designers: what do you value?

ACKNOWLEDGMENTS

This work was supported by the National Science Foundation (awards: IIS-0725223, IIS-0855995, and IIS-0968561) and was influenced by all of the participants at the Studying Professional Software Design Workshop in 2010.

REFERENCES

Baldwin, C. Y. and Clark, K. B. 2000. *Design Rules*, MIT Press, Cambridge, MA.

Benekohal, R. F. and Treiterer, J. 1988. CARSIM: Car-following model for simulation of traffic in normal and stop-and-go conditions, *Transportation Research Record* (1194), 99–111.

Birtwhistle, G. M., Dahl, O. J., Myhrhaug, B., and Nygaard, K. 1979. *Simula Begin*, Chartwell-Bratt, Bromley.

Brooks, F. P. 2010. *The Design of Design: Essays from a Computer Scientist*, Addison-Wesley Professional, Boston, MA.

Case, P. and Pineiro, E. 2006. Aesthetics, performativity and resistance in the narratives of a computer programming community, *Human Relations* 59(6), 753–782.

Chopra, S. and Dexter, S. 2007. Free software and the aesthetics of code. In *Decoding Liberation*, Chopra, S. and Dexter, S. (eds), Routledge, New York, pp. 73–110.

Chung, L. and do Prado Leite, J. 2009. On non-functional requirements in software engineering. In *Conceptual Modeling: Foundations and Applications*, Borgida, A. T., Chaudhri, V. K., Girogini, P., and Yu, E. S. (eds), Springer-Verlag, Berlin, pp. 363–379.

Deb, K., Pratap, A., Agarwal, S., and Meyarivan, T. 2002. A fast and elitist multiobjective genetic algorithm: NSGA-II, *Evolutionary Computation, IEEE Transactions* 6(2), 182–197.

Emmorey, K., Tversky, B., and Taylor, H. A. 2000. Using space to describe space: Perspective in speech, sign, and gesture, *Spatial Cognition and Computation* 2, 157–180.

Gero, J. S. and Kannengiesser, U. 2006. A framework for situated design optimization. In *Innovations in Design Decision Support Systems in Architecture and Urban Planning*, van Leeuwen, J. and Timmermans, H. (eds), Springer-Verlag, Berlin, pp. 309–324.

Gero, J. S. and Kazakov, V. A. 1996. An exploration-based evolutionary model of a generative design process, *Computer-Aided Civil and Infrastructure Engineering* 11(3), 211–218.
Keeney, R. L. and Raiffa, H. 1993. *Decisions with Multiple Objectives: Preferences and Value Tradeoffs*, Cambridge University Press, Cambridge.
Lee, J. and Lai, K. Y. 1991. What's in design rationale?, *Human–Computer Interaction* 6(3–4), 251–280.
Maeda, J. 2006. *Laws of Simplicity*, MIT Press, Cambridge, MA.
McDonnell, J. 2012. Accommodating disagreement: A study of effective design collaboration, *Design Studies*, 33(1), 44–63.
Metcalfe, J. E. and Shimamura, A. P. 1994. *Metacognition: Knowing about Knowing.* MIT Press, Cambridge, MA.
Moore, M., Kaman, R., Klein, M., and Asundi, J. 2003. Quantifying the value of architecture design decisions: Lessons from the field. In *Proceedings of 25th International Conference on Software Engineering*, IEEE Computer Society, Los Alamitos, CA, pp. 557–562.
Nadin, M. 1996. Computational design. Design in the age of a knowledge society, *Journal of Design and Design Theory* 2(1), 40–60.
Nickerson, J. V. and zur Muehlen, M. 2006. The ecology of standards processes: Insights from Internet standard making, *Mis Quarterly* 30(3), 467–488.
Osborn, A. F. 1963. *Applied Imagination: Principles and Procedures of Creative Problem Solving*, 3rd revised edition, Charles Scribner's, New York.
Purcell, A. T. and Gero, J. S. 1998. Drawings and the design process: A review of protocol studies in design and other disciplines and related research in cognitive psychology, *Design Studies* 19(4), 389–430.
Resnick, M. 1997. *Turtles, Termites, and Traffic Jams: Explorations in Massively Parallel Microworlds*, MIT Press, Cambridge, MA.
Rittel, H. W. 1980. *APIS, A Concept for an Argumentative Planning Information System*, Institute of Urban & Regional Development, University of California, Berkeley, CA.
Rittel, H. W. J. and Webber, M. M. 1973. Dilemmas in a general theory of planning, *Policy Sciences* 4, 155–169.
Sacks, H., Schegloff, E. A., and Jefferson, G. 1974. A simplest systematics for the organization of turn-taking for conversation. *Language* 50, 696–735.
Schön, D. A. 1983. *The Reflective Practitioner: How Professionals Think in Action*, Basic Books, New York.
Schummer, J., MacLennan, B., and Taylor, N. 2009. Aesthetic values in technology and engineering design. In *Philosophy of Technology & Engineering Sciences*, Vol. 9, part V: *Handbook of the Philosophy of Science*, Gabbay, D., Thagard, P., and Woods, J. (eds), Elsevier Science, pp. 1031–1068.
Searle, J. R. 1969. *Speech Acts*, Cambridge University Press, Cambridge.
Simon, H. A. 1996. *The Sciences of the Artificial*, MIT Press, Cambridge, MA.
Tversky, B., Lee, P., and Mainwaring, S. 1999. Why do speakers mix perspectives?, *Spatial Cognition and Computation*, 1, 312–399.
Tversky, B., Heiser, J., Lee, P., and Daniel, M.-P. 2009. Explanations in gesture, diagram, and word. In *Spatial Language and Dialogue*, Coventry, K. R., Tenbrink, T., and Bateman, J. (eds), Oxford University Press, Oxford, pp. 119–131.
van der Hoek, A. and Lopez, N. 2011. A design perspective on modularity. In *Proceedings of the Tenth International Conference on Aspect-Oriented Software Development*, ACM Press, New York, pp. 265–280.
Visser, W. 2006. *The Cognitive Artifacts of Designing*, Lawrence Erlbaum, Mahwah, NJ.
Weick, K. E. 1995. *Sensemaking in Organizations*, Sage, London.
Winograd, T. 1987. Language/action perspective, *Human–Computer Interaction* 3(1), 3–30.

CHAPTER 20

Accommodating Disagreement

*A Study of Effective Design Collaboration**

Janet McDonnell

University of the Arts London

CONTENTS

20.1 INTRODUCTION

A better understanding of how design, design innovation, and design collaboration come about is a prerequisite for constructive intervention in design education and design practice. Efficiency and effectiveness in design collaboration rely on making it possible for collaborators both to exercise their skills and knowledge relevant to the design task and, at the same time, to be able to exercise their abilities to work collaboratively with others to achieve a shared goal. Alongside a scholarly interest in knowing more about what it is we do when we collaborate to design, there are commercial interests at stake in understanding how to improve design processes and their outcomes.

* Reprinted from *Design Studies* 33 (1), McDonnell, J., Accommodating disagreement: A study of effective design collaboration, 44–63, Copyright 2012, with permission from Elsevier.

This investigation focuses on the verbal interaction between two experienced software designers collaborating to develop initial design ideas for a new software application. The study examines the negotiation of the design—the ways that it is brought into being as the designers work on the design task. Attention is centered on the interactional strategies that support their design collaboration. The exchanges facilitate and enact movement around both the "problem space" suggested by the design brief and the "space" of design possibilities, what we might refer to collectively as the design situation. Putting it more prosaically, this particular study focuses on what stops the collaborators getting stuck.

In contrast to work that focuses on breakdowns and failures, the work presented here pursues an interest in fine-grained inspection of how collaboration flows to successfully and productively move a design forward. In a recent study of a conversation between a designer and his clients (McDonnell, 2009), we focused on the talk-in-interaction to identify ways in which design decisions and their justifications are supported by the subtle orientation of the speaker to (appeal to) his/her audience (Perelman and Olbrechts-Tyteca, 1971). Working with the performative aspects of language, and from the view that design conversation brings design into being, we are interested in the skills for progressing the design—for negotiating what to do (next)—that are performed through talk. Access to a shared data set intended to support the study of professional software design (Petre et al., 2010) presented an opportunity to further these interests by inspecting the conversation between two experienced software designers working on a design brief together.

20.2 MATERIALS AND APPROACH

20.2.1 Data

The data studied comprised material collected during a 2-hour software design exercise carried out by two software engineers, Carla and Joe, working together to develop initial design proposals to meet a written brief. The designers were asked to collaborate on this task by working at a whiteboard while being observed and videotaped. Figure 20.1 shows what the camera captured—the work evolving on the whiteboard and the participant(s) at the whiteboard. For most of the time in the session, Joe was at the whiteboard, while Carla was off camera but audible.

The task was to design a road traffic simulator with interactive functions to support learning about traffic patterns in an educational setting.* The two software engineers were asked to come up with some initial design ideas sufficient to brief a team of software (prototype) developers. At the end of the design session, the participants were asked to briefly present the state of the design to the observers (for a few minutes). These data, which will be referred to here as the AmberPoint session, were one of three similar data sets based on design sessions that were recorded and initially distributed for an international workshop on Studying Professional Software Design, which took place in 2010 (see Acknowledgments for further details). The pattern of this larger research project, to which an earlier version of the study presented here made a contribution, followed that of using shared data sets to

* The brief for the design task was published in full in *Design Studies* 31, 2010, 542–544.

FIGURE 20.1 Snapshots from the video recording at 0:08:55 (left) and 1:11:39 (right).

advance our understanding of design through research (Cross et al., 1996; McDonnell and Lloyd, 2009).

The study presented is based on the following materials from the AmberPoint session:

- The video recording of the session, taken from a single view, static camera centered on the whiteboard. The extracts presented in this chapter are the author's own transcriptions from the video recording.
- A transcript supplied by the workshop organizers which records what was talked about by the two designers (what ten Have describes as a "secretarial transcript" [ten Have, 2007: p. 94]). The transcript includes time stamps to provide anchor points to moments in the video recording. These also serve as common reference points for researchers who are analyzing the common data set in different ways for a variety of research purposes.
- A transcription of a short post-session debriefing interview with the designers was also made available; some reference is made to this in the following text.

It appears from the AmberPoint data that the collaborators know each other; there is an explicit reference in the talk during the design session (at time stamp [1:47:33]) which indicates that they know each other in an institutional context; this is something that was also confirmed by those who collected and provided the data for analysis (Petre et al., 2010). The post-session interview adds support to the belief that the designers, Carla and Joe, do work together from time to time and it discloses that they have, respectively, 26 and 20 years experience as software designers. Despite the ease with each other, and perhaps the confidence in each other's integrity, that this disclosure may account for, it is important to be clear that the AmberPoint data are not design occurring in a natural setting, they are researcher-provoked data (Silverman, 2001). Nevertheless, those observed are experienced software engineers who are colleagues. This and the "laboratory" aspects of the setting are a pervasive influence on what we see. Although the interaction is contrived for research purposes, it has elements of institutional talk, particularly that the participants' institutional identities will be relevant to their interaction (Heritage, 2005). Nevertheless, the design task, the time constraint, the presence of observers, the physical restrictions of the static single camera position, and the requirement to work standing at a whiteboard, among other factors, frame the event and inevitably have effects *on* the collaboration, as does the fact that the participants know each other; some of the more subtle aspects of these issues are returned to later.

20.2.2 Method

The study focuses on the talk-in-interaction during the design session. The term *talk-in-interaction* covers both informal and formal conversation, the latter being talk subject to functionally specific or context-specific restrictions, or specialized practices or arrangements (Schegloff, 1999). Talk-in-interaction includes forms of structured talk associated with all manner of talk at work where institutions, roles, and so on have impacts not only on what is talked about but also on the ways in which talk is organized.

The study presented pays attention to the turns-at-talk. It draws on the approach of conversation analysis (Sacks, 1995; Schegloff, 1997). From this perspective, a turn-at-talk is considered as an action, its meaning is approached by considering the local context, in particular by examining the next action, that is, the turn-at-talk that follows it. This approach, in focusing on the performative, thus dispenses with a reliance on making claims about what is in a speaker's or a hearer's mind. The approach is described succinctly by Matthews (2009: p. 36) in the introduction to his study of the interference of the rules governing social order and the rules of brainstorming as follows: "[i]n conversation analytic studies, the point is not to build theory but to unpack how events are organised and ordered—to see how social actions are structured and accomplished. This is not done by employing e.g. theoretically-derived concepts or definitions, but rather it is attempted through analysis of the data on the basis that each turn at talk displays *the speaker's understanding* of what is going on."

In the context of working with qualitative data, a turn-at-talk is a relatively objective unit of analysis for spoken interaction. For example, it is usually accepted that although it may be difficult in some circumstances to distinguish turns-at-talk (such as where there are many participants, several of whom speak at the same time), it is not necessary to test for inter-coder reliability of unit identification. The study presented here does not interpret what is said to apply or derive coding categories to the data. The transcript of the session was used as a "noticing device"; that is, the transcript supplied by the workshop organizers served as a way of noticing and discovering the features that are drawn attention to later in this chapter (ten Have, 2007: p. 95). The extracts presented in this chapter are more detailed retranscriptions of particular sequences of exchanges between the two participants; they serve as examples of what has been noticed in the recording of the spoken material. Time stamps for the start and end of each abstract are provided to give the reader a sense of when, in the course of the session, each example is taken from and these time stamps also serve as anchor points to the video material. This notation also supports comparative studies and analyses of a common data set by other researchers so that comparisons of interpretations can be made. "Generations" of design researchers have benefited from the ability to study prior analyses of the Delft protocols (from which the first collection of studies was published in Cross et al. [1996]) and explicit examples of how comparisons can be made when a common data set is available have been published for the DTRS7 data set (McDonnell and Lloyd, 2009).

Throughout the extracts in this chapter, italicized talk and added italicized material in square brackets are used to draw the reader's attention to parts of what is said and not to indicate intonation. Neither intonation nor overlapping speech is indicated by notation, pauses are indicated by the symbol "+," and long pauses and other relevant paralinguistics are indicated by plain text in square brackets, for example, [laughs]. Text in parentheses indicates a contribution that is unclear on the recording.

20.3 CHARACTERIZATION OF THE DESIGN SESSION AS FLUID EXPERT DESIGN PRACTICE

The AmberPoint session was selected for attention in this study because the participants' strategies demonstrate some of the characteristics associated with expert design performance; in particular, breadth first, depth next progression (e.g., Akın, 2001), counterplay of raising issues and dealing with them (Rowe, 1987), and solution orientation (Lawson, 1980). An independently conducted study of the AmberPoint session, which uses a coding scheme to analyze a number of characteristics of the decision making, describes Carla and Joe's decision mode as predominantly cooperative (Christiaans and Almendra, 2010*) within a scheme that allows for the classification of decision modes into cooperative, autocratic, or autonomic. Their analysis recognizes different roles, at different times being undertaken to this cooperative end, at times, for example, seeing Carla as challenging Joe's suggestions and at other times as a more constructive engagement—however, both to the same cooperative ends (Christiaans and Almendra, 2010).

The AmberPoint designers start the recorded session in silence as they read through the design brief. After about 6 minutes, following an invitation from the observer, Joe gets the designing underway [0:06:08]. Carla responds by immediately bringing the design into being verbally, as she says, "so it's like a drawing tool ..." [0:06:27], "*it*" is understood by Joe to be a reference to the virtual design (Medway, 1996) as he colludes with this in his reply by suggesting that they start to draw something ("*it*") on the whiteboard. Thus, immediately at the start of the interaction, the designed software becomes real conversationally and thence via the marks on the whiteboard as well as via what is spoken. The remainder of this section briefly characterizes the design session in terms of accomplished, rapid, opportunistic movement of topic focus, that is, it looks at what is talked *about* before turning to how the talk is organized to accomplish fluid collaboration. Prior studies of designing provide us with a rich collection of concepts for characterizing this flowing of attention, we could talk in terms of moving between the problem space and solution space; problem setting and problem solving; the counterplay of raising issues and dealing with them; and the making and critiquing of design moves. Three axes of attention flow are drawn out in the following sections.

20.3.1 Moving between Requirements and Design

From the outset and throughout the session, the collaborators move seamlessly between talking about the emerging software design itself and the requirements stated in, or inferred

* The aim of Christiaans and Almendra's study is to produce a description of the AmberPoint session for comparison with two other recordings of pairs of software engineers addressing the same software engineering task; and, further, to make a comparison of these studies of software engineering with studies of decision making in product design.

from, the design brief. Here are two examples of exchanges. Extract 20.1 is from early on in the session and Extract 20.2 is from about halfway through the session. Joe describes the software (design) and Carla switches to summarizing the requirements before describing part of the design and then returning to the requirements.

EXTRACT 20.1 Seamlessly Moving between Requirements and Design

[0:08:01–0:08:55]
Joe: there's *a* kinda *drawing area* here which kinda doubles as the kinda
Carla: uhuh
Joe: simulation
Carla: uhuh
Joe: of that part when they're actually running it now I don't know if there'd be two modes *an editing mode* and a + *simulation mode*
Carla: simulation exactly
Joe: a simulation mode
Carla: exactly
Joe: or it could be *you can draw while you're simulating*
Carla: well so so one is one is you want to *change the layout of the map*
Joe: okay
Carla: two is you want to *change the parameters* you gave it *in terms of speeds and timings* right and three you wanna + *run it* meaning *little dots are moving um showing you how the traffic is flowing* and what does that mean how do you + what kind of metric do you get back *to tell you that this is working* you know how
Joe: yeah
Carla: do you *assess*
Joe: yeah it kinda feels
Carla: the success of the timing

In Extract 20.2, the talk is about a design detail, a table, which will feature in the interface. Here, as for Extract 20.1, the conversation slips easily between talk about the design feature (the table) and its appraisal and refinement by Carla in two turns wrapped around a turn from Joe in which he brings in a requirement (to have up to six intersections) relevant to the current design focus of attention.

EXTRACT 20.2 Seamlessly Moving between Requirements and Design

[1:11:39–1:11:52]
Joe: this *table* [interrogatively]
Carla: yeah
Joe: yeah +
Carla: so *it wouldn't scale very well* if you added a lot of streets you couldn't do that so maybe

Joe: we only have *to have up to six intersections* [with amusement]
Carla: [laughs] but I'm saying having *it* just *show up when you're dragging it* is fine
Joe: yeah

20.3.2 Moving between Design Context and Use Context

The collaborators set about their task by working from what they suppose the user experience needs to be. This leads them to develop ideas about the components of the design such as what information is displayed and when, and design features to allow the user to interact with the traffic simulator. Only later do they turn to set down explicitly the object model that will be needed to support the system behavior that the users are to interact with and the functionality with which the users are to be provided. As they work on the design, we see them flip to and fro between talking as system designers and talking as though they are the system users. Glock (2009) has pointed out that if we pay attention to the indexical expressions in design conversation, we can investigate the contexts that are addressed. We can distinguish between the speech event—the negotiation of the design that the designers are engaged with, and the narrated event—the users' interactions with the imagined future system that they are engaged in designing. In the talk we see "*I*," "*you*," and "*we*" standing in, in close proximity, for the designers qua designers, the system users, and sometimes for both at the same time, when the design feature being discussed coincides—in the way it can be expressed—with a user requirement. Extract 20.3, an exchange about an hour and a half into the session, exemplifies this movement between contexts.

EXTRACT 20.3 Moving between Design Context and Use Context

[1:27:17–1:27:48]
Joe: [earlier part of turn omitted] and then there's building the simulation between + which *we* [*designers, users*] might use the map to do because *you* [*users*] can click on things to fill in details about how the simulation is gonna work and *you* [*user*] wanna save you wanna name the simulation and then +
Carla: yeah and in some ways *I* [*designer*] would say within the simulation *you* [*designers, users*] may wanna to keep again separate the intersection from the cars because *you* [*user*] wanna have this map with this kind of in ... intersections and pump different kinds of
Joe: through it
Carla: cars
Joe: yeah
Carla: through it so *you* [*user*]
Joe: right
Carla: (turn unclear)
Joe: so *you* [*user*] might want to leave one of these constant and change the other
Carla: exactly exactly

Carla signals a flip at the start of her turn from user to designer using "*I* would say," whereas Joe's last turn in Extract 20.3 confirms that he has understood Carla's last "*you*" to refer to the user. For the most part, this flipping between perspectives goes on without being explicitly marked by saying something like "from the user's perspective"—although explicit marking does occur occasionally [e.g., at 1:40:33]. There are only eight uses of the terms "user" or "users" in the entire session, but if we attend to the switches between the speech event and the narrated events, we see that the user context is a pervasive reference signaled by many of the uses of "*you*," "*we*," and "*I*," such as the ones shown in Extract 20.2.

20.3.3 Moving between Breadth and Depth

Carla and Joe start their discussion by establishing their shared understanding of what it is, in general, that they have been asked to deal with and by raising some of the larger questions that they need to address, as we see in Extract 20.1. The exchange occurs in the ninth minute, the first six minutes having been occupied with reading through the brief independently.

This takes them, after some further exchanges, quickly (after about four more minutes) to considering some detail concerning the nature of road intersections and how much detail about intersections they might need their design to accommodate. This first excursion into detail leads back (up) to the level of the representation of (street patterns on) the map—visualization of the map being the part of the task that they first started to address just prior to the exchanges shown in Extract 20.1. The whole process closely fits the description of "the inventive core of the process" of interaction design characterized by Crampton Smith and Tabor (1996) as comprising five activities: understanding, abstracting, structuring, representing, and detailing. In particular, their description that, "[t]hese five processes are not executed sequentially. Designers circle among them as activity in one throws light on another. When considering the structure of the information, for instance, a designer might also be thinking about how much text can be put on the screen, or how this particular audience might most intuitively interact with the information" (Crampton Smith and Tabor, 1996: pp. 46–47). Crampton Smith and Tabor's description particularizes for interaction design what we know from decades of studies of expert design, that there is a movement between the consideration of broad issues—the "architectural" level—and detailing issues, and that in the movement between depth and breadth, each consideration informs progress with the other. Recently, for example, Atman and colleagues have conducted a series of detailed studies of the patterns of this movement and its consequences for design progression (e.g., Atman et al., 2009).

20.4 SIGNALING TENTATIVENESS SUPPORTS CONSTRUCTIVE COLLABORATION

Having briefly characterized some of the ways that the AmberPoint session suggests that expert designers are at work, in this section we inspect more closely *how* the conversation signals of tentativeness serve to keep the design process itself moving along. We look at the way that the designers talk to each other serves to prevent them from getting held up or stuck. In short, we take notice of what they do to maximize the progress they make with the designing in the time available.

Tentativeness is signaled in a number of ways. As well as explicitly enumerating a number of alternative possibilities, hedging and other pragmatic devices are used to reinforce the provisional nature of propositions. In a study of concept formation in design conversations, Dong et al. (2005) label this use of linguistic cues the *projection of possibilities*, an offering of design ideas marked as negotiable through the speaker's way of introducing them, which indicates that the proposer's own commitment to them is tempered. "The projection of possibilities allows the members of the team to signal that they are proposing something and at the same time signal that they are not fully committed to what they are proposing" (Dong et al., 2005: p. 7). This phenomenon has also been remarked upon in earlier studies of design protocols (e.g., Brereton et al., 1996). In the design conversation we are studying here, signaling tentativeness leads to several types of moves in the design process that are important for progression, including the following:

- Enabling the interlocutor to express agreement and confirm that a conjecture is shared and therefore to contribute to consolidating a shared view. Dong et al. (2005) refer to this as the method of *naming*, the concretization of a proposed idea generated through a projection of possibilities.
- Prompting the refinement of potential design details.
- Making it acceptable to backtrack to unravel earlier design conjectures.

Examples are now given of each of these in turn. We first return to Extract 20.1, shown again as Extract 20.4 with signals of tentativeness italicized, in which we see that Carla responds to Joe by confirming and extending his tentatively couched proposition and extending it herself also tentatively.

EXTRACT 20.4 Tentativeness as an Opening for Expressing Agreement

[0:08:01–0:08:55]
Joe: there's a *kinda* drawing area here which *kinda* doubles as the *kinda*
Carla: uhuh
Joe: simulation
Carla: uhuh
Joe: of that part when they're actually running it now *I don't know if there'd be two modes* an editing mode and a + simulation mode
Carla: simulation exactly
Joe: a simulation mode
Carla: exactly
Joe: or *it could be* you can draw while you're simulating
Carla: well so so one is one is you want to change the layout of the map
Joe: okay
Carla: two is you want to change the parameters you gave it in terms of speeds and timings right and three you wanna + run it meaning little dots are

moving um showing you how the traffic is flowing and what does that mean how do you + *what kind of metric* do you get back to tell you that this is working *you know* how

Joe: yeah

Carla: do you assess

Joe: yeah it kinda feels

Carla: the success of the timing

In Extracts 20.5 and 20.6, we see examples of two occasions where tentativeness elicits an exchange in which a possible detail of the design is explored. Immediately after the exchanges shown in Extract 20.5, attention moves back to the design of the representation of the road pattern (now) informed by the excursion into exploring the detail of aspects of road intersections (and exemplifying a movement from depth back to breadth as outlined in Section 20.3.3).

EXTRACT 20.5 Tentativeness Prompts Refinement of a Design Detail

[0:15:11–0:16:17]

Carla: yeah because you know it has to be adjusted to however the fast you can possibly be going + um + also one more time with the streets *I don't know exactly how to capture this* but with traffic flow the the distances between the intersections and the speed

Joe: oh

Carla: right the distances between each intersection um influence + so how far you know is between one and two

Joe: so this

Carla: and three

Joe: is *kind of* a segment

Carla: so each leg yeah

Joe: so there's also a leg

Carla: yeah

Joe: or a segment it's another object

Carla: so what would be the easy way to address to name each leg *kind of* it's so I guess it's a +

Joe: I'd *probably* points on coordinate right so like if this is A1 and this is A2 then the leg would be A1 to A2

Carla: okay but there has to be *some* mechanism where you can reference each

Joe: yeah yeah yeah yeah

In Extract 20.6, we see Joe at "I *almost* wonder…" seeking permission to backtrack: to revoke a design detail he has put in place earlier. Carla's response acknowledges his reasons for finding the element troublesome but—again with a hedge—she shows that it might be possible to retain the feature, fixing the apparent "problem" with it by other means.

EXTRACT 20.6 Tentativeness as Seeking Agreement for Backtracking

[0:35:51–0:37:10]

Joe: I understand what you mean I understand what you mean but I think part of the traffic light problem is figuring out how long you should have the overlapping red lights +

Carla: so *maybe*

Joe: to avoid accident

Carla: that's a that's a separate setting and I know it's the duration of each light you set it on I haven't completely thought of it

Joe: yeah

Carla: whether it's *maybe* one or two but the other one it's implied because if this is green then you cannot tell the system make other one green as well it's just not happen so you only set one but the overlap of the red and the duration of the yellow is something that you can +

Joe: (turn unclear)

Carla: you need to set

Joe: I *almost* wonder if *I don't know I don't know* if I like the pop-up window anymore'cos there's a lot of

Carla: lot of

Joe: complexity it it it's *almost* like you've got this thing selected and then you've got a form down

Carla: and then I mean that

Joe: here where you can fill out details

Carla: if you go away from it yeah or you yeah it seems almost like you +

Joe: what if you need to see what I've got on this one and it's underneath there I can't go look at it +

Carla: that's true

[5 seconds pause]

Carla: but *I think that that's it it's*

Joe: we can (remainder of turn unclear possibly: move it)

Carla: that's presumably

Joe: (turn unclear possibly: it's one of those details)

Carla: we can place it anywhere

The signaling of tentativeness also supports two of the most powerful design strategies that Carla and Joe use regularly throughout the session to enable them to keep going. These strategies are as follows:

- To simplify their task by various means
- To set aside issues to be addressed elsewhere

Simplification and deferral support their need to overcome uncertainties and ambiguities in the design brief and their current lack of understanding arising from these or other causes. Christiaans and Almendra's study of the AmberPoint session, which we have already referred to, identifies deferral by Carla and Joe as a strategy they use for key decisions. In Christiaans and Almendra's scheme, a key decision is one made at moments when the product creation occurs (Christiaans and Almendra, 2010). According to these researchers, Carla and Joe recognize that some areas of the design will need to remain unspecified in their own design work; they recommend, for a later stage, referring (back) to the client for further and more precise information on some matters. They see Carla and Joe as making trade-offs in dealing with the constraints that they face in terms of time, their own expertise in relation to the task, and the level of information they have at their disposal.

A further study of simplification and deferral would be interesting and there is no doubt that both play important roles in authentic design activity taking place in natural settings. Texts intended to educate inexperienced designers make reference to just such strategies to assist progression with a design task. The engaging series of books by Gordon Glegg from the beginning of the 1970s, uses a direct style now uncommon. He wrote, "I want to give a little help to someone whose mind is more or less blank ... My suggestion is that, to de-blank your mind, you should simplify matters still further and try solving the primary problem by listing the ways in which it could be done if one of the design requirements was omitted. Then go back through the list and find which could most easily be adapted so as to re-instate the missing condition. The same technique can also be used at any point in the design" (Glegg, 1972).

However, a fine-grained analysis of the several occasions in the AmberPoint session where Carla and Joe defer further development of the design in a particular direction or the (again several) occasions where they simplify their task might not yield any new insights generalizable to authentic design practice as the constraints of the experimental setting seem likely to have been a strong influence on how, and particularly why, Carla and Joe can both defer and simplify when faced with obstacles here. They do not need to consider professional, commercial, or any other consequences of what they do in this design *exercise*. There is, however, one fundamental dilemma that the collaborators deal with very successfully, although never by confronting it head-on. Section 20.5 looks at how they do this.

20.5 ACCOMMODATING DISAGREEMENT

We know from studies of the way in which design progresses that design ideas and therefore the details of any particular design are not developed monotonically. In the journey to arrive at a final design, the progression accommodates parallel lines of enquiry, even in the case of design by a solitary designer. The parallel lines of development include the pursuit of conjectures that can be mutually inconsistent; therefore, during design, designers are able to maintain simultaneous sets of "truths" to support these different assumptions or lines of enquiry.* It appears that, even to be modestly successful at designing, a designer must be at ease with uncertainties, contradictions, and partial knowledge. To be

* Twenty years ago, considerable efforts were made to develop automated systems to support design that could accommodate parallel, alternative, but nevertheless mutually incompatible, truth maintenance systems. One of the most notable design assistance systems of this kind was the Edinburgh Designer System (see, e.g., Logan et al., 1992).

able to *create* anything, a creator, here a designer, must be in possession of Keat's negative capability (Gittings and Mee, 2002), that is, the ability to be at ease with working in a state of uncertainty, since it is such a state of partial knowledge that makes creation possible at all. It is intrinsic to any nontrivial design task that there must be a state in designing when not everything is resolved, even if, unlike Keat's context (poetry and literature), we might not be willing, for some design domains, to accept that there is no point at which this state must be transformed to a state of resolution: the concretization of the design, when establishing fact becomes a priority and a necessity.

In the AmberPoint data, we see that throughout the session Carla and Joe maintain different beliefs about how the traffic should be generated in the traffic simulator from the underlying object model. They never resolve their differences on this matter. The evidence for this comes from the debriefing session. In describing their proposals for the simulation configuration object, Joe describes both possibilities (his and Carla's) within the description of this object, "and so simulation configuration really consists of a map the name of the map the traffic configuration so where *the traffic patterns how is the traffic turning at different intersections or how do cars choose whether to turn* and then the configuration for the lights" [2:09:19]. Carla's view is that the traffic patterns should be generated from the combined behaviors of individual cars or types of car, each of which is to "know" where it wants to go and how to get there. Joe favors generating traffic densities from intelligence held by the intersection and other road-related objects, such as the traffic lights. Neither designer tries to persuade the other in any serious way—for example, we might expect Carla to appeal to real life! Instead, the two collaborators manage to work around this *fundamental* difference in views about how to resolve this design requirement in the form of software architecture. It may well be the case that once again the nature of the experimental setup discourages a confrontation and resolution of this issue.* There are no professional matters or principles at stake. It may also be the case that as Carla and Joe *are* actually professional colleagues in real life, there is a good working relationship between them, a trust that a technical "disagreement," if we can even call it that, will be resolvable at the right moment in the design process through some means that both parties will consider reasonable.

An analysis that looked at the software engineering strategy the two have adopted, of working principally from the user model toward the object model that will support it, might account for the collaborators' ability to keep on designing without resolving their different views on this issue. In a study of these same AmberPoint data, Tang et al. (2010) characterize Joe and Carla's approach as a high-level design, dropping into low-level (detail) visualization to validate the developing design; an approach characterized by a good deal of inductive reasoning and a high level of explicit reasoning—that is, reasoning where the reasons for a choice or a suggestion are stated explicitly. The inductive reasoning supports the generation of new ideas as design options and the sharing of reasons for proposals or

* After all, it is *only* a 2-hour session, "to get just a high-level sense of what you think the major points are" as the interviewer says in the feedback session at [1:59:02], having already said, "we don't expect you to completely solve it you know 2 hours with no access you know" [1:58:17].

choices "encourages mutual understanding between the designers," creating opportunities to explore the design problem and solution further (Tang et al., 2010: p. 636). The effect of this on the way that they work together is to keep them asking questions to challenge their own individual ideas and those of each other.

In the study presented here, we want to draw attention to *how* Carla and Joe move on with the design while maintaining incompatible beliefs about the object model that will generate traffic behavior in the simulator. They keep moving (past) this potential obstacle by

- Signaling contributions as hypothetical (Extract 20.7)
- Including both design possibilities within the (larger) unfolding design in two ways:
  - Explicitly—by enumerating the two alternatives in what they say (Extracts 20.7 and 20.8)
  - By using distinct terms that encapsulate each belief set

Extract 20.7 shows Carla signaling a conjectural excursion (italicized) after it is clear that she is aware that their views on how traffic will be "pumped" differ. Her signaling of a conjecture projects the possibility (Heritage, 2005) for Joe to encourage the development of the conversation in that direction. Joe gives permission at his second turn in the extract, and thus, in turn, projects the possibility for Carla to continue along this line. This is what she does in the final turn shown.

EXTRACT 20.7 Handling Disagreement by Signaling Design Moves as Hypothetical

[0:44:41–0:46:04]
Joe: are you saying there should be a car out there +
Carla: hah + *I'm entertaining this idea* yeah whether there should be a car out and there are different types of car you know +
Joe: yeah put it up there [laughing]
[Carla draws a cartoon of a car on the whiteboard]
Joe: there you go + see
[Carla and Joe laugh]
Joe: (turn unclear)
Carla: um + I don't know ah + *talk through this with me* because on one hand as you say you can generate these again but you still need to be given some params you need to characterise the cars either you characterise it as you say and you know I want you have a palette like that… [she continues to speak, referring to Joe's concept, pointing to his representation of it on the board, and she summarises in her own words, redrawing it as she does so] … or you have a sort of categories of cars

Enumeration within a contribution avoids the need to resolve the issue at that particular moment. An example from Carla is in the turn to which we have just referred to in Extract 20.7. Extract 20.8 shows an example of Joe using the same strategy to head off dissent by inviting Carla to stay with the issue he is summarizing, that is, working with the notion of percentages of traffic, leaving aside how these will be arrived at.

EXTRACT 20.8 Handling Disagreement by Explicitly Acknowledging Alternatives

[1:25:53–1:26:32]

Joe: so then that there's the er for for each + for each er + approach + so um A1 east percentage of traffic that's going left right and forward + A1 west + A1 north [reads from board, writes on it during pauses]

Carla: this is cars [interrogatively]

Joe: yeah it's percentage of

Carla: (turn unclear)

Joe: cars that are going

Carla: uhuh

Joe: left

Carla: uhuh

Joe: right if we do it at the in yeah I'm still not this is where we need to check with the professor again about + but that is it's is it part of the simulation

Carla: uhuh

Joe: it's we build up a whole set of cars

Carla: yeah

Joe: or we say at each intersection this is the percentage of

Carla: yeah yeah the level of control we need on the traffic how it's it's going it's it's not clear to me

Explicit enumeration is relatively easy to spot once we have noticed it as a strategy for accommodating disagreement and moving on around it. More subtle is the use of encapsulation in the service of accommodating disagreement. Carla and Joe use encapsulating terms to set up a conceptual barrier (Wirfs-Brock et al., 1990) around the collection of design decisions they cannot agree upon. The term *encapsulating* is chosen because although this linguistic strategy successfully fulfills the function of enabling further negotiation of the design, the term does not seem to serve as functional or data abstractions of *agreed* elements of the design solution. Two terms play this encapsulation role: "density" and "traffic patterns." Both Joe and Carla use "density" (three times each) until Joe makes it serve to encapsulate his proposition for the part of the design that the two are in disagreement about. He does this at [0:46:32] with, "you would have to do that with a combination of the density at certain hours and for a given intersection a percentage what you expect to go left right or straight for that hour." From this point on only Joe uses the term "density" (five more times), but while Carla does not use it again, now it is "tainted" by Joe's definition of it which embodies his preferred solution—it does do a disagreement encapsulation job and serves to allow this contentious part of the design to be referred to without reopening it as an issue to be resolved, at that moment, between the designers.

To serve the same disagreement encapsulation function, Carla uses the term *traffic pattern*. She introduces its specific meaning at [1:27:55] with, "your map is a named object your intersection is a named object and your traffic patterns are named object" (so far so good!) until at [1:44:05] she elaborates, "so the traffic pattern object the big guy owns all the cars and as it manufactures them … it creates them in some flavor and says you shall go in A and turn left whenever possible …." Carla uses "traffic pattern" (five more times) much as Joe uses "density" to serve as a way of talking around the contentious element of the design by naming her proposition for it. Joe, in turn, does not use Carla's term except when he is speaking (narrating) as the system's client.* What we seem to be seeing here is that the collaborators share a coherent framing of their task at some level, within which there is disagreement, that is, no coherent frame about one particular area of the design at a more detailed level. They handle this by naming the contentious area; each gives it a different name, as their frames are different.

To clarify, Carla and Joe agree that there is a functional requirement that the design should allow the traffic pattern to be adjusted and controlled in the simulator. They have not reached agreement on the means of achieving this in the software architecture. By naming the two solution forms differently, they both acknowledge and preserve the fact that there is an unresolved disagreement. This is not a case of glossing over the issue, "papering over a crack"—that would have serious consequences further along in the design process—they are not using a single term for the problematic element, which they can choose to think of differently. There is no evidence that there is any misunderstanding between them over the fact that there are two alternatives—the examples of explicit enumeration by both parties are some evidence for this claim but the overwhelming evidence comes from the description of the design that Joe gives during the debriefing session, which has already been referred to and quoted earlier.

Dong et al. (2005), in the work referred to previously, describe the process of "linguistic technicalizing" or naming, "whereby descriptions of actions and objects are outlined and then captured or represented by a name… this allow for the generation of new and unique meaning" through which the group (here the two participants) in the design conversation "accumulates design content" (p. 8). Participants in a conversation to develop a design, build up an understanding about something—perhaps a design detail or a solution idea—and technicalize it through talk. What we see here is the same phenomenon; however, although both participants may share a common understanding of the meanings of what they are separately naming, they do not agree about the status of the accumulated knowledge that each name represents in terms of its part in the design solution (specifically whether the named content will feature as a component of the design).

Here, the encapsulation via different terms serves well enough to let the designing continue, but we do not see the accumulation we would expect from a single term (frame),

* At [1:34:14] "… so it could be the professor says you know during these hours this is what the traffic pattern is like …."

which would be a signal of collective agreement about what is contributing to the aggregation. In a bigger corpus, we might look to a lexical chain analysis to differentiate discontinuities in agreement (Dong, 2009). What we would expect to see is a very different flow of the concept from one speaker to another; signaling that named concepts may be understood qua options but agreement over their status in the design is not in place. Thus, we might see no connection between "traffic pattern" and "density" in terms of lexical flow or lexical cohesion, indicating no semantic relationship between one speaker's use of "density" and the other speaker's use of "traffic pattern."* Although such discontinuities might signal potential problems with design collaboration from the point of view that ascribes positive values in design teams to common framing, we need other means to inspect the *negotiation-serving* function of encapsulation terms. That is to say, the technicalizing may well be serving a different critical function, namely, supporting collaboration by enabling progress with design to take place despite the lack of shared frames at some level. This is an important distinction that has implications for studying and attempting to measure, or assess, collaboration quality (McDonnell, 2010).

20.6 SUMMARY AND IMPLICATIONS

This study has attended to some of the features of the collaboration between two software design professionals working on the development of the conceptual design stage of a software design task under laboratory conditions. The study sought to draw notice to how fluid collaboration is taking place through the talk-in-interaction. Using cues from the spoken exchanges, the design session is characterized as an expert design performance. The data have been analyzed by other researchers for other purposes, and reference to two other studies is made to support and extend the general characterization of the design session. In this study, the effectiveness of the collaboration is closely inspected to draw out a number of the features that might contribute to an account of how it is effective. An account of the interaction as fluid expert design practice is followed first by an inspection of the ways in which tentativeness supports the negotiation of design progression. Secondly, by focusing on a part of the design where the designers' separate and different ideas on how a fundamental component of the design should be approached are acknowledge by them and "worked past," we have drawn attention to how the disagreement is bracketed so that it does not impede progress. Three conversational devices are highlighted which support working past the impediment. These are signaling the propositional status of a conversational turn, explicitly acknowledging that there are design alternatives yet to be resolved by enumerating them, and, finally, marking off different solution concepts about which there is disagreement (over which to choose), by naming (technicalizing) them. This conversational device serves to encapsulate the disagreement without losing sight of the fact that it exists. The encapsulation does not support the design of the software by data abstraction very well, but it does serve the *collaboration* itself by providing a way of avoiding having to resolve a disagreement sufficiently for designing to continue.

* A. Dong, personal communication.

For more than 50 years there has been a debate in the design studies arena over the extent to which designers are born or made; in particular, whether creativity is innate or if it can be developed through education and/or experience. In parallel, but in far more diffuse forums, there has been discussion, sometimes fuelled by research, into the ability to collaborate on shared tasks, including design tasks. The relevant issues here are whether the social skills for effective collaboration come about naturally and whether there are skills applied in social interaction for the purposes of designing that are different from the social skills needed to negotiate everyday life. Are there strategies at work in effective collaboration that we can benefit from noticing, perhaps to make a difference to the educational experiences we provide for designers, or to influence what is paid attention to in developing as reflective collaborative practitioners?

An analysis that focuses on the evolving design itself often treats phenomena such as vagueness, hesitation, and delay with negative connotations *from the point of view of the evolving design.* In the last decade or so, studies that pay close attention to how design comes about through conversations between collaborators give a different account of these phenomena (when they are construed as an openness to possibilities and/or an invitation to negotiation) in terms of the way that they serve designing and the making of collaboration possible at all (McDonnell, 2010). The study presented here is a contribution to the perspective that has moved beyond seeing design as an instrumental process and that tries to better understand the reality of design as a situated social process by paying close attention to what the mechanisms are that contribute to making design collaboration effective.

ACKNOWLEDGMENTS

The author thanks the organizers of the workshop, Marian Petre, Andre van der Hoek, and Alex Baker, and the NSF for the opportunity to take part in the NSF-sponsored project Studying Professional Software Design (grant CCF-0845840) and for access to its shared data set. In addition, the author thanks Dr. Andy Dong from the University of Sydney for commenting on an earlier analysis of the data set and Sunil Bhambra from Central Saint Martins for technical assistance.

REFERENCES

Akın, O. (2001) Variants in design cognition. In C. Eastman, W. Newstetter, and M. McCracken (Eds), *Knowing and Learning to Design* (pp. 105–124). Amsterdam: Elsevier.

Atman, C., Borgford-Parnell, J., Deibel, K., Kang, A., Ng, W., Kilgore, D., and Turns, J. (2009) Matters of context in design. In J. McDonnell and P. Lloyd (Eds), *About Designing: Analysing Design Meetings* (pp. 399–416). London: Taylor & Francis.

Brereton, M., Cannon, D., Mabogunge, A., and Leifer, L. (1996) Collaboration in design teams: Mediating design progress through social interaction. In N. Cross, K. Dorst, and H. Christiaans (Eds), *Analysing Design Activity* (pp. 319–341). Chichester: Wiley.

Christiaans, H. and Almendra, R. (2010) Accessing decision-making in software design. *Design Studies* 31, 641–662.

Crampton Smith, G. and Tabor, P. (1996) The role of the artist-designer. In T. Winograd (Ed.), *Bringing Design to Software* (pp. 37–57). New York: ACM Press.

Cross, N., Dorst, K., and Christiaans, H. (Eds) (1996) *Analysing Design Activity.* Chichester: Wiley.

Dong, A. (2009) *The Language of Design: Theory and Computation.* London: Springer.

Dong, A., Davies, K., and McInnes, D. (2005) Exploring the relationship between lexical behavior and concert formation in design conversations. In *Proceedings of ASME International Design Engineering Technical Conference and Computers and Information in Engineering Conference* (pp. 61–71). New York: ASME.

Gittings, R. and Mee, J. (Eds) (2009) *John Keats Selected Letters*. Oxford: Oxford University Press.

Glegg, G. (1972) *The Selection of Design*. Cambridge: Cambridge University Press.

Glock, F. (2009) Aspects of language use in design conversation. In J. McDonnell and P. Lloyd (Eds), *About Designing: Analysing Design Meetings* (pp. 285–301). London: Taylor & Francis.

Heritage, J. (2005) Conversation analysis and institutional talk. In K. L. Fitch and R. E. Saunders (Eds), *Handbook of Language and Social Interaction* (pp. 103–148). Mahwah, NJ: Lawrence Erlbaum.

Lawson, B. (1980) *How Designers Think*. London: Architectural Press.

Logan, B., Corne, D., and Smithers, T. (1992) Enduring support: On defeasible reasoning in design support systems. In J. Gero (Ed.), *Artificial Intelligence in Design '92* (pp. 433–454). Dordrecht: Kluwer Academic.

Matthews, B. (2009) Intersection of brainstorming rules and social order. In J. McDonnell and P. Lloyd (Eds), *About Designing: Analysing Design Meetings* (pp. 33–47). London: Taylor & Francis.

McDonnell, J. (2009) Collaborative negotiation in design: A study of design conversations between architect and building users. In J. McDonnell and P. Lloyd (Eds), *About Designing: Analysing Design Meetings* (pp. 251–267). London: Taylor & Francis.

McDonnell, J. (2010) "Slow" collaboration: Some uses of vagueness, hesitation and delay in design collaborations. *International Reports on Socio-Informatics* 7(1), 49–56, (www.iisi.de), International Institute for Socio-Informatics.

McDonnell, J. and Lloyd, P. (Eds) (2009) *About Designing: Analysing Design Meetings*. London: Taylor & Francis.

Medway, P. (1996) Virtual and material buildings: Construction and constructivism in architecture and writing. *Written Communication* 13, 473–514.

Perelman, C. and Olbrechts-Tyteca, L. (1971) *The New Rhetoric: A Treatise on Argumentation*. Notre Dame: University of Notre Dame Press.

Petre, M., van der Hoek, A., and Baker, A. (2010) Editorial to special issue on studying professional software design. *Design Studies* 31, 533–544.

Rowe, P. (1987) *Design Thinking*. Cambridge, MA: MIT Press.

Sacks, H. (1995) *Lectures on Conversation*. Oxford: Blackwell.

Schegloff, E. A. (1997) Whose text? whose context? *Discourse and Society* 8, 165–187.

Schegloff, E. A. (1999) Discourse, pragmatics, conversation, analysis. *Discourse Studies* 1, 405–436.

Silverman, D. (2001) *Interpreting Qualitative Data* (2nd edn). London: Sage.

Tang, A., Aleti, A., Burge, J., and van Vliet, H. (2010) What makes software design effective? *Design Studies* 31, 614–640.

ten Have, P. (2007) *Doing Conversation Analysis* (2nd edn). London: Sage.

Wirfs-Brock, R., Wilkerson, B., and Wiener, L. (1990) *Designing Object-Oriented Software*. Upper Saddle River, NJ: Prentice-Hall.

CHAPTER 21

Application of Network Analysis to Conversations of Professional Software Designers

An Exploratory Study

Sian Joel-Edgar

University of Exeter Business School

Paul Rodgers

Northumbria University

CONTENTS

21.1 INTRODUCTION

The concept of network analysis "encompasses theories, models, and applications that are expressed in terms of relational concepts or processes. The unit of analysis in network analysis is not the individual, but an entity consisting of a collection of individuals and the linkages" (Wasserman and Faust, 1994). Network analysis looks at the connections between actors and how such connections form into a network. Network analysis helps to interpret group data such as communities of practice. It can identify cliques, trace how information flows through networks, and holistically understand what is going on with a connected number of individuals. Network analysis can also be used to test hypotheses for groups or clusters of people; for instance, the idea that boys socialize more with other boys or the idea that people with weak ties are useful for learning about new ideas or jobs (Granovetter, 1973, 1982).

In comparison with other types of methodological approaches, network analysis looks at the relational approach rather than at the attribute. It also looks at the structure and the composition of the connections that make a group rather than looking at individuals and their characteristics.

In this chapter, we explore the use of network analysis in the understanding of conversations. Network analysis has often focused on the social structure of groups and on how individual people connect to others (otherwise known as social network analysis). In this exploratory study, we do not look at a person's social network, instead we look at the interrelationships of objects (within conversations) with each other. To do this with professional software design activities, the actors in question are not the designers themselves but the objects that are referred to (such as the "car" or the "intersection").

Essentially, we have taken the techniques of social network analysis and applied them to nonhuman objects that are the references that software designers make. We do this with the same intent with which a social network study would try to understand social group behaviors, as we aim to appreciate the groupings of conversational objects. We attempt this approach because of an underlying belief that communication patterns are intertwined with that which is produced (Herbsleb and Grinter, 1999) and that dialog is embedded within code (Mahendran, 2002).

Previous studies of design practice as it takes place have shown that conversations that revolve around objects during the early stages of design reveal insights that help the participants' understanding of the design problem (Luck and McDonnell, 2006). Moreover, the analysis of designer-to-designer conversations has been shown to be an appropriate method of analysis for revealing spoken interactional behaviors around the use of objects, as well as revealing knowledge that is embedded in the objects themselves (Luck, 2007). Dong et al. (2005) have shown that designers articulate their individual knowledge and design perspectives through language when they are expressing their concepts that enable the design team to bridge the relations among object ideas stored in each designer's mind. Lawson and Loke (1997) go further when they state that more work needs to be done on how we hold conversations about design. They suggest that we should concentrate less on pictures and more on words. Subsequently, Lawson (2004) has suggested that to understand design expertise, we need to recognize that design practice includes the roles of teams, communication, and shared experiences and understandings, and research needs to concentrate on conversations and memories as much as on drawings.

In looking at the connections between software objects, the following analysis aims at defining the important objects and groups of objects that the software designers repeatedly refer to in their conversations as interconnected. If objects are grouped together, then it is proposed that these objects be coded together. Moreover, it may imply that any central and connected object, with a relationship to many other objects, should be a priority in the design project, whereas any object that is weaker than other objects should not be the first priority in the overall design project as there are no other objects dependent upon it. Also, any object groups within the network can be tested together. A network component that contains a few objects can be tested together in the first instance and then all objects can be tested as a whole. Components with a collection of objects that solely interrelate to one another (a clique) may not need to work with other objects.

21.2 METHODOLOGY

This section outlines the steps and methods utilized by the authors in the analysis of the three software design case studies. Essentially, the steps taken to complete the task at hand involved:

- Understanding the objects (nodes) and their relationships (edges)
- Producing a network map of interactions between the objects
- Analyzing the network characteristics that can be beneficial or problematic

In order to carry out the first of these steps and to understand the three different software design processes, the transcription of each case was examined and objects were identified from the conversations. Objects in the text were defined as objects in a manner akin to those within object-oriented programming. For example, the software designers were working and using descriptive language (i.e., nouns) such as "Roads" and "Maps." In this analysis, therefore, objects are nouns—words used to identify any class of people, places, or things—and each of these terms constitutes an object for subsequent analyses. Additionally, attributes to objects were included such as "Time." For example, "… My initial thinking was you'd have *intersection* object that contained four *roads* they're all incoming *roads*."

Two individuals decided on the list of both the objects and the attributes of the objects, and one individual researcher carried out the analysis of the three case studies using this lexicon. Ideally, the analysis should be carried out by more than one person so that the results and the findings can be cross-referenced. Next, each of the three design sessions was analyzed, and the objects and the attributes of the objects were highlighted in the transcribed text as shown in Figure 21.1.

It should be noted that there were many nuances to the conversation that were not captured using this exploratory technique. For instance, if the software designers referred to an object for half of their discussion, but then decided to exclude it, this was not reflected in the overall analysis.

For the sake of this analysis, words similar to those highlighted in the design session transcriptions were defined as being the same. For example, objects such as "Time" were defined to include other similar words such as "Clock." There are other issues surrounding object attributes such as attributes that can be gleaned from other attribute references, for example, "Rate" can equate to "Number" in relation to "Time," both of which are already attributes. References to "Rate" could therefore be included in the count of other attributes

Intuit Designers

[0:27:29.2]
Male 1: So what are the aspects of the speed that are important? It's how many - the number of cars that can flow through a light, or a signal. And so the signal has the properties on signal is this time here and that time is going to you know cars.

[0:28:08.0]
Male 2: Yeah, and then there's also the - well there's time at the signal, and then there's just shear travel time down the street. Just moving from point A to point B is going to take time.

AmberPoint Designers

[1:26:19.5]
Male: Yeah, percentage of cars we're doing it off, rather than do it at the intersection. This is why we need to check with the professor again. This is part of the simulation. It's either we build up a whole set of cars or we say at each intersection, this is the percentage of . . .

[1:26:32.0]
Female: Yeah, yeah I don't know the level of control we need on the traffic - how it's going, its not clear to me.

Adobe Designers

[0:38:19.8]
Male 1: Yeah, so what we have in our model so far is intersection that has the light. And maybe the time and the current state. The intersection contains basically four roads.

[0:38:39.1]
Male 2: Right, so maybe the[inaudible] here is we're saying this has some one to n, or one to 2n roads, right? So the intersection has light gating a pair of roads, light gating another pair of roads. There's the light, I guess per pair of roads.

FIGURE 21.1 Objects identification in transcribed design sessions.

and depending on how the reference was made in the transcript, the "Rate" reference could be ignored, or placed in either (or both) the "Time" or the "Number" category. Moreover, some strings were grouped together although their impact on a resulting program would be different. For example, "Red Light" or "Green Light" was defined as being part of the object "Light." However, the relationship of the "Red Light" object to the car object would result in the car stopping, whereas the "Green Light" object would refer to the car object moving. For example, "So basically we need an option to say—so there needs to be a toggle to say if this has a *sensor* then—if there's no *traffic*, never turn it *green*." In this case, the reference to "turn it green" relates to turning the traffic light green, and thus contributes as a reference to the "Light/Signal" object. This relationship is based on that of one broadly defined object to another, but it does not reflect the meaning of the relationship to the object. The connections between objects and attributes included both positive and negative connections (if a negative action involved the software responding in some way). For example, if the traffic lights turned from red to green when there were no cars on the roads, this would be a negative relationship that still required a connection to be counted as there would be an action in the software code. Further analysis could allow for objects to be contextualized, resulting in far more objects and a more intricate analysis.

In order to carry out the analysis, a social network tool, UCINet (Borgatti et al., 2002), was used to calculate go-between, weaker, and more dominant objects. Weaker and dominant objects are categorized using Freeman's degree centrality algorithm, which states that the number of ties or neighbors adjacent to a given node in a symmetric graph is the degree of that node (Freeman, 1979). Nodes with lots of ties and the neighboring nodes connected to them are considered dominant objects, while the inverse is true for those node objects considered weaker. In order to carry out this analysis, the valued data object matrix was dichotomized, which resulted in a binary matrix. The go-between value was calculated using eigenvector centrality (Bonacich, 1972), where it is the strategic position of the node object that is of importance.

21.3 RESULTS

This section details the results of the three software design sessions. That is, the software design activities with the three teams of designers—Intuit designers, AmberPoint designers, and Adobe designers. As background to this study, the following information is given:

Intuit Designers

- Intuit designer 1 is male. He has a BA in computer science and 24 years of industrial experience.
- Intuit designer 2 is male. He has a BA in computer science and 2 years of industrial experience. Combined, the Intuit designers have 26 years of experience. Both of these designers are considered among the strongest designers in the company.

AmberPoint Designers

- AmberPoint designer 1 is female. She has a BA in computer science and 26 years of industrial experience.
- AmberPoint designer 2 is male. He has a BA in computer science and 20 years of industrial experience. Combined, the AmberPoint designers have 46 years of experience.

Adobe Designers

- Adobe designer 1 is male. He has a BA and a master's degree in computer science and 11 years of industrial experience.
- Adobe designer 2 is male. He has 16 years of industrial experience. Combined, the Adobe designers have 27 years of experience.

The full list of objects stated by the three design teams during the three separate design activity case studies is shown, case by case, in Table 21.1.

It is worth mentioning that there are some objects that are not expected from the conversations, as can be seen in Table 21.1. For instance, the Adobe designers refer to the object "Queue" whereas the Intuit designers and the AmberPoint designers do not. Moreover, as can be seen in Figure 21.4 and Table 21.4, the "Queue" object is a dominant one in the conversations between the Adobe designers. The next stage involved making connections between objects and observing if a conversational link was made between two objects. For instance, if the designers stated that "an *intersection* should contain *lights*," then that statement would show a connection between the *intersection* object and the *lights* object. The resulting analyses in Figures 21.2 through 21.4 show the full series of three network diagrams of connections between objects identified in each of the three design session case studies. It should be pointed out that the weight of the line denotes the level of connection between two objects. So, for example, in Figure 21.2,

TABLE 21.1 Objects Lists Comparison by Design Team

Intuit Designers	AmberPoint Designers	Adobe Designers
Intersections	Intersections	Intersections
Lights/signal	Lights/signal	Signal/lights
Cars	Cars	Cars
Traffic	Traffic	Roads/streets
Lanes	Leg/segway/lane	Time
Roads	Roads/streets	Simulation
Sensors	Sensors/data recording points	Direction
Time	Time	Queue
Direction	Controller	Speed
Controller	Distance/dimensions	Number [random]
Lengths	Speed	Controller
Speed	Map	Traffic
Map	Density/number/percentages	Turn
	Block	Lane
	Widget/palette	On-ramp/interior road
	Drawing/dragging	Sensors
	Grid	Map
	Simulation	Network
	Direction/turning	Geometrics/dimensions
	Feedback/results	

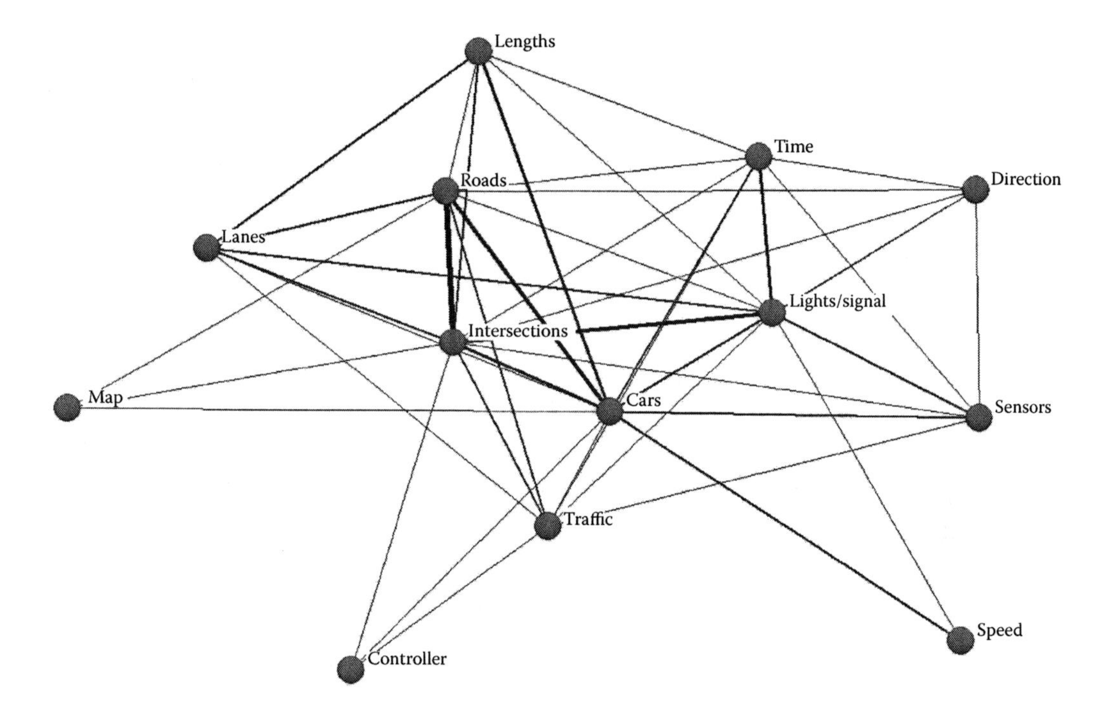

FIGURE 21.2 Intuit designers' objects network.

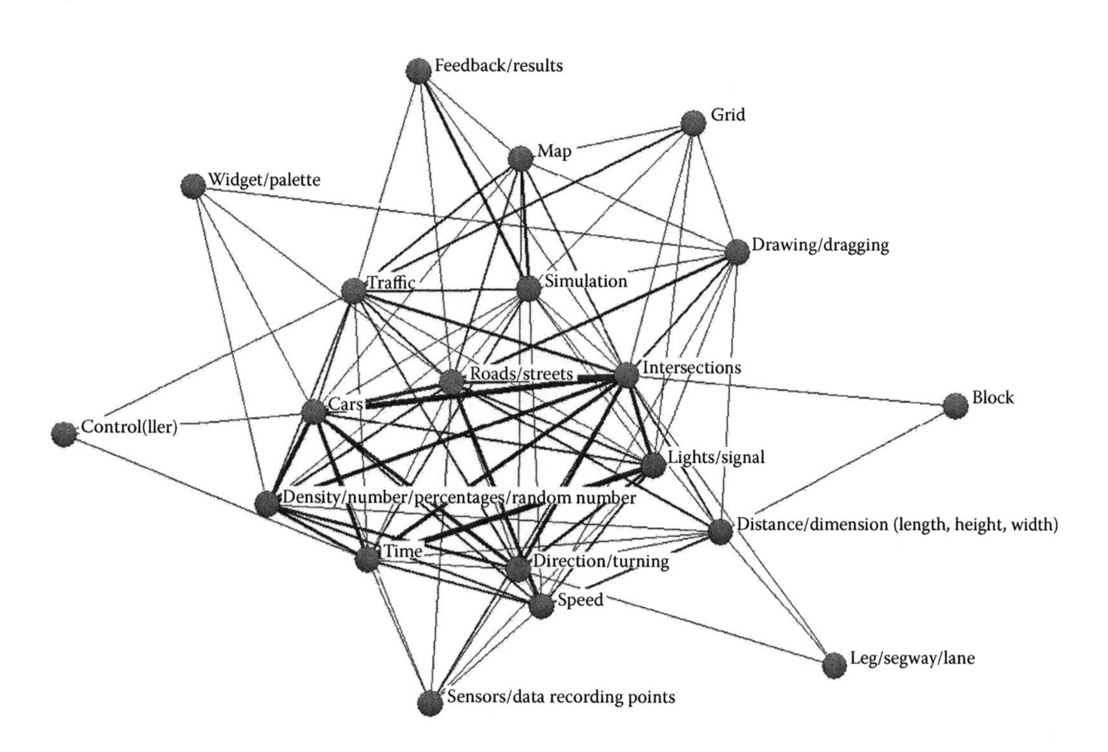

FIGURE 21.3 AmberPoint designers' objects network.

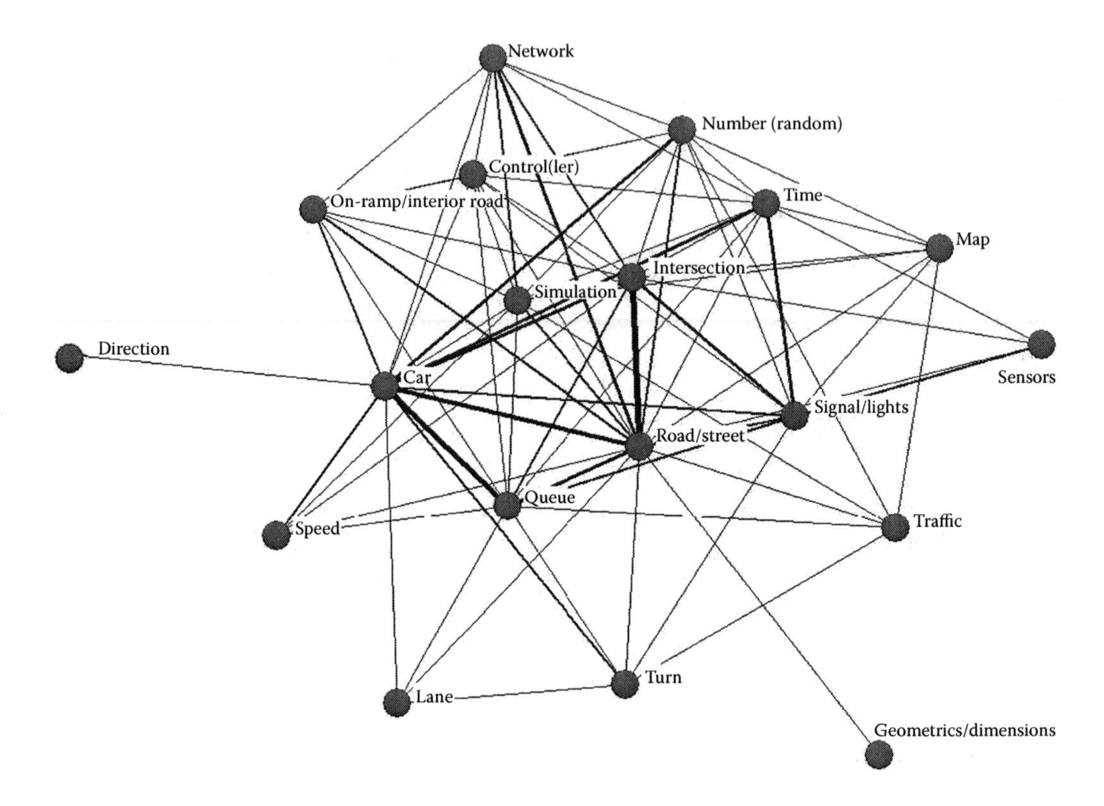

FIGURE 21.4 Adobe designers' objects network.

"Roads" and "Intersections" have more connections between them than say "Roads" and "Time."

It is envisaged that this methodology should highlight those objects that are pivotal and have many connections to other objects, and other objects that are weaker and less significant to the task. As part of a reflective design process, this approach will provide the designers with key information about which objects are important in the overall design of the traffic system.

21.4 TRENDS

The utilization of network theory in this chapter has facilitated the identification of a number of trends in each of the three software design case studies. The network diagrams show that there are categories of objects that can be classed into one of three types—dominant objects, go-between objects, and weaker objects as well as objects that are grouped together. This enables designers and other stakeholders to identify important and less important objects in the overall design process and also to focus their resources and efforts on particular objects in the design task.

21.5 DOMINANT, GO-BETWEEN, AND WEAKER OBJECTS

Table 21.2 highlights that the Intuit designers concentrate their attention on the core or dominant objects such as "Roads" and "Intersections" and treat the broader concepts such as "Map" and "Controller" as somewhat less significant. The AmberPoint designers, on the other hand, appear to concentrate their efforts not only on the core object-oriented objects but also on their attributes such as "Time" (Table 21.3).

This is also the case for the Adobe designers but to a lesser extent (Table 21.4). Figure 21.5 shows the common objects shared by all three software design teams across all three categories of objects. That is, across the dominant (most important), go-between (medium importance), and weaker (least important) objects.

As can be seen from Figure 21.5, the recurring dominant objects are "Intersections," "Roads/Streets," "Cars," and "Lights/Signal." All three design teams also found the objects "Sensors," "Speed," and "Simulation" to be moderately significant. The objects "Controller" and "Map" were found to be weaker and thus of lesser importance among all three design team conversations.

TABLE 21.2 Intuit Designers' Objects Classification

Dominant Objects	Go-Between Objects	Weaker Objects
Roads	Lengths	Controller
Intersections	Time	Speed
Cars	Direction	Map
Lights/Signal	Sensors	
	Traffic	
	Lanes	

TABLE 21.3 AmberPoint Designers' Objects Classification

Dominant Objects	Go-Between Objects	Weaker Objects
Traffic	Speed	Widget/palette
Cars	Distance/dimension	Controller
Density/number/percentages	Sensors/data recording points	Leg/segway/lane
Time	Simulation	Block
Direction/turning	Map	Grid
Lights/signal	Feedback/results	
Intersections	Drawing/dragging	
Roads/streets		

TABLE 21.4 Adobe Designers' Objects Classification

Dominant Objects	Go-Between Objects	Weaker Objects
Queue	Sensors	Direction
Cars	Turn	Lane
Lights/signal	Speed	Geometrics/dimensions
Intersections	Simulation	Traffic
Time	Controller	Map
Roads/street	Network	
	On-ramp/interior	
	Road	
	Number	

	Dominant	Go-between	Weaker
Intuit designers	Roads, intersections, cars, lights/ signal	Lengths, time, direction, sensors, traffic, lanes	Controller, speed, map
AmberPoint designers	Traffic, cars, density/ number/ percentage, time, direction/ turning, lights/ signal, intersections, roads/ streets	Speed, distance/ dimension, sensors/ data, recording points, simulation, map, feedback/ results, drawing/ dragging	Widget/ palette, controller, leg/ segway/ lane, block, grid
Adobe designers	Queue, cars, lights/ signal, intersections, time, roads/ streets	Sensors, turn, speed, simulation, controller, network, on-ramp/ interior road, number (random)	Direction, lane, geometrics/ dimensions, traffic, map

FIGURE 21.5 Comparison of dominant, go-between, and weaker objects.

Tables 21.2 through 21.4 illustrate which objects are dominant (most important), go-between (medium importance), and weaker (least important) in each of the three software design case studies. Furthermore, the object occurrences in the design team conversations can be quantified. Figures 21.6 through 21.8 show the number of object occurrences for each software design process case study.

Next, we can compare the conversational use of objects across the three different software design sessions. As can be seen in Figure 21.9, each software design session has been broken down into the total number of objects mentioned, the total word count of the conversation, and the combined experience of the designers (in years).

The implications of the network analysis results are twofold—code related and designer related. The code-related implications first refer to the approach a designer may take when starting the coding process. The dominant objects shown through the network analysis have many links to other objects. It would make sense, therefore, for the designer to begin with these dominant objects that connect to many others. The designer then moves on to the go-between objects that are more strategically placed within the network. These objects interconnect two other objects and, it can be argued, should be coded after the more dominant ones. It is noted that other objects fall between the cracks of being neither dominant nor go-between. Although these objects may have more conversational references made to

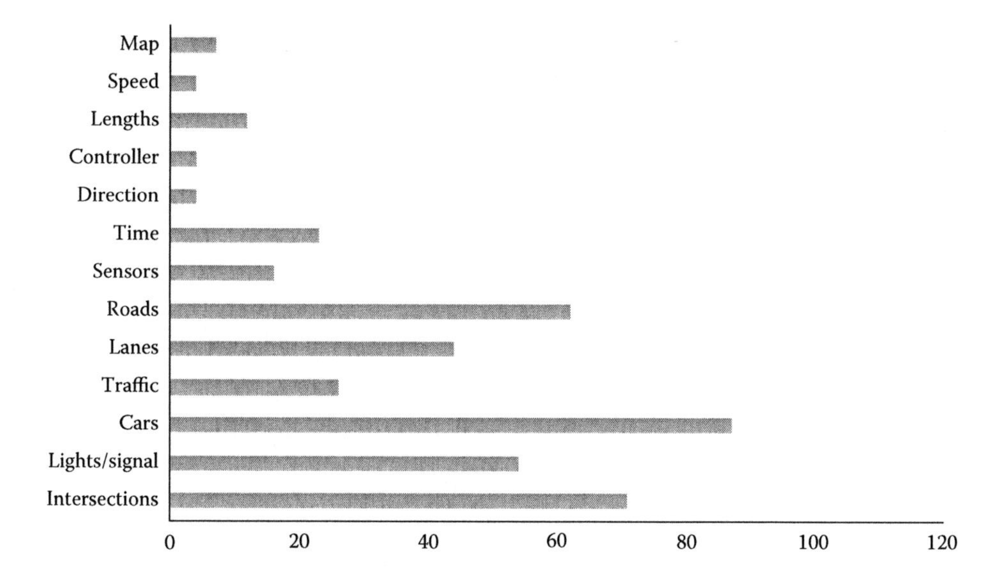

FIGURE 21.6 Intuit designers' object occurrences.

them than some go-between objects, they should be coded after the go-between objects, as they are not as well placed. Weaker objects that do not connect to other objects should be coded last. In the same manner in which the designer approaches the coding process, the testing process follows the same staged method. Dominant objects tested in the first instance and weaker objects being last to be tested. Additionally, component groups that are identified can be tested together. Components can be logically split from the rest of the network without implication. The tester can therefore test a component in isolation, theoretically without needing other objects in order to perform a function.

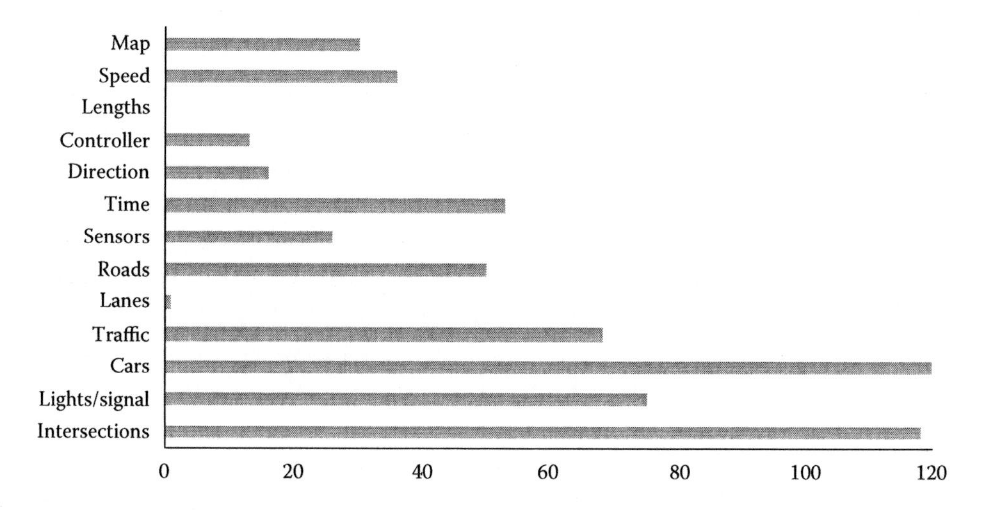

FIGURE 21.7 AmberPoint designers' object occurrences.

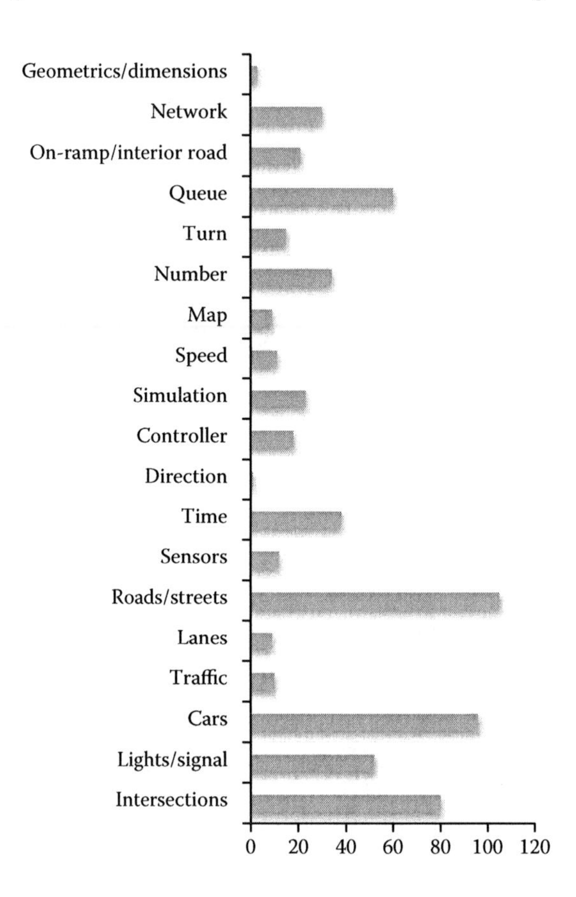

FIGURE 21.8 Adobe designers' object occurrences.

Designer-related implications can be seen if the design process is somehow judged. If the program produced by one set of designers performs better or is deemed more successful, their network will be evaluated as somehow a better example than the network produced by the design team who receive a poorer result. If repeated patterns occurred in the network that correlate with a more successful program, these network trends can be encouraged. If, for example, an entire course was set the same task, the network patterns can be revealed to each design team and may form a constructive approach to offering advice on how to develop the program, particularly if this was done in hindsight and helped the design team to improve their work. The network analysis results can therefore be used as a tool for the designer to reflect on his or her work.

Reflection is a conscious and rational action that leads to reframing a problem or attending to new issues. Reflection is crucial in designing, because by reflecting on their behavior, a team of designers can rationally make decisions to start new activities (Valkenburg and Dorst, 1998). Personal and group reflection is widely acknowledged as a critical tool in design practice (Adams et al., 2003). Indeed, some research claims that reflection, for both individuals and teams, is a key driver in successful design practice (Stempfle and Badke-Schaub, 2002). It may possibly be the case that the Adobe designers were unaware that they

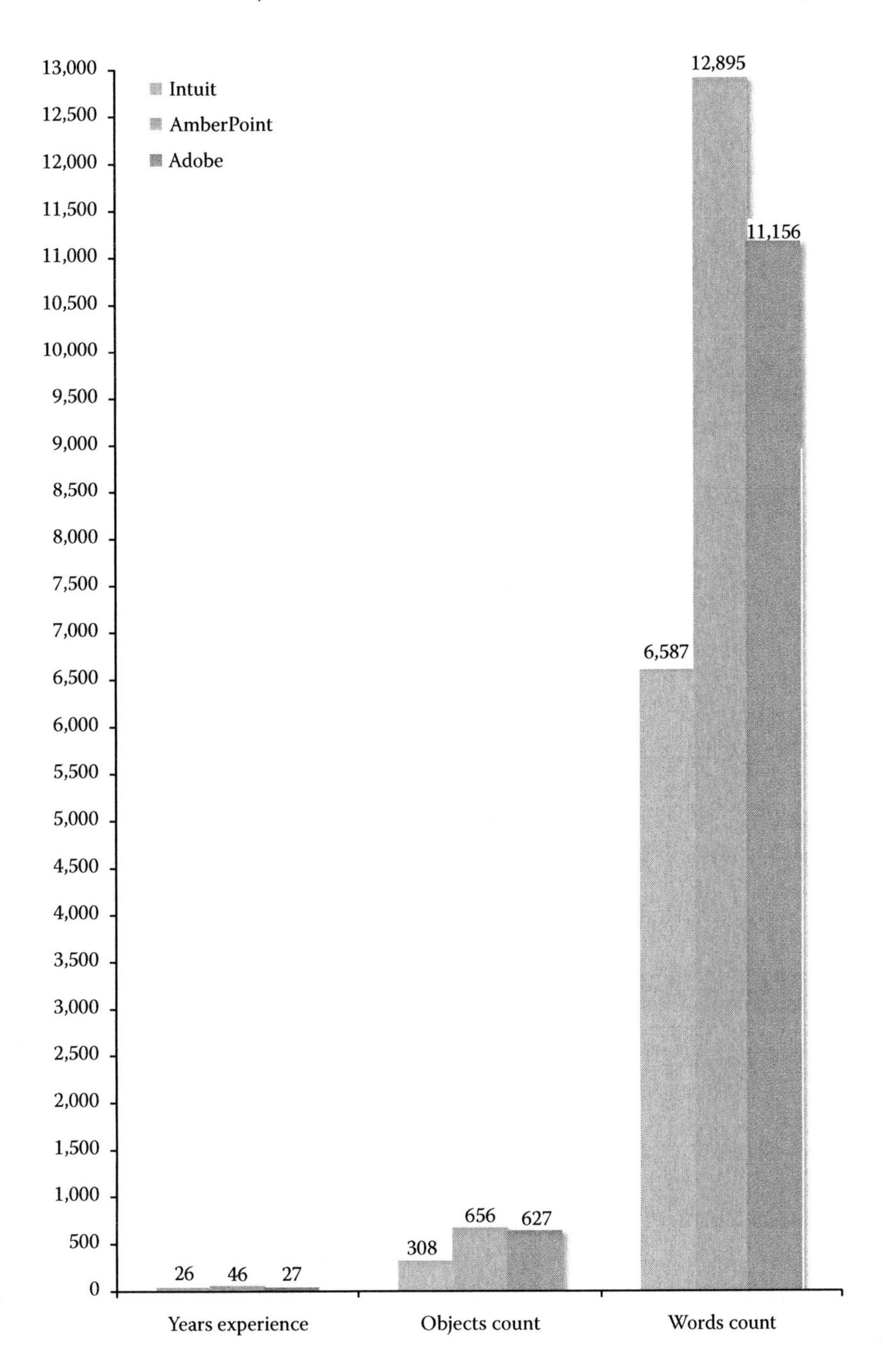

FIGURE 21.9 Combined experience (years), objects, and word count across all three software design sessions.

frequently referred to the "Queue," an object that the other design teams did not refer to. The Adobe designers, perhaps, in hindsight, preferred not to place so much attention on this aspect of the program. Or, perhaps, the other design teams had not considered this aspect of the software.

21.6 CONCLUSIONS AND FUTURE WORK

This chapter has presented a network analysis of three different software design process case studies. The chapter has outlined the steps and methods in the network analysis approach and has illustrated how it was successfully applied to the conversations between the software designers across all three cases. The methodology behind the identification of network objects, the network results produced for each case study, and the trends that are revealed within each of the three case studies have also been presented. The aim of each analysis was to highlight the dominant, go-between, and weaker objects in the software designers' conversations that would highlight any similarities and/or differences between the individual designers involved in the three cases and their particular ways of working through a common design problem.

As a result, the chapter has highlighted the crucial, moderate, and less significant objects within the three design sessions that may help the designers and other stakeholders to identify and focus on during their design activities. So, from the results produced thus far, the main objects of importance for the designers to focus on are roads/streets, intersections, cars, and lights/signal. These objects may be unsurprising given that the design problem here relates to a traffic system. However, it is interesting that all three design teams concur with one another with respect to the most important design objects.

It is acknowledged that the mapping of networks has previously been applied to social, human, and nonhuman collaborations within software design projects (Bijker and Law, 1992; Callon, 1987; Latour, 1993, 1996; McMaster et al., 1997; Vidgen and McMaster, 1996; Sack et al., 2006). Sack et al. (2006), for example, combined ethnography, text mining, and sociotechnical network analysis and visualization to understand open software systems holistically. They used a form of "computer-aided ethnography" similar to Teil and Latour's (1995) notion of "computer-aided sociology" to follow participants' roles and statuses. Their approach to social network analysis was also influenced by actor network theory and used social network analysis but they extended it to include nonhuman actors in the networks. Furthermore, Sonnenwald and Iivonen (1999) applied Ranganathan's (1957) framework for knowledge organization (personality, matter, energy, space, and time) to various case studies. They used a mixture of methodologies to each of those human information facets and used social network analysis to understand participants and their social networks. It can be argued that, in this chapter, we have used social network analysis to understand Ranganathan's energy facet: the action, the factors that cause the action, the problems, and the goals that the participant faces.

The research carried out here is specific to one aspect of the software design process. Further analysis is required into other areas of the design process (such as Ranganathan's other framework facets) to strengthen the claims put forward here and also to understand the human behavior of software designing in its totality. The authors believe that this conversational analysis, object-based approach will add positively to the field, particularly if it is part of a larger holistic comprehension of software design in general. It is envisaged that it can also add to previous network analysis research that looks to map the structure of program and class hierarchies (Borgatti, 2002) such as those using unified modeling language.* Comparisons can then be made with the network produced from the program, with its class structures, and with that described by the designers prior to developing the software.

Future work will concentrate on which designer from each team of two refers to which object. If one designer, for example, has certain object preferences over other objects to those of his design partner, then this might reveal significant results. Further research will also include the references that the designers make to other people involved in the project (i.e., "the Professor"), who were responsible for setting the design task, the brief that was proposed, and how the brief was interpreted by the design team. Comparisons can also be made with each network produced at the end of the project and the network produced when the designers were planning the program at the beginning of the project.

REFERENCES

Adams, R.S., Turns, J., and Atman, C.J. (2003). Educating effective engineering designers: The role of reflective practice, *Design Studies*, 24(3), 275–294.

Bijker, W. and Law, J. (Eds) (1992). *Shaping Technology, Building Society: Studies in Sociotechnical Change*, MIT Press, Cambridge, MA.

Bonacich, P. (1972). Factoring and weighting approaches to status scores and clique identification, *Journal of Mathematical Sociology*, 2, 112–120.

Borgatti, S.P. (2002). *NetDraw: Graph Visualization Software*, Analytic Technologies, Harvard, MA.

Borgatti, S.P., Everett, M.G., and Freeman, L.C. (2002). *UCINET for Windows: Software for Social Network Analysis*, Analytic Technologies, Harvard, MA.

Callon, M. (1987). Society in the making: The study of technology as a tool for sociological analysis, *in* W.E. Bijker, T.P. Hughes, and T.J. Pinch (Eds), *The Social Construction of Technical Systems: New Directions in the Sociology and History of Technology*, pp. 83–103, MIT Press, Cambridge, MA.

Dong, A., Davies, K., and McInnes, D. (2005). Exploring the relationship between lexical behavior and concept formation in design conversations, in *Proceedings of ASME 2005 International Design Engineering Technical Conferences & Computers and Information in Engineering Conference (IDETC/CIE 2005)*, September 24–28, Long Beach, CA.

Freeman, L.C. (1979). Centrality in social networks I: Conceptual clarification, *Social Networks*, 1, 215–239.

Granovetter, M.S. (1973). The strength of weak ties, *American Journal of Sociology*, 78, 1360–1380.

Granovetter, M. (1982). The strength of weak ties: A network theory revisited, *in* P. Marsden and N. Lin (Eds), *Social Structure and Network Analysis*, pp. 105–130, Sage, Beverly Hills, CA.

Herbsleb, J. and Grinter, R.E. (1999). Architectures, coordination, and distance: Conway's law and beyond, *IEEE Software*, 63–70.

Latour, B. (1993). Ethnography of a "high-tech" case: About Aramis, *in* P. Lemonnier (Ed.), *Technological Choices: Transformation in Material Cultures Since the Neolithic*, pp. 372–398, Routledge, London.

Latour, B. (1996). *Aramis, or the Love of Technology*, MIT Press, Cambridge, MA.

* Diomidis Spinellis. Automated drawing of UML diagrams. http://www.umlgraph.org/.

Lawson, B. (2004). Schemata, gambits and precedent: Some factors in design expertise, *Design Studies*, 25(5), 443–457.

Lawson, B. and Loke, S.M. (1997). Computers, words and pictures, *Design Studies*, 18(2), 171–183.

Luck, R. (2007). Using artefacts to mediate understanding in design conversations, *Building Research and Information*, 35(1), 28–41.

Luck, R. and McDonnell, J. (2006). Architect and user interaction: The spoken representation of form and functional meaning in early design conversations, *Design Studies*, 27(2), 141–166.

Mahendran, D. (2002). Serpents and primitives: An ethnographic excursion into an open source community, unpublished master's thesis, University of California.

McMaster, T., Vidgen, R.T., and Wastell, D.G. (1997). Towards an understanding of technology in transition: Two conflicting theories. In *Proceedings of the Information Systems Research in Scandinavia*, IRIS20 Conference, University of Oslo, Hanko, Norway.

Ranganathan, S.R. (1957). *Prolegomena to Library Classification* (2nd edn), Library Association, London.

Sack, W., Détienne, F., Ducheneaut, N., Burkhardt, J.-M., Mahendran, D., and Barcellini, F. (2006). A methodological framework for socio-cognitive analyses of collaborative design of open source software, *Computer Supported Cooperative Work (CSCW)*, 15(2), 229–250.

Sonnenwald, D.H. and Iivonen, M. (1999). An integrated human information behavior research framework for information studies, *Library & Information Science Research*, 21(4), 429–457.

Stempfle, J. and Badke-Schaub, P. (2002). Thinking in design teams—An analysis of team communication, *Design Studies*, 23(5), 473–496.

Teil, G. and Latour, B. (1995). The Hume machine: Can association networks do more than formal rules? *Stanford Humanities Review*, 4(2), 47–65.

Valkenburg, R. and Dorst, K. (1998). The reflective practice of design teams, *Design Studies*, 19(3), 249–271.

Vidgen, R. and McMaster, T. (1996). Black boxes, non-human stakeholders, and the translation of IT through mediation, *in* W.J. Orlikowski, G. Walsham, M.R. Jones, and J.I. DeGross (Eds), *Information Technology and Changes in Organizational Work*, pp. 250–271, Chapman & Hall, London.

Wasserman, S. and Faust, K. (1994). *Social Network Analysis: Methods and Applications*, Cambridge University Press, Cambridge, MA.

CHAPTER 22

Conjectures on How Designers Interact with Representations in the Early Stages of Software Design

Kumiyo Nakakoji

Software Research Associates, Inc.

Yasuhiro Yamamoto

Tokyo Institute of Technology

CONTENTS

22.1 INTRODUCTION

The three design sessions being studied in the Studying Professional Software Design (SPSD) Workshop (Petre et al. 2010) demonstrate a wide variety of activities, with participants given the same design prompt with the same physical setting in which two designers work on the task in front of a whiteboard. The differences are not in the approaches that were taken, but are in the aspects of software design that the participants worked on. The variety of the three design processes seems to reflect the wide range of activities that software design involves and the different levels of detail that need to be covered in software design.

Studies have been published about the three design sessions, including those that focus on how the design process proceeds through a fine-grained analysis (Baker & van der Hoek 2010), how a pair of designers collaborate with each other through an ethnographic analyses (Rooksby & Ikeya 2012), and how designers make decisions through a design protocol analysis (Christiaans & Almendra 2010).

In contrast, this chapter focuses on how designers interact with representations in the early stages of software design. Each team worked on the design problem by talking to each other and using a whiteboard as a means of externalization (Yamamoto & Nakakoji 2005). The drawings on the whiteboard that were captured in the video data and the transcript of the conversations provide a rich source of representations with which the designers interacted.

The focus of the study presented here is similar to that of Schoen (1983), who observed how two architectural designers engaged in sketching and talking about a design problem. We are interested in exploring the dynamism of the representations externalized by the designers in a practical setting, viewed as the designers' conversations with the material (Schoen 1983). Our work is distinctive from those that analyze sketched diagrams (e.g., Do 2005), or those that analyze the cognitive aspects of designers in controlled study settings (e.g., Suwa & Tversky 2001).

Our study primarily focuses on whiteboard usage together with word usage in terms of which of the following were generated and then used: what diagrams were drawn on the whiteboard, what text was put on the whiteboard, what phrases were uttered in the transcript, as well as how these communications were referred to, revisited, modified, and reappropriated in the subsequent process. By contrast, Ju and colleagues (2006) reported on how people use a whiteboard in a collaborative setting and Walny and colleagues (2011) analyzed what they called spontaneous visualizations on the whiteboard.

This chapter analyzes the use of a whiteboard along with textual representations uttered and written down on the whiteboard during the course of software design practice.

22.2 DATA ANALYSES SETTING

In analyzing data that consist of meeting videos and their transcripts, we use design practice streams (DPS) tools (Nakakoji et al. 2012). The DPS tool has three components: MovieViewer, StrokeViewer, and TranscriptViewer (Figure 22.1). It helps us to browse segments of the video data relevant to the focused topic by specifying a region on the whiteboard or choosing a few terms used in the transcript.

The DPS tool uses a simple time-stamp mechanism to relate the video data of a design meeting with the digital whiteboard drawing data (i.e., a set of time-stamped strokes drawn during the design meeting) and the textually transcribed data (i.e., a set of time-stamped utterances). It does not require any manual tagging, annotations, or semantic interpretations. The detailed mechanism of the DPS tool is described in Nakakoji et al. (2012).

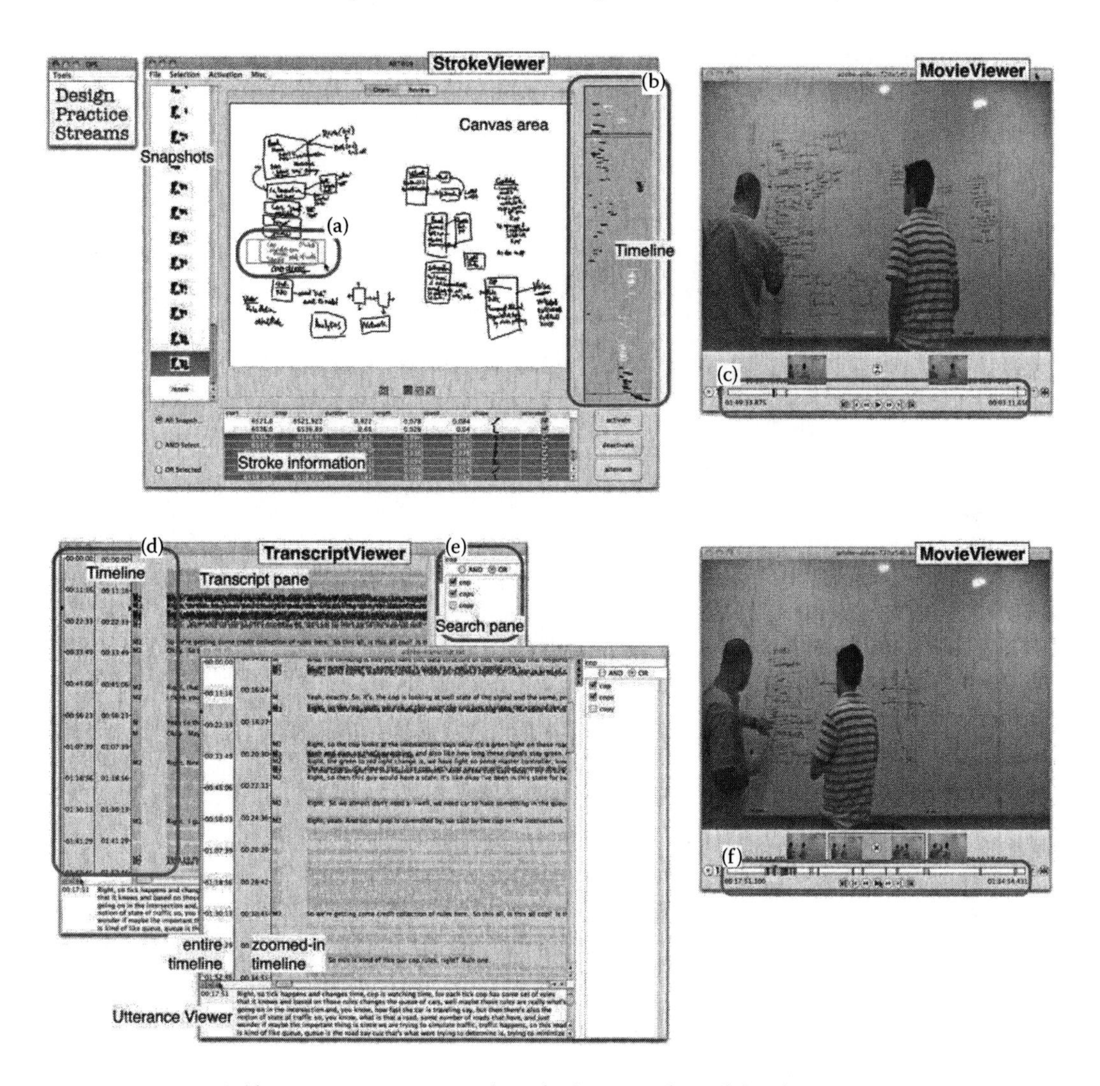

FIGURE 22.1 DPS (design practice streams) tools showing the Adobe data.

The original data set did not include the whiteboard stroke data, so we manually generated a set of stroke data for the videotaped whiteboard drawing of one of the three teams (Adobe) in sync with the video data. By using the DPS tool to browse the Adobe team design process, we found that the diagrams drawn on the whiteboard evolved over time—not necessarily in a consecutive manner but intermittently, for example, by adding labels to verbally refer to some parts of the diagram, or by adding more graphic representations to reappropriate the diagram for a purpose that was different from the original one (Nakakoji et al. 2012). We used the DPS tool for the video and the transcript data to analyze the design processes of the other two teams.

In what follows, we list a set of observations that we made in analyzing the data. We first identified distinctive elements of how the designers interact with the representations by viewing the three sets of design session data with the DPS tool and recording the time stamp of each element. We then textually annotated each of these elements with particular

FIGURE 22.2 Data analysis setting.

issues and concerns. Finally, we categorized the elements into groups by spatially arranging them while applying a variation of the KJ (Kawakita Jiro) method (Kawakita 1968; Figure 22.2).

The results consist of the following topics:

- Observation A: using the whiteboard as a medium for design
- Observation B: working with an emerging concept that did not exist in the given specification
- Observation C: using the whiteboard for temporal drawing
- Observation D: producing a variety of notations on the whiteboard
- Observation E: reappropriating the representations on the whiteboard
- Observation F: detailing user interface representations
- Observation G: recognizing a "eureka" moment
- Observation H: experiencing a temporal gap between talking and drawing
- Observation I: observing courses of design
- Observation J: redrawing on the whiteboard

We do not claim here that these observations are an exhaustive list of the features of the software design. Rather, we argue that the list gives us a partial preunderstanding about how designers interact with externalized representations in the early stages of software design. Some may be cases peculiar to the three design teams and the given environmental settings, some may be specific to the domain of software design, and some may be applicable to other design practices.

22.3 OBSERVATIONS

The following scenarios provide examples of each type of observation. The three teams we observed are the Adobe, AmberPoint, and Intuit teams.

Observation A: Using the whiteboard as a medium for design

The designers used a whiteboard drawing as a snapshot to hold their intermediate design ideas.

Jim of the AmberPoint team wanted to take a picture of the whiteboard (AmberPoint: 00:48:35) by saying, "But I don't want to [erase it], that's always one of my concerns of writing on a whiteboard is that I've got some great ideas on here but now I need to reshuffle a little bit, and so I would just pull out a camera" (00:48:47). He actually did not erase anything right away. The purpose of taking a picture of the whiteboard seemed to be to record a snapshot of the important artifact that he had been working on so that he could reshuffle the ideas in the subsequent process.

The designers looked back at what they had written on the whiteboard.

The designers articulated the process they had used. Male 1 of the Intuit team summarized what they had just done and commented on the process by saying, "I like jumping into details and then coming up and saying, 'Okay, now what do we really need here, what are we seeing here?'" (Intuit: 00:14:51).

The designers used "designing" and "drawing" as synonyms.

Facing the whiteboard, Ania of the AmberPoint team asked a question: "What else are we going to be drawing?" (AmberPoint: 00:09:33). Jim, also facing the whiteboard, posed a question: "Are there other objects?" (00:10:52). They seem to be saying the same thing but stating it differently—one in terms of things to draw in a graphical user interface (GUI) window, and the other in terms of objects to list.

The designers walked through the user interaction steps by interacting with a drawn GUI on the whiteboard.

The designers simulated a short interaction flow with the drawn GUI window on the whiteboard by using gestures. When Jim of the AmberPoint team was saying, "So the user can change where the intersections are, the distance between them, by moving them around," he moved both of his hands and showed a gesture expanding the road drawn in the user interface window on the whiteboard.

While sketching a GUI, M of the Adobe team kept talking about how a user would interact with the system, such as, "So once you drag this one here you have to specify how that one's connected, right? So as you drag, as you drag each intersection on it will snap and say, okay, I'm connected to here and to here, and it snaps. And here, and what not. And so then drag it and then it automatically connects and as you move this back and forth it increases the capacity for this road so now it ..." (Adobe: 01:05:45). While doing so, M made gestures pointing to and touching the GUI window drawn on the whiteboard.

The designers occasionally acted as if there had been something drawn on the whiteboard. When discussing how a student would interact with the system, Ania of the AmberPoint team used gestures interacting with the whiteboard, and relayed possible interaction steps by saying, "So there is some sort of drawing palette, right, that says okay, I have this thing I drag something, I'm drawing a road and I call it something and I draw and I call it B ..." (AmberPoint: 00:20:10). Ania used the object listed on the left side of the whiteboard as the palette, although it was just the list of objects and was not drawn as the palette.

Observation B: Working with an emerging concept that did not exist in the given specification

A new concept that was not a part of the given specification became a major design element. When trying to understand how cars go through an intersection, M of the Adobe team mentioned "cop" for the first time (Adobe: 00:14:06) by saying, "It's almost like you need [a] traffic cop, right, traffic cop controller." The term "cop" does not appear in the given task prompt.

After M mentioned the term, M2 drew a larger square around the model and its components on the whiteboard (00:14:16). He added a box underneath the square he had just drawn as if he was ready to add an "external" entry to the "model." M2 wrote "Event (e)" in the new box, saying, "So, something happens as an external controller is ticking the model, is that what you're saying?" (00:15:07). Then, M2 wrote "cop" on the whiteboard near the "model" label (00:15:15), and then M said, "Well, I don't know, is the cop really a model or is that a controller? A cop is kind of controlling the state of the model" (00:15:18). M2 then immediately erased it (00:15:26). He then deleted the square frame surrounding the model field.

After pondering for a few seconds, M2 started boxing each of "intersection," "car," and "time" by saying, "These are objects, right?" (00:15:29). He deleted "Event" that he kept unboxed.

When M2 wrote "cop" again on the whiteboard (00:16:37), he wrote "responds to time changes" and "changing state of model" by verbalizing them. He then drew a square around the "cop" and its description, and asked M if M thought it was reasonable.

This resulted in the list of boxed terms under the model field, and "cops" had become a boxed object under the "model" category. Thus, "cop" had become an object in their model (see Figure 22.3a).

In the subsequent design process, the "cop" object served as an important player in their model. The terms "cop" and "cops" were mentioned in 28 utterances in the transcript, ranging over the entire design process (see Figure 22.3b).

A new concept was introduced to help the designers understand the problem better, but it did not become a design element.

When trying to describe how cars come and go through an intersection in the traffic simulation, the Intuit team came up with the term "faucet." When wondering what would be the "start" and "stop" of the simulation, Male 1 of the Intuit team used "faucet" as a

(a) (b)

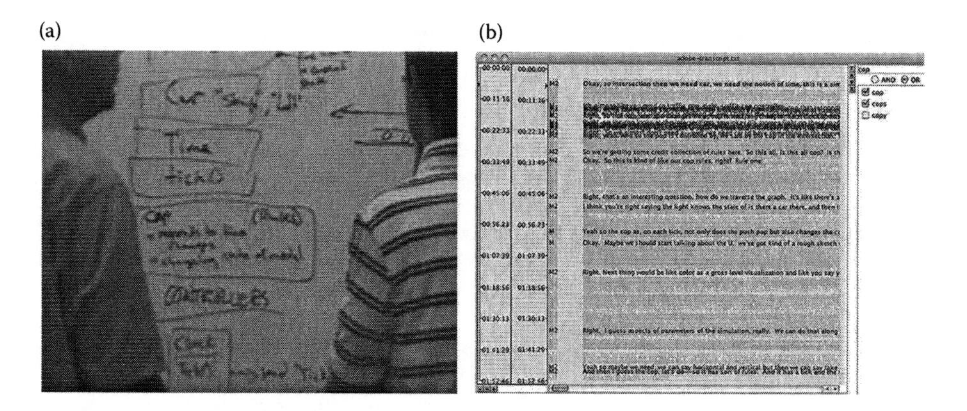

FIGURE 22.3 "Cop" drawn on the whiteboard (a) and uttered in the transcript (b).

metaphor to describe the behavior of the controller, as in the following: "I'd imagine some guy turning on a faucet. When the sink's full, turn it off, let it drain down, or tell me there's more room, I'll turn it on again" (Intuit: 00:41:35). This metaphor seemed to be a comprehensive one, and Male 2 agreed with Male 1 by saying, "Perfect, okay, yeah, that makes sense" (00:41:58). In contrast to the "cop" object with the Adobe team, the Intuit team did not make "faucet" a part of their model.

The designers spent a long time deciding to make a new concept as a design element. When Male 1 of the Intuit team drew a line between "Roads" and "Lane" by saying, "Roads have lanes" (Intuit: 00:11:01), Male 2 asked whether they cared about such details as lanes. Then Male 2 concluded by saying, "I don't think we need to make assumption of lanes, period" (00:12:11).

Despite this decision, the notion of "Lane" kept emerging in the subsequent discussions. When they finished discussing the length of the road in front of the pictures of the two intersections on the whiteboard, Male 1 walked toward the left of the whiteboard, where "Roads" was listed as "Data." He then annotated "Lane" connected to "Roads" with a cross in red, by saying, "So we've decided that lanes may be not so significant but length is going to be significant" (00:19:38). He continued by saying, "And it's significant because it holds a number of cars" (00:19:40), as he drew a red line starting from "Roads" and wrote "Length" and "# of cars."

Male 1 then said, "I'm still wondering about the left-turn lane" (00:20:21). After describing why having lanes affects the complexity of the problem, he claimed to have "a protected left" to "go with the minimum" (00:21:03) in taking the simplest assumption. The two continued discussing how having lanes affects the traffic simulation for a few more minutes. Male 1 then expressed his concern by saying, "I hate to bog down in that detail."

Finally, after about 30 minutes, when Male 1 went back to where he had written "Roads" under "Data," he added "s" to "Lane" connected to "Roads" to make "Lanes" (00:48:59). He then wrote "L/S/R" near the "Lanes," with Male 2 saying, "More paths, kind of" (00:49:08). This way, they seemed to have agreed that lanes are like "paths" for each of the directions (left, straight, and right).

Observation C: Using the whiteboard for temporal drawing

The designers casually drew simple diagrams on the whiteboard to quickly establish shared understanding.

M of the Adobe team drew a graph diagram when explaining the relationship between "capacity" and "wait time" (01:33:30). He erased the graph immediately after he finished his explanation (Adobe: 01:34:02).

When M2 of the Intuit team asked whether "all the opposing lanes [were] always the same state" (Intuit: 01:48:25), M drew a picture of a simplified intersection in a corner space, trying to illustrate the real situation, and described that "it [what M2 said] is a simplified assumption" (01:48:29). M then immediately erased the intersection that he had just drawn.

The designers used diagrams to understand a real-world phenomenon.

M2 of the Adobe team drew a picture of an intersection on the whiteboard after a few minutes of talking about the model (Adobe: 00:06:41). This picture was used not so much as an image of the map visualization, but as an object-to-think-with in understanding the role and behavior of an intersection and its relationships with cars, lanes, and signals.

The Intuit team drew diagrams as a representation of the world to talk about a model for the world. Male 1 drew a picture of an intersection (i.e., crossing roads) and added visual marks corresponding to the components of "Data" that he had just written, such as cars and signals, while talking about how they were related to one another (Intuit: 00:01:00). Then they added arrows and little marks on the intersection diagram that they had drawn when talking about Data. While discussing, they wrote "Abstract," underlined it, wrote "rules" beneath "Abstract" (Intuit: 00:13:00), and drew a line between "rules" and "signals." As they talked about the objects listed under "Data" and "Abstracts" and their relationships by adding lines, they tried to understand the world of the traffic system; as Male 1 said, "So we [are] kind of tinkering around with this real-world model" (00:14:41).

Observation D: Producing a variety of notations on the whiteboard

The designers used specific notations to represent the type or function of a diagram drawn on a whiteboard.

The designers used boxes, lines, and arrows as specific notations to represent objects and hierarchical relationships. M2 of the Adobe team wrote down "clock controller" under the label "controllers," and boxed the words. Then, M2 drew an arrow and wrote "send 'tick' event to model" (Adobe: 00:13:40). This seems to become the description of what a "clock controller" would do.

In talking about the "hierarchy," Male 1 of the Intuit team drew a line between "interaction" and "signal," indicating that there was a hierarchical relationship between the two.

The designers used a pseudo-code expression to describe the system to be designed.

In summarizing what they had designed, M2 of the Adobe team wrote in a programming-language-like notation, "main() {n = new network, c = new clock. c.run(n)}" and described it as "at a very gross level" (Adobe: 01:37:50).

When M2 of the Adobe team wrote down each rule for the "cop rules," he wrote them in detail by using a pseudo programming-language-like notation (Adobe: 00:34:32), such as, "1 car per tick passes if car is the head of the road" when "car = = straight and I.light = = greenlight R.pop" (00:36:15). When he wrote "I.light = green" on the white-board, he was actually saying "And green, and intersection I.light." He then put "(1)" in the head of the line he had just written, saying "So that's case one, that's case one. Case two would be ..." (00:36:06) and he put "(2)" in the new line.

The designers used colors on the whiteboard to match the GUI colors and to distinguish different concepts or phases.
While drawing the time line, Jim of the AmberPoint team used a green pen for a green light, a red pen for a red light, and a yellow pen for a yellow light. He asked for another color by saying, "That represents green arrow if we need to. Left-turn is orange, I don't know." He then used a purple pen (AmberPoint: 00:39:58).

When they started thinking about "container," Male 1 of the Intuit team wrote "map" and underlined it with a green pen in the space in the top part of the whiteboard (Intuit: 00:30:50). He then circled "interactions" and "Roads" connected from "map" (00:31:07) also in green. Male 1 kept using a green pen and drew two circles and a connecting line, and labeled them "I1," "I2," and "P1," respectively (00:33:50). He then asked, "How do you define the behavior of the traffic?" (00:34:30). He then wrote "behavior" at the bottom part of the whiteboard in green as if it would serve as a reminder to rethink later the things written in green (00:34:50).

Observation E: Reappropriating the representations on the whiteboard
The designers reappropriated a drawn diagram on the whiteboard.
A diagram of two intersections was first drawn when the two designers of the Adobe team were discussing how the state of an intersection changes from time T0 to time T1 (Adobe: 00:09:57; see Figure 22.4a). About 30 minutes later, when the designers were talking about how cars flow from one intersection to another, they started pointing to the drawn intersections as if they were adjacent to each other. M2 then drew dotted lines to connect the two

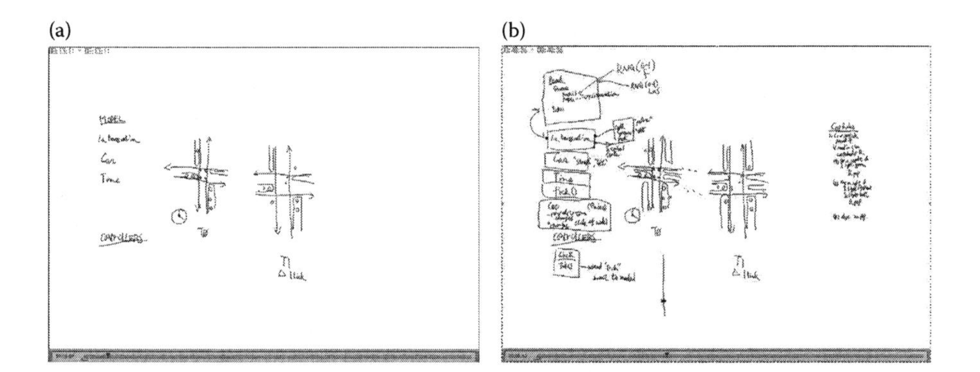

FIGURE 22.4 Intersections reappropriated—initial drawing (a) and later drawing (b).

intersections (00:40:33; see Figure 22.4b). The drawing resulted in a discrepancy between the labels (T0 and T1) and the dotted lines.

The designers reappropriated the written text on the whiteboard.
When the designers were reexamining the "rules" of signals, the two designers of the Intuit team discussed the relation between "time" and "red/yellow/green," both of which had been written under "rules" (Intuit: 00:51:32). Male 2 said, "There should be a time component attached to each one of these things," pointing to "red/yellow/green." Then, Male 1 added a "t-" to each of "red/yellow/green" to create "t-red/t-yellow/t-green," and said, "There's a time for each of those. t-red, t-yellow, t-green" (00:52:09). Thus, "red/yellow/green," which was originally written to mean something to do with the rules related to the three options, had come to mean the three time durations for the red, yellow, and green lights when each was prefixed with "t-."

Observation F: Detailing user interface representations
The designers went into extreme detail in sketching GUI components on the whiteboard.
Jim of the AmberPoint team started drawing how the "summary area" would look by writing down what information they would like to see in the area (AmberPoint 00:14:00). He listed all the road names and intersections that were drawn on the left side of the map (00:14:50).

The designers spent a few minutes just redrawing the details. In response to the discussion on how a user would specify the duration of the cycle of a signal, Jim of the AmberPoint team erased all the inside area of the pop-up window where he had spent several minutes listing the conditions, and started to redraw them on the whiteboard (AmberPoint: 00:36:30).

The designers detailed not only in drawing but also in talking.
In sketching the pop-up window for an intersection to specify the signal behavior, the AmberPoint team discussed details such as "2 minutes" for the default cycle duration of a signal (AmberPoint: 00:29:59), and "on average 40 seconds because with the green arrow" (00:30:42), and "so it would be 50 seconds at the top of the cycle followed by, I don't know, 10 seconds of yellow, followed by how about 55 here" (00:31:00). Jim used a green pen to represent a green light, a red pen for a red light, and so on (see Figure 22.5).

The designers prioritized the completeness of a textual list to the exact layout of GUI components in drawing a GUI.
When Jim of the AmberPoint team added the "speed" column in the drawing of the "summary area" window, the section did not fit within the drawn square area and stuck out from the area toward the right, but he left it there (AmberPoint: 00:14:50).

Observation G: Recognizing a "eureka" moment
The designers came up with a new understanding of the problem through talking.
After talking about making "Roads" own traffic, Male 1 of the Intuit team said, "so the road is an object" (Intuit: 00:36:08) and wrote as operations "add cars" in green next to "Roads" (00:36:25). Male 2 initially nodded for a few seconds, expressing his agreement.

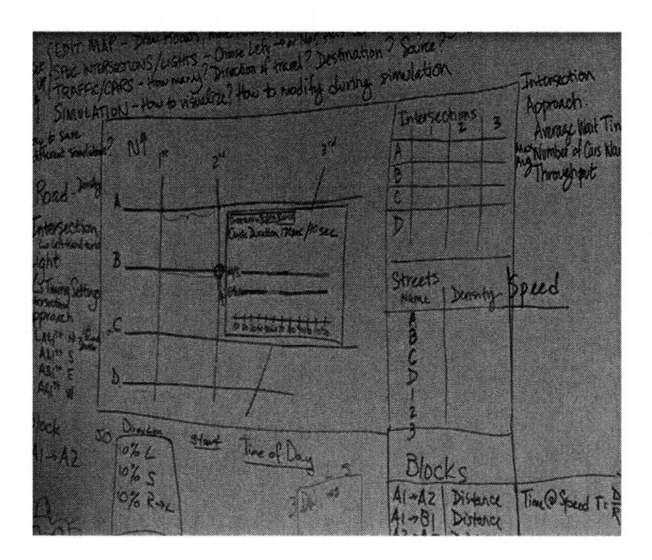

FIGURE 22.5 Detailed drawing on the whiteboard.

But then Male 2 murmured, "Would it add a car?" (00:36:28). Immediately after that, Male 1 echoed Male 2 by saying, "Yeah, I'm trying to get how do we [add a car]?" (00:36:29). This finding led them to talk about "input" to a road in the subsequent discussions.

The AmberPoint team became aware of a problematic situation by walking through a user's interaction with the currently drawn interface. While Jim was talking about a student's walk-through interacting with the current user interface and how the information in the summary area would change (AmberPoint: 00:25:35), he had become aware of a possible problem with the summary window by saying, "So it's funny because some of these things are editable and some of them just kind of give you details about how your edits made a change," and he said, "So I think it's a little confusing" (00:26:25).

Jim was redrawing a map window and a window for setting up the intersection, and was about to draw a window for a car/traffic configuration while saying that these were "editing and setting modes." Ania then wanted to have "the map drawing as one thing" but the intersection specification and the car specification "as a part of the simulation" (01:18:49). The two modes were then renamed to "building map" and "building simulation" (01:21:35).

The designers refined a partial design solution based on a new understanding that had emerged through the detailed drawing.

Jim continued listing all possible patterns of signals in the pop-up window by first finishing the N/S options and then the E/W options (AmberPoint: 00:34:06). While listing, he had become aware that the signal has one state when both directions are red (00:34:30).

Observation H: Experiencing a temporal gap between talking and drawing

The designers did not immediately record seemingly important terms on the whiteboard.

The Adobe team first mentioned the notion of "interior" road (Adobe: 00:40:25). It was some time later that they added the "interior" and "onramp" properties to the "Road" on the whiteboard (Adobe: 00:42:55).

The designers did not record on the whiteboard a number of GUI ideas verbally discussed.
When the Adobe team discussed the user interface for the system, a number of ideas such as the following were verbally described, but they were never written on the whiteboard. "Right, so this is now a UI piece, and the dot brings up your editor with frequency of, your frequency value. Maybe if you select a car, I mean a road, it allows you to edit. I wonder if we make the capacity of the geometry and you could sort of simulate a highway or something ..." (01:11:52).

The designers wrote down text on the whiteboard for possible later use.
The designers wrote down a not-so-certain object on the whiteboard so that they could come back to it later. When the designers of the Intuit team went back to working on the "rules" of signals under "Abstract," Male 2 decided to write "left/straight/right" (00:51:25) under "time," "sensor," and "red/yellow/green" after wondering a little bit. Male 1 agreed with what Male 2 had just done by saying, "Yeah let's put it in there and we'll capture that" (Intuit: 00:51:17).

Observation I: Observing courses of design
The designers started the design process by listing names as objects on a whiteboard.
Male 1 of the Intuit team first wrote "Data" on the whiteboard, and underlined it. He then added "signals," "Roads," and "cars (traffic)" (Intuit: 00:08:59).

M and M2 of the Adobe team started by working on "basic data structures" (Adobe: 00:05:56). During the conversation, terms that seemed to play important roles in the model were mentioned, and M2 started writing them down on the left side of the whiteboard, for example, "intersection," "car," and "time." Other terms that also seemed important, such as "queues" (00:08:28), were mentioned but were not written on the whiteboard then.

The AmberPoint team listed object names in the thin left area of the whiteboard. The remaining part consisted of the visual map area and the summary window area, which listed intersections, streets, and blocks, each with a complete list of roads displayed in the map area (AmberPoint: 00:17:33).

The discussions of a user interface and those of a model were intertwined.
When talking about how "push" and "pull" were done by "cop," M of the Adobe team wondered, "Well, it does say that something that the user could control" (Adobe: 00:25:32), and the designers went back to the problem description sheet to better understand how the simulation is controlled by a user (e.g., a student) (00:25:47).

When Male 2 of the Intuit team was talking about how input traffic would flow cars into the road, he wrote "input traffic" and drew a circle on the right side of the intersection picture that Male 1 had previously drawn. Then, Male 2 drew an arrow from the circle into a road, which was a part of the intersection (00:36:40). Male 1 then wrote "R1" as a label to the road, and he used "R1" in saying, "Let's say there's, I'm thinking, the box they're using they click, they're going to define a flow into R1 and then they're going to click 'go'" (00:36:55).

When talking about "queuing and de-queuing of cars" for visualizing cars in the user interface (Adobe: 01:25:34), the Adobe team started talking about the notion of "queue" and went back to talking about the model, such as, "Would that be a property of the queue or a property of the car?" (01:26:44). The subsequent conversation for the next several minutes kept going back and forth between two topics, the user interface visualization and the model.

Observation J: Redrawing on the whiteboard

The designers added labels to drawn objects on the whiteboard to talk about them.

When the AmberPoint team talked about signals and left-hand turns, Jim labeled each horizontal road with "A," "B," "C," and "D," and each vertical road with "1st," "2nd," and "3rd" (AmberPoint: 00:12:06). Jim started naming an intersection by using these labels, for instance, "At the corner of A and 1st, there's a north approach" (00:12:31).

The designers added a little context to a diagram later in the process.

Jim of the AmberPoint team added the North-up sign in the top-left corner of the sketched map GUI window drawn on the whiteboard (AmberPoint: 00:28:17). At that point, the designers had been working on the map from the very beginning of the meeting for about 20 minutes (00:08:21).

Labeling took place not only during the meeting but also after the meeting.

During a break when M2 of the Adobe team stepped out of the room, M looked at the whiteboard and added "visitor" (Adobe: 01:21:03) as a label to the two lines of text, "rule/actions" and "objects/rules," which had been written down about 20 minutes before the break. The term "visitor" was mentioned at the time M2 put the two lines of text on the whiteboard but it had not been written down. M seemed to have remembered the context and recorded the context to the two lines.

The designers redrew diagrams on the whiteboard to afford more space.

When drawn objects on the whiteboard evolved, the designers redrew the same diagram in a different space to afford more space. They did not seem to be aware of how the diagrams would evolve when they began to draw them.

M2 of the Adobe team erased the "model" label and wrote "Road" on the whiteboard where "model" had been (Adobe: 00:18:30). When adding properties to the "Road," M2 deleted "Road" and wrote it again in an upper position on the whiteboard so that there was more space below to write more properties (Adobe: 00:24:05).

M of the Adobe team copied a picture of the clock toward the right to have more space (Adobe: 01:22:03).

Jim of the AmberPoint team enlarged the pop-up window for the intersection toward the right since the area got quite cramped with its contents. (AmberPoint: 00:33:40).

The designers redrew a great many details in copying a GUI window on the whiteboard before erasing them to afford more space.

In response to the discussion on how a user would specify the duration of the cycle of a signal, Jim erased all the inside area of the pop-up window where he had spent several minutes listing the conditions, and started rewriting them (AmberPoint: 00:36:30).

The sectioning of the whiteboard evolved over time.
The designers also "moved" drawn objects on the whiteboard by copying and erasing them when a grouping of objects emerged. Different types of groups were identified in different sections of the whiteboard.

When M mentioned the "controller" (Adobe: 00:09:37), M2 went back to the place on the whiteboard where he had written a list of terms, and put "model" on top of the list and underlined it (00:10:05). Immediately after that, he wrote "controllers" in the space below and underlined it. Now the whiteboard had two areas, one labeled "model" and the other labeled "controllers." The initial list of terms became a part of the "model."

A topic change triggered the designers to erase whiteboard drawings.
When the Adobe team started talking about sketching user interfaces, they first tried to find things that they could erase on the whiteboard, but then decided they would not, and started sketching a GUI window in the small space at the bottom of the whiteboard (Adobe: 01:04:46). When they later changed the topic to discuss a user story, they erased the pictures of the two intersections on the whiteboard to create space (01:19:09).

The designers sought more appropriate wording and expressions when writing on the whiteboard.
When Jim of the AmberPoint team started listing "objects that they need to deal with" (AmberPoint: 00:09:20) on the left side of the whiteboard, he asked Ania, "Are we going to call them streets or roads?" Ania responded by saying, "I guess roads, intersections" (00:09:31). Jim then wrote down "Roads" on the whiteboard.

When the notion of "leg" emerged (AmberPoint: 00:15:46), Ania first asked, "What would be an easy way to name each leg?" (00:15:51). Jim said, "If this is A1 and this is A2, the leg would be A1 to A2" (00:16:01) and Jim drew "A1 → A2" on the left side of the whiteboard under the list of objects (00:16:20).

Jim added "legs" above "A1 → A2" on the left side of the whiteboard, and added "legs" to the interface summary area on the right side of the whiteboard. But then he did not seem to like the term "legs" and changed it to "blocks" on the left side of the whiteboard (AmberPoint: 00:16:30).

While working on a user story, M of the Adobe team kept writing natural language sentences by verbalizing—for instance, "I want to evaluate success of each simulation" (Adobe: 01:31:56). M then pondered what expression he should use by saying, "… average capacity on the road or something? Is it capacity or average umm or umm.. capacity means like …" (01:32:04). M2 then said, "Well, capacity is a good one to have because then you can evaluate wait time per, for a given capacity" (01:32:31). M kept the expression "average capacity" on the whiteboard. A few minutes later when M wrote "I want to control the aspects of the intersection" on the whiteboard (01:35:20), M2 rephrased it by saying "I guess 'aspects of parameters of the simulation,' really" (01:35:28). M changed the writing into "I want to control the parameters of simulation" (01:35:50).

The designers kept consistencies among the drawings on the whiteboard when refining labels and layouts.
When Jim of the AmberPoint team decided to use "blocks" instead of "legs," he also changed "legs" to "blocks" on the right side beneath the summary area, drew a window-like square around the "blocks" that he had just written, and listed possible block names, such as A1 → A2, A1 → B2, and so on (00:16:40).

The designers did not keep consistencies among the drawings on the whiteboard when refining labels and layouts.
When discussing how an intersection behaves with roads, Male 1 of the Intuit team labeled the four roads of the intersection he initially drew with "R1," "R2," "R3," and "R4." He then used the labels to describe the behavior of the intersection, such as in "The intersection is going to say, okay, my light is green, R1, I mean it's green for 30 seconds – R1, give me 10 cars, we can calculate that, and put them in R3" (Intuit: 00:46:56). He had already written "R1" on a different road of the adjacent intersection at 00:36:55, so there now existed two R1s in the single drawing.

22.4 DISCUSSION

This section discusses the observational findings listed in the previous section. We believe that the observations described in Section 22.3 can help us formulate hypotheses about how the software designers interact with representations, as well as inform the design of tools for software design, such as whiteboarding tools or diagramming editors for software designers.

22.4.1 Whiteboard Usage in Software Design

Just as Dekel and Herbsleb (2007) observed with object-oriented design teams, the three design teams demonstrated the use of the whiteboard as a medium to represent the artifacts that they had worked on (see Observation A). As Ania of the AmberPoint team pointed out in the video that summarized their task, Jim was said to have strong expertise in using a whiteboard. He often seemed to have planned which part of the whiteboard to use for what purposes.

As Petre (2009) argued, designers "used juxtaposition and annotation deliberately and expressively" by sketching on the whiteboard, using both text and diagrams, as well as formal representations (see Observation D). The notations used on the whiteboard partially adhered to unified modeling language (UML) class diagram notations, but were not limited to the UML, and incorporated many informal representations (Dekel & Herbsleb 2007). The designers used pseudo-code expressions to specify rules, conditions, and timings on the whiteboard rather than using natural language textual representations. What the designer uttered was transformed into a programming-language-like expression when being written down on the whiteboard.

The designers deliberately chose colors when sketching a GUI window on the whiteboard. The choice sometimes depended on the choice of colors of the GUI components, but other times it was merely to distinguish different objects and concepts (Observation D).

22.4.2 How Objects Evolve over Time

Two of the three observed teams started the course of design by listing nouns on the whiteboard as object names (see Observation I), as observed by Dekel and Herbsleb (2007). The designers started the design process by first identifying the names of things that would serve as components for the software to be built (see Observation B). Such components are likely to become class and object names in software programs. The teams tried not to erase them throughout the design process.

Some objects had been straightforwardly identified (e.g., "Signal" "Road," and "Car"), and others had been debated for quite a while (e.g., "Lane" in the Intuit team and "Cop" in the Adobe team). The former are seemingly objects and class names that are commonly used by the three teams; the latter appear to be those that are used by a specific team.

The Intuit team spent a long time deciding whether they should list "Lane" as an object, and they did not write it down on the whiteboard (see Observation B). Their concern was around a conflicting need between introducing a seemingly necessary concept and keeping the model simple. The Intuit team initially thought that they might need "Lanes" as an object but then they worried that the notion was too detailed and would complicate the model. They wanted to avoid it and did not immediately write it down as an object on the whiteboard. But the notion kept coming up during the subsequent process. Later, in discussing lanes, the team first identified that the notion of "Lane" is related to the notion of "length" in terms of the number of cars a road segment is able to hold between the two intersections. Toward the end of the design session, the designers also identified the notions of left-turn, right-turn, and straightforward paths, related to lanes. In the end, the notion of "Lane" did not appear in the resultant model, but "the length of a road" and "left/right/straight paths" did.

22.4.3 Sharing Mental Imagery of Software to Be Designed

We observed that the design teams brought up new concepts that were not part of the given requirements specification. Such concepts as "cop" and "faucet" explained in Observation B seem to be the result of externalizing the designer's mental imagery about the software to be designed (Petre 2009).

The Adobe team and the Intuit team demonstrated the effective use of such concepts as metaphors in understanding the complex behavior of an "intersection." The Adobe team used the term "cop" to illustrate how an intersection handles cars flowing from one road to another. The Intuit team used the term "faucet" to understand the incoming flow of traffic into a road.

It is interesting to note that "cop" was made into an object in the Adobe team's design, but "faucet" was not in the Intuit team's design.

22.4.4 Dynamism of Design Process

The use of the DPS tool allowed us to observe and identify how the designers refined the whiteboard drawings over the course of the design process (see Observation J). These observations would inform whiteboard tool designers about how a whiteboarding tool could transcend the traditional ways of using a whiteboard.

We found that the designers added labels to the drawn GUI window not as a GUI label component, but as names for the drawn objects so that they could talk about them. It was also observed that the two designers of the AmberPoint team sought more appropriate wording in listing object names, naming a portion of the user interface, and writing down a user story. One team changed a noun from singular to plural by adding an "s" to a word written on the whiteboard probably because it would correspond to a one-to-one or one-to-many relationship between the objects.

As shown in Observation J, the designers did not always keep consistencies between drawings on the whiteboard when changing names and labels. We also observed that there were temporal gaps between the time when the designers uttered a name and the time when the designers wrote it down (see Observation H).

It was interesting to note that the Adobe team went into extreme detail in sketching GUIs on the whiteboard (see Observation F). When the Adobe team sketched GUI windows in the study, they almost always enumerated all the possible elements and seldom omitted details. They did not seem to mind redrawing different versions of the GUI ideas repeatedly (see Observation F), similar to how an interaction designer repeatedly draws detailed sketches in a sketchbook (Nakakoji & Yamamoto 2010). This team used the term "draw" almost as a synonym for "design" (see Observation A).

The degree of detail may not necessarily be to complete the design detail but to invoke reflections. For instance, when Jim of the AmberPoint team started another attempt at drawing by saying, "So there's 120 seconds we want broken into increments" in specifying the duration of a cycle of a signal, he had come up with the idea of the "time line" (AmberPoint: 00:37:26) representation.

At the same time, sketching a GUI window on the whiteboard does not afford the intended resolution for the designers. Jim of the AmberPoint team put "5 sec" as an increment and started putting dots in a time line drawn in the pop-up window, but it became so small and congested that he changed to a 10-second increment (AmberPoint: 00:37:50). It was not clear whether this was a design decision or just a way to accommodate his hand drawing.

Reappropriation was observed for both graphical representations and textual representations (see Observation E). We need more studies to distinguish the type of reappropriation that is demanded because of the nature of the design from those that are merely for the sake of convenience in saving time and effort to generate similar drawings.

22.4.5 Program Design and Interaction Design

The AmberPoint team worked on the interaction design of the system, focusing on how a user would interact with the system through what user interface components in great detail. They started the task by drawing a GUI sketch for "laying out the roads" and a "visual map" (AmberPoint: 00:07:44), and then talked about what role this window would play in terms of what kind of functionality. For instance, "You need kind of a summary area to kind of tell you what your settings are for the individual intersections, and what kind of effect, like how much is the traffic backup at this light, or what's the average wait time at this light" (00:08:53).

This was a typical process of interaction design (Cooper 1999), which starts by focusing on how a user interacts with the system to be designed, and then leads to identifying the necessary functionality. As such, the AmberPoint team detailed user interface components and walked through user interaction steps by interacting with a drawn GUI on a whiteboard through gestures (Observation A).

The Intuit team worked on the underlying substrate models of the simulation mechanism, primarily focusing on the program design aspect of software. Given the design prompt saying that the client wants to have "traffic simulation," the team spent a long time trying to understand how traffic operates in the real world, and discussing what assumptions to make to create a model that is simple to handle but covers the necessary functionality. Their struggle in dealing with "Lanes" of a "Road" (as explained in Observation B) typifies the issue.

The Adobe team covered both aspects, ranging from simulation engines to user stories. They started by working on the object model, then they specified control flows and rules, depicted user interface layouts and how a user interacts with the system, wrote user scenarios, outlined the main program structure, and drew entity–relationship (ER) diagrams. The level of detail they worked on was different from those of the other two teams, probably due to the limited amount of time assigned to the teams.

The Adobe team demonstrated how the discussions of a user interface and those of a model were intertwined (see Observation I). Thinking about controls over the model may trigger the designers to think about user interactions; conversely, thinking about user interactions through a user interface may trigger the designers to think about the model.

22.5 FINAL THOUGHTS

This chapter presents a list of our observations on what representations designers externalize and interact with during the early stages of software design. The representations we focused on include diagrams and text drawn on a whiteboard and uttered in transcripts. In analyzing the data with the DPS tool, we were particularly interested in when and how the representations were referred to, revisited, modified, and reappropriated over time.

By comparing the process of textual and graphical representations produced by the Adobe team and Intuit team, we found that the two teams seemed to be engaged in two completely different design tasks.

As discussed in Section 22.4, the Adobe team focused on the interaction design of the simulation software, and the Intuit team focused on the program design of its engine. We are aware of few studies that focus on how the interaction design (i.e., the design of the world of using) should be related to the program design (i.e., the design of the world of making). Software engineering research has traditionally looked into the design of program code. How a user would interact with the system, which is sometimes called experience design, has primarily been studied in the field of human–computer interaction (HCI).

Interaction design and program design are strongly tied together, as demonstrated by the Adobe team's process (Observation I), but it is only so in the current practice. It is not clear whether the two design aspects should be done in parallel, whether one should precede the other, or whether they should be carried out by different teams of expertise or by a single team. More understanding is necessary to develop theories and models of the two aspects of software design.

We would like to further investigate qualitative and quantitative studies to identify the relationship between the design of the world of using and that of making.

ACKNOWLEDGMENTS

We deeply appreciate the editors and the reviewers for valuable comments and encouragement.

REFERENCES

Baker, A., A. van der Hoek, Ideas, subjects, and cycles as lenses for understanding the software design process, *Design Studies*, 31(6), 590–613, 2010.

Christiaans, H., R.A. Almendra, Accessing decision-making in software design, *Design Studies*, 31(6), 641–662, 2010.

Cooper, A., *The Inmates Are Running the Asylum: Why High-Tech Products Drive Us Crazy and How to Restore the Sanity*, SAMS, Indianapolis, IN, 1999.

Dekel, U., J.D. Herbsleb, Notation and representation in collaborative object-oriented design: An observational study, in *Proceedings of OOPSLA '07*, pp. 261–280, ACM Press, New York, 2007.

Do, E.Y-L., Design sketches and sketch design tools, *Knowledge-Based Systems Journal*, 18(8), 383–405, 2005.

Ju, W., W.L. Neeley, T. Winograd, L. Leifer, Thinking with erasable ink: Ad-hoc whiteboard use in collaborative design, CDR Technical Report #20060928, Stanford University, 2006.

Kawakita, J., *An Idea Development Method, Chuuko Shinsho, Chuuo Kouron-sha*, Tokyo, Japan, 1968 (in Japanese).

Nakakoji, K., Y. Yamamoto, A study of sketches for interaction design in the ARTware project, in *Design Symposium 2010*, JSPE, Tokyo, Japan, 2010 (in Japanese).

Nakakoji, K., Y. Yamamoto, N. Matsubara, Y. Shirai, Toward unweaving streams of thought for reflection in early stages of software design, *IEEE Software, Special Issue on Studying Professional Software Design*, 29(1), 34–38, 2012.

Petre, M., Insights from expert software design practice, in *Proceedings of ESEC/FSE '09*, pp. 233–242, ACM Press, New York, 2009.

Petre, M., A. van der Hoek, A. Baker, Editorial, *Design Studies*, 31(6), 533–544, 2010.

Rooksby, J., N. Ikeya, Collaboration in formative design: Working together at a whiteboard, *IEEE Software*, special issue on *Studying Professional Software Design*, 29(1), 56–60, 2012.

Schoen, D.A., *The Reflective Practitioner: How Professionals Think in Action*, Basic Books, New York, 1983.

Suwa, M., B. Tversky, Constructive perception in design, in J.S. Gero, M.L. Maher (Eds), in *Computational and Cognitive Models of Creative Design V*, pp. 227–239. University of Sydney, Sydney, 2001.

Walny, J., S. Carpendale, N.J. Riche, G. Venolia, P. Fawcett, Visual thinking in action: Visualizations as used on whiteboards, *IEEE Transactions on Visualization and Computer Graphics*, 17(12), 2508–2517, 2011.

Yamamoto, Y., K. Nakakoji, Interaction design of tools for fostering creativity in the early stages of information design, *International Journal of Human–Computer Studies (IJHCS)*, special issue on *Creativity*, L. Candy, E. Edmonds (Eds), 63(4–5), 513–535, 2005.

CHAPTER 23

Postscript

CONTENTS

23.1 INTRODUCTION

Although this book captures the broad variety of analyses that the workshop participants conducted, and the reflective pieces are meant to convey some of the flavor of the cross-perspective discussions that took place at the workshop, this volume cannot truly convey its "buzz"—the intensity and vividness of the discussions that happened in the room. Fueled by the range of disciplines represented by the participants, the discussions identified conflicts, challenged assumptions, looked for synergies, raised temperatures, and, quite frequently, prompted peals of laughter.

We had designed considerable discussion time into the agenda, and this time was filled enthusiastically. Memorable examples include: Nigel Cross pressing Dewayne Perry about whether software design really is a design discipline and the two trading characteristic examples from product design, architecture, and software design in a set of challenges about similarity and difference. Mary Shaw, Michael Jackson, and Jeff Nickerson thrashing through theoretical perspectives on problems, solutions, and evaluations—and striving to integrate them. Jim Dibble, at the whiteboard, reiterating a current design problem from his firm to a small group of onlookers, in order to illustrate which parts of the video were typical of his practice, but also which were different. David Budgen poring over the Wall of notes with one of the professional designers, pointing out interesting cards, and discussing and questioning points they found. Dewayne Perry, Andrew Ko, and Sol Greenspan discussing the impact of the organizational context on the software design process. Fred Brooks challenging the workshop as a whole: the speakers are all analysts and not synthesists—what should we be synthesizing for practice and teaching?

Involving the designers in the workshop was instrumental—and revelatory, both for the researchers and for the designers themselves. They were impressively candid and undefensive. It would have been easy for them to take an apologetic stance concerning their design work on the videos, or to dismiss the researchers altogether. To the contrary, and to their credit, the opposite occurred. They quickly became full participants in the inquiry into their own practices, explaining, questioning, and hypothesizing as much as the researchers. As one of them captured on a card on the Wall:

> Thank you for including us. It is not often that the biologists invite the lab rats to their conferences. Amazed by the level of interest in the process of our professional work. In 2 hours we learned what we do not know:
>
> - About the domain (traffic simulation, traffic light control, educational software)
> - About the user (professor, students)
>
> What I have learned:
>
> - We lack a meta-language to talk about design collaboration;
> - Yes, tools for documenting exploration, design thoughts and issues/assumptions would be incredibly helpful;
> - Idea of backlog;
>
> ...

Having the designers on hand gave the discussions a reference to current practice, it gave the researchers a chance to ask clarifying questions, and it gave the designers a chance to call attention to phenomena of interest to them, or to redress misconceptions. This, in turn, reemphasized the importance of the dialog between research and practice.

The discussions at the workshop were wide ranging, including debates about fundamental definitions, close comparisons of different analyses, disciplinary comparisons, challenges about whether software design is a design discipline at all, and attempts at theory generation. What follows merely hints at the richness of the discussions, highlighting a sample of the key themes that emerged.

23.2 REPRESENTATIVENESS AND EARLY SOFTWARE DESIGN

Michael Jackson characterized the videos as "more like brainstorming than design," and Mary Shaw identified a "dissonance" between the content of the videos and the practices and models that the discipline advocates, such as separating the descriptions of problems from the discussion of a solution, selecting appropriate abstractions, generating design alternatives, or thoroughly understanding the requirements. The participating designers themselves made it clear that the context affected their behavior during the recorded sessions, constraining it from some familiar practices, such as the separation of front-end (interaction) design from the design of back-end processing.

Nigel Cross questioned whether the videos really represent software design; in his view, they contained a lot of problem analysis, but not much design. In contrast, Andrew Ko asserted that: "Almost everything people would call 'design' is problem formulations," and Paul Grisham drew attention to the balance between discovery and subsequent evolution in more general practice: "What do you have to know to evolve something? 75% of the time is spent in discovery."

Dewayne Perry divided the design space into architecture (information and processing), low level (interfaces, algorithms, data structures), and code (representations of those). He asserted that: "The design is only done if/when the code is done." Andrew Ko, too, noted that software design has a relationship to software manufacture, because the medium of design is the same as the medium for implementing design.

Clearly, these discussions revolved around the notion of representativeness: to what degree did the practices captured by the videos align with the participants' own views of what software design is and the role that it plays in the software development life cycle? This, however, is where disagreement immediately arises: to date, no consensual view of software design exists, and the aforementioned perceptions and opinions all stem from the differing views of the participants. Indeed, our choice to focus on early design was fueled by our own beliefs that design is a broad activity that spans the software life cycle, and that early design had received too little attention to date, overshadowed by design that focused solely on devising the code structure.

Ania Dilmaghani provided an interesting parallel regarding the limitations of the data—the artificial situation, the short time scale—in observing: "It's like taking a biopsy. The needle hits where it hits." While acknowledging the need to start somewhere, she wondered how to improve the sample "without killing the patient." Nigel Cross summarized the impact on the designers' performance: "You have to cut your suit in the cloth available." There were repeated calls for studies of software design that capture a broader view of the process, a topic that we return to later.

23.3 DESIGNING CHANGES TO SOFTWARE

Another recurring theme of discussion was the software product's life cycle, and the way that it extends beyond the delivery of the completed product. Almost any product can be changed, in some way, after it is delivered. What makes software unusual is the expectation of the customer, the user, and other stakeholders, that it *will* change. As Dewayne Perry pointed out, software is held to a different standard of quality than other disciplines. In software, something faulty is sometimes still worth having, partly because of the idea that bugs can be fixed, patches can be issued, and new versions can be released. The software issued is, more often than not, simply a precursor to the "final product."

This persistence of design activity after software release gives a different complexion to the notions of "evaluation" and "reflection" in software design. Dewayne Perry asserted that design and evaluation are "two halves of the same problem." In general, the discussion did not distinguish clearly between reflection and evaluation, although a distinction was implied between reflection-on-process and reflection-on-product. Vesna Popovic noted that reflection takes a narrative form in software design, because the system is dynamic. She contrasted that with physical design, in which reflection is not narrative but summative, and is used for evaluation. Some discussion concerned criteria, for example, "'Goodness' evaluated in terms of internal quality—or in terms of suitability for the user?" and "How do software designers decide when a design is 'done enough'?"

Unfortunately, making changes to software can be surprisingly costly in terms of human resources, not to mention risky. For one, changes to one aspect of a program can readily cascade and affect other aspects of the system in unexpected ways; this effect is multiplied as the system gets more complicated, and if it is poorly designed in the first place. Furthermore, making changes can be especially difficult if one does not know the decisions that led to the current design (Andrew Ko), and effective tracking of design rationale is notoriously difficult. Because software is so complex, it can be very difficult to map a desired change in the world back to the code segments that would enact it. Understanding the ways that the expectation of change affects the software process and treating this change-making process as a design process represent outstanding intellectual challenges.

While the videos captured greenfield design, the topic of designing *changes* to software persisted throughout the workshop, both in terms of how the designers did and did not capture their assumptions and decisions, and in terms of representativeness when so much day-to-day practice in software design concerns evolving an existing application.

23.4 REPRESENTATION

A recurring theme throughout the workshop was the slipperiness of the idea of "a software design." Emily Navarro reported the lack of good notations for software design as an overarching theme during the discussions, questioning: "Is a good (set of) notation(s) really the solution, or is there more to it than that?" In any design field, the design itself exists in a variety of states: in the minds of the designers, in sketches, in prototypes, and in a variety of other forms. Gerhard Fisher captured the challenge of representation in software design as: "Making visible the computation processes you want." Paul Grisham observed: "The product is not code, but behavior." These comments point to the fact that the final design of a program is its source code, but that it would be a mistake to focus solely on the design of the structure of this source code. Eventually, the software is executed, and the behavior that is exhibited by the executing program is what must be designed, first and foremost.

Rachael Luck expressed surprise at the forms of representation, in particular that there was "... nothing analogous to a plan in the build world, a recognized notation/representation, something equivalent to sketching as an abstraction of the real world." Clayton Lewis, too, in comparing software design and product design, commented that there is "... so little use of holistic representations." He wondered if software is *necessarily* complicated, or if we can address the difficulty of representing software artifacts by conceiving of an approach with fewer degrees of freedom? The nature of partial designs in software seems to challenge boundaries established by other domains (such as the distinction between design and manufacture), and requires further exploration to understand.

Jim Dibble observed that we do not have a metalanguage to talk about the design *process*. His examples included an explicit agreement to leave part of the whiteboard for assumptions, and an explicit role assignment (proposer, devil's advocate). He discussed the value of documenting exploration, design thoughts, issues, and assumptions as design proceeds. One wall card noted a "lack of language for *quality* of the solution." Jeff Nickerson agreed that software designers are not trained in the metalanguage of design. He noted that there is little support for conceptual design. Others referred to the checkered history of efforts in design rationale.

23.5 COLLABORATIVE CONTEXT

Early software design was broadly acknowledged as a collaborative process. In addition to the observations made earlier, much of the discussion about representation highlighted the role of the *design dialogs* in refining and supplementing the *design representations*. Michael Jackson remarked: "Design is very verbal." The importance of the collaborative context, its impact on representation, and the importance of narrative in representing and reasoning about software were remarked upon repeatedly. Nevertheless, Paul Rodgers was concerned that: "... the discussions have overlooked the socio-cultural aspects of the design sessions and also the socio-cultural elements of the design problem." Harold Ossher observed that, in practice, much of the *problem* is worked out collaboratively, but the *solution* finding is done independently. In such cases, the idea of a shared design concept by a team takes on a new character, and the issue of coordinating the interpretations of this design concept by members of a team becomes crucial.

Ania Dilmaghani spoke about the dialogs associated with the creative process, drawing attention to the "bedside manner" needed for effective discussion. Paul Rodgers observed that: "Greater familiarity with your partner brings greater comfort, trust ...." Yet, Irina Solovyova questioned: "Does good collaboration necessarily lead to good design?" Rachael Luck wondered how, given the reliance on description and semantic interpretation, software developers change jobs and do productive work. She raised issues of loss of social capital when someone leaves an organization, and the delay for cultural induction when someone joins. Others drew attention to the different structures of evolving collaborative contexts, such as distributed teams and open source development.

The discussion of collaboration also broadened into a discussion about elements of social and organizational contexts that affect design practice. Jeff Nickerson observed that "any design is done under pressure," and a number of conversations drew attention to the impact on design of organizational structure, time to market, and competition for resources. Ania Dilmaghani emphasized that "insights are welcome," but that they must fit into the context and pressure characterized by money, time to market, milestones, and releases. Rachael Luck drew attention to pressures arising from the client or stakeholders. Sol Greenspan drew attention to "the adoption problem for tools," explaining that one reason why tools fail is organizational: in any large company there is always a battle about what tools to buy and whether they should standardize. The implications in terms of tool support, formal representations, and design practice demand further exploration.

23.6 EXPERTISE AND INDIVIDUAL DIFFERENCES

Paul Rodgers posted a note on the Wall: "Designers are part of the problem. Discuss." The impact of individual differences is one of the most robust effects observed in the psychology of programming literature, and it has also been documented in the software engineering literature. It featured in the workshop discussion as well, in the context of both individual design performance and interaction within teams. Acknowledgment was made of *abstraction* as a key ability for software designers.

Irina Solovyova, noting that assumptions and cultural expectations are based on personal experience, queried: “How big a role do those personal assumptions play in structuring a design problem?” For example, both members of the AmberPoint team were avid user interface designers, which pulled that team’s design toward that aspect of the problem. Joanne Atlee, saying that “Design is strongly influenced by designers’ expertise,” identified the need to diversify teams. There was a discussion about the impact of the background and experience of the different teams on software design. For example, the Intuit session had two designers who had very different experience levels, and the more experienced designer tended to take the lead at the whiteboard. Yet, informally, the less experienced designer seemed to coin quite a few important advances. Reference was made to Wenger’s notion of “communities of practice,” that designers profit from interaction with experienced colleagues. Gerhard Fisher urged that: “We do not just need reflective practitioners, we need reflective communities.”

It is worth noting that the six designers recorded in the videos were regarded as being among the strongest at their respective companies, but there remain open questions about how these designers’ particular skill sets affected their approach to the shared design problem. The answers might lead to insights into the nature of expertise and experience in software design.

23.7 LIMITATIONS AND THE DESIGN OF FUTURE DATA COLLECTION

The data were sufficient to inspire a variety of analyses and to support rich discussion about the nature of software design, as well as to evoke participants’ perspectives on, assumptions about, and experiences with software and other design. It served its purpose.

Yet, the limitations of the data set were clear, and one of the outcomes of the workshop was a wish list for future data collection. Fundamentally, there is a tension between gathering data that are realistic and creating a data set that is tractable for study in a workshop of this kind. The ideal case involves studying real software designers, on a real software project, through the entirety of the product’s life cycle, and possibly beyond. But, of course, achieving each of these elements of realism presents serious challenges: recording work on real projects presents risks to the participating organization, gathering data in real-world settings is difficult, and, as the corpus grows in size, the feasibility of studying it effectively diminishes sharply. As an initial foray, this workshop’s data set erred on the side of simplicity, but future data sets will need to explore a much richer space.

Of course, the question of how to study software designers is not as simple as choosing a level of realism. There is no canonical software design activity, and studies are needed along a variety of dimensions, if a broader understanding of the discipline is to be achieved. For example:

Parts of the software process: As we have observed, design is needed throughout the software process, from the earliest discussions of the problem, through the development of the code itself, and even after its release. This workshop was concerned primarily with conceptual design early in the process, which has been insufficiently studied, but design and redesign persist throughout, in a variety of settings: elicitations with clients or stakeholders, architectural design discussions, individual solution finding, archaeological forays and rationale reconstructions during maintenance, and so on.

Modes of working: Software is designed by people, but the number of people and the way that they work together on a project can vary greatly. Sometimes, design is a solitary activity, one of introspective thought and reflection. Other times, design is highly collaborative, with much conversation and explanatory sketches in support. Furthermore, as technology advances, distributed software development projects are increasingly common, and present yet another mode of software design.

Different roles and expertise: Software design is performed by a variety of individuals, with wildly varying attitudes and experiences. Some design is performed by entry-level programmers, and other design is performed by architects with decades of experience. Furthermore, experienced software designers are not a homogeneous group; there are myriad ways that designers might specialize their skills, whether intentionally by seeking expertise or simply as a matter of exposure to particular domains.

In addition to these dimensions of consideration, the discussions at the workshop led to an understanding of some essential challenges to understanding software design. As we reflect on how to move forward with future studies, the following are some factors of which it would be prudent to be mindful:

Design decisions are tightly interwoven: As we have already highlighted, design takes place across the life cycle. By the time code is being written, a great deal of conceptual design and system architecture is likely to have preceded it. Yet, writing the code will lead to a reflection on those parts of the design process, and quite likely call for a reconsideration of certain decisions. As a result, a complex network of design decisions emerges with newer decisions often implicitly relying on previous decisions. This presents a challenge as we focus our studies on the kinds of design that take place later in the process—how to account for the rich context that has already been built?

Reuse: In practice, software is rarely built from scratch: code from an organization's previous projects might be reused or commercial components might need to be integrated. Even when existing code is not being used directly, designers must increasingly engage in "brownfield development," where the system being designed must work harmoniously with other software systems already in place. Finally, organizations increasingly create software product lines, where new versions are built from the designs and artifacts of existing products. Integrating these realities into studies of software designers will be far from trivial.

Nature of software: Much of the early research in software design, for example, in the psychology of programming community, focused on "programming in the small," from a conception of programs as discrete artifacts of commensurable size. Yet, in a few short decades, software has evolved into massive, distributed, evolving, cooperating systems. Currently, a major shift toward mobile apps and cloud computing is taking place. The question for the discipline is: what next? The concept of "software design" will remain a moving target, and our study of it will need to remain agile.

In the midst of these challenges and considerations, no one solution for a next study emerged from the workshop, but one concrete proposal was to capture videos of actual collaborative design meetings held by small teams in industry. These videos would be collected at various points during development, and would serve as snapshots of the larger process. The argument was that the discussions in these meetings would provide sufficient background to allow the researchers to understand the major decisions made in between the recorded sessions. Another proposal was to record software design competitions at a software engineering conference, which would provide a controlled setting, albeit one that remains disconnected from some of the realities discussed earlier.

23.8 IN CLOSING

The workshop ended on a clear consensus: despite the limitations of the data set, the analyses and dialog were valuable, and the exercise would be worth doing again. The participating designers reiterated the imperative to bring industry and academia together for research and discussion.

Numerous questions arise as to how we ought to be using the findings from this workshop to improve the design of software. What tools and methodologies might be devised to aid the current generation of software designers, and what educational innovations could be devised to better prepare the next generation? What are the implications, for example, of different modes and contexts of designing for tool development? In distributed projects, for example, there is no discussion around the coffee machine, no sketching on the whiteboard; ongoing design work is conducted mainly on the telephone or independently. On the one hand, it may be easier to capture the progress of design via the interactions; on the other hand, new issues of communication and coordination may require attention. As another example, how might we cope with the increasing scale and complexity of software? Should designers perhaps intentionally restrict the degrees of freedom they allow themselves? The ability to create software that can do anything is a mixed blessing, and by finding appropriate ways to restrict its design, we may actually facilitate more effective design processes. As a final example, how can we better capture the rationale for a design? Particularly, how can rationale be captured in the moment, rather than through an explicit process of design rationale capture? How might designers be better able to record their thoughts, assumptions, and decisions?

In addition to focusing on understanding software design, this workshop attempted to provide a cross-disciplinary forum for considering design in general. We would like to reiterate our goal of encouraging communication between software design researchers and those from other design disciplines. As Fred Brooks observed in the plenary discussion, there is reason to believe that the older design disciplines have better-formed disciplines, more common practices, and more mature ways of communicating about design. There is a need to make progress on the big questions about the nature of software design. What is it about software that makes it unusual, and how does that affect our ability to transfer lessons from other fields? Conversely, what insights from the software design community might be leveraged in other disciplines? Finding ways to answer these questions, and otherwise enhance communication across disciplinary boundaries, remains our most prominent goal.

Appendix I Design Prompt

Traffic Signal Simulator

AI.1 PROBLEM DESCRIPTION

For the next two hours, you will be tasked with designing a traffic flow simulation program.

Your client for this project is Professor E, who teaches civil engineering at UCI. One of the courses she teaches has a section on traffic signal timing, and according to her, this is a particularly challenging subject for her students. In short, traffic signal timing involves determining the amount of time that each of an intersection's traffic lights spend being green, yellow, and red, in order to allow cars in to flow through the intersection from each direction in a fluid manner. In the ideal case, the amount of time that people spend waiting is minimized by the chosen settings for a given intersection's traffic lights. This can be a very subtle matter: changing the timing at a single intersection by a couple of seconds can have far-reaching effects on the traffic in the surrounding areas.

There is a great deal of theory on this subject, but Professor E. has found that her students find the topic quite abstract. She wants to provide them with some software that they can use to "play" with different traffic signal timing schemes, in different scenarios. She anticipates that this will allow her students to learn from practice, by seeing firsthand some of the patterns that govern the subject.

AI.2 REQUIREMENTS

The following broad requirements should be followed when designing this system:

1. Students must be able to create a visual map of an area, laying out roads in a pattern of their choosing. The resulting map need not be complex, but should allow for roads of varying length to be placed, and different arrangements of intersections to be created. Your approach should readily accommodate at least six intersections, if not more.

2. Students must be able to describe the behavior of the traffic lights at each of the intersections. It is up to you to determine what the exact interaction will be, but a variety of sequences and timing schemes should be allowed. Your approach should also be able to accommodate left-hand turns protected by left-hand green arrow lights. In addition:

    a. Combinations of individual signals that would result in crashes should not be allowed.

    b. Every intersection on the map must have traffic lights (there are not any stop signs, overpasses, or other variations). All intersections will be 4-way: there are no "T" intersections, nor one-way roads.

    c. Students must be able to design each intersection with or without the option to have sensors that detect whether any cars are present in a given lane. The intersection's lights' behavior should be able to change based on the input from these sensors, though the exact behavior of this feature is up to you.

3. Based on the map created, and the intersection timing schemes, the students must be able to simulate traffic flows on the map. The traffic levels should be conveyed visually to the user in a real-time manner, as they emerge in the simulation. The current state of the intersections' traffic lights should also be depicted visually, and updated when they change. It is up to you how to present this information to the students using your program. For example, you may choose to depict individual cars, or to use a more abstract representation.

4. Students should be able to change the traffic density that enters the map on a given road. For example, it should be possible to create a busy road, or a seldom used one, and any variation in between. How exactly this is declared by the user and depicted by the system is up to you.

Broadly, the tool should be easy to use, and should encourage students to explore multiple alternative approaches. Students should be able to observe any problems with their map's timing scheme, alter it, and see the results of their changes on the traffic patterns.

This program is not meant to be an exact, scientific simulation, but aims to simply illustrate the basic effect that traffic signal timing has on traffic. If you wish, you may assume that you will be able to reuse an existing software package that provides relevant mathematical functionality such as statistical distributions, random number generators, and queuing theory.

You may add additional features and details to the simulation, if you think that they would support these goals.

Your design will primarily be evaluated based on its elegance and clarity—both in its overall solution and envisioned implementation structure.

AI.3 DESIRED OUTCOMES

Your work on this design should focus on two main issues:

1. You must design the interaction that the students will have with the system. You should design the basic appearance of the program, as well as the means by which the user creates a map, sets traffic timing schemes, and views traffic simulations.
2. You must design the basic structure of the code that will be used to implement this system. You should focus on the important design decisions that form the foundation of the implementation, and work those out to the depth you believe is needed.

The result of this session should be: the ability to present your design to a team of software developers who will be tasked with actually implementing it. The level of competency you can expect is that of students who just completed a basic computer science or software engineering undergraduate degree. You do not need to create a complete, final diagram to be handed off to an implementation team. But you should have an understanding that is sufficient to explain how to implement the system to competent developers, without requiring them to make many high-level design decisions on their own.

To simulate this hand-off, you will be asked to briefly explain the above two aspects of your design after the design session is over.

AI.4 TIMELINE

- 1 hour and 50 minutes: Design session
- 10 minutes: Break/collect thoughts
- 10 minutes: Explanation of your design
- 10 minutes: Exit questionnaire

Appendix II SPSD 2010 Original List of Workshop Participants

Name	E-mail	Affiliation	Country
Aldeida Aleti	aaleti@swin.edu.au	Swinburne University of Technology	Australia
Rita Almendra	almendra@fa.utl.pt	Technical University of Lisbon	Portugal
Joanne Atlee	jmatlee@uwaterloo.ca	University of Waterloo	Canada
Alex Baker	abaker@ics.uci.edu	University of California, Irvine	United States
Linden Ball	l.ball@lancaster.ac.uk	Lancaster University	United Kingdom
Werner Beuschel	beuschel@fh-brandenburg.de	Brandenburg University of Applied Sciences	Germany
Fred Brooks	brooks@cs.unc.edu	University of North Carolina at Chapel Hill	United States
David Budgen	david.budgen@durham.ac.uk	Durham University	United Kingdom
Janet Burge	burgeje@muohio.edu	Miami University	United States
Henri Christiaans	h.h.c.m.christiaans@tudelft.nl	Delft University of Technology	The Netherlands
Nigel Cross	n.g.cross@open.ac.uk	The Open University	United Kingdom
Michael Desmond	mdesmond@us.ibm.com	IBM Research	United States
Francoise Detienne	francoise.detienne@telecom-paristech.fr	CNRS—Telecom ParisTech	France
Jim Dibble	jdibble@amberpoint.com	AmberPoint	United States
Ania Dilmaghani	adilmaghani@amberpoint.com	AmberPoint	United States
Natalia Dragan	ndragan@cs.kent.edu	Kent State University	United States
Gerhard Fischer	gerhard@colorado.edu	University of Colorado	United States
John Gero	john@johngero.com	Krasnow Institute for Advanced Study	United States
Mike Godfrey	migod@uwaterloo.ca	University of Waterloo	Canada
Sol Greenspan	sgreensp@nsf.gov	National Science Foundation	United States
Paul Grisham	grisham@mail.utexas.edu	University of Texas at Austin	United States

(*continued*)

Name	E-mail	Affiliation	Country
Mark D. Gross	mdgross@cmu.edu	Carnegie Mellon University	United States
Christopher Han	christopher.han@stanford.edu	Stanford University	United States
André van der Hoek	andre@ics.uci.edu	University of California, Irvine	United States
David Holloway	davidh@slugworth.com	Google Inc.	United States
Nozomi Ikeya	nozomi.ikeya@parc.com	Palo Alto Research Center	Japan
Michael Jackson	jacksonma@acm.org	The Open University	England
Sian Joel	s.joel@napier.ac.uk	Napier University	Scotland
Malte Jung	mjung@stanford.edu	Center for Design Research, Stanford University	United States
Andrew Ko	ajko@uw.edu	University of Washington	United States
Ben Kraal	b.kraal@qut.edu.au	Queensland University of Technology	Australia
Micah Lande	micah@stanford.edu	Center for Design Research, Stanford University	United States
Clayton Lewis	clayton.lewis@colorado.edu	University of Colorado	United States
Rachael Luck	luck621@btinternet.com	University of Reading	United Kingdom
Jonathan Maletic	jmaletic@cs.kent.edu	Kent State University	United States
Ben Matthews	matthews@mci.sdu.dk	University of Queensland	Australia
Janet McDonnell	j.mcdonnell@csm.arts.ac.uk	Central Saint Martins, University of the Arts London	United Kingdom
Leonardo Murta	leomurta@ic.uff.br	Universidade Federal Fluminense	Brazil
Kumiyo Nakakoji	kumiyo@kid.rcast.u-tokyo.ac.jp	University of Tokyo/Software Research Associates Inc.	Japan
Jeffrey Nickerson	jnickerson@stevens.edu	Stevens Institute of Technology	United States
Harold Ossher	ossher@us.ibm.com	IBM Research	United States
Dewayne Perry	perry@mail.utexas.edu	University of Texas at Austin	United States
Marian Petre	m.petre@open.ac.uk	The Open University	United Kingdom
Vesna Popovic	v.popovic@qut.edu.au	Queensland University of Technology	Australia
Paul Rodgers	paul.rodgers@northumbria.ac.uk	Northumbria University	United Kingdom
John Rooksby	rooksby@googlemail.com	University of St Andrews	United Kingdom
Mary Shaw	mary.shaw@cs.cmu.edu	Carnegie Mellon University	United States
Irina Solovyova	irina.solovyova@utsa.edu	University of Texas at San Antonio	United States
Neeraj Sonalkar	sonalkar@stanford.edu	Center for Design Research, Stanford University	United States
Antony Tang	atang@swin.edu.au	Swinburne University of Technology	Australia
Barbara Tversky	btversky@stanford.edu	Columbia University/Stanford University	United States
Hans van Vliet	hans@cs.vu.nl	VU University Amsterdam	The Netherlands
Willemien Visser	willemien.visser@telecom-paristech.fr	CNRS, Telecom ParisTech—INRIA	France
Claudia Werner	werner@cos.ufrj.br	Federal University of Rio de Janeiro	Brazil

Index

D

E

F

G

R

S

Y